Therapeutic Strategies in Cancer Biology and Pathology

Therapeutic Strategies in Cancer Biology and Pathology

Gajanan V. Sherbet

School of Electrical Electronic and Computer Engineering, Newcastle University, Newcastle upon Tyne, UK

The Institute for Molecular Medicine, Huntington Beach, CA, USA

ELSEVIER

AMSTERDAM • BOSTON • HEIDELBERG • LONDON • NEW YORK • OXFORD
PARIS • SAN DIEGO • SAN FRANCISCO • SINGAPORE • SYDNEY • TOKYO

Elsevier
32 Jamestown Road, London NW1 7BY
225 Wyman Street, Waltham, MA 02451, USA

British Library Cataloguing-in-Publication Data
A catalogue record for this book is available from the British Library

Library of Congress Cataloging-in-Publication Data
A catalog record for this book is available from the Library of Congress

ISBN: 978-0-12-416570-0

For information on all Elsevier publications
visit our website at store.elsevier.com

This book has been manufactured using Print On Demand technology. Each copy is produced to order and is limited to black ink. The online version of this book will show color figures where appropriate.

Dedication

To

Shri Sathya Sai Baba

Contents

Preface

தாளாண்மை என்னும் தகைமைக்கண் தங்கிற்றே
வேளாண்மை என்னுஞ் செருக்கு.

(Noble legitimate pride adorns the benevolence of untiring effort and perseverance)
Thiruvalluvar (Tamil poet second century India)
(Thirukkural Chapter 62, verse 613)

Recent years have seen intensive effort directed towards the identification of molecular targets that can provide approaches to the development of novel therapeutic strategies in cancer management. The focus here is on tumour and metastasis-associated genes, viz. those identified with the ability to promote tumour growth and metastasis or suppress metastatic spread, those linked with cellular changes that relate specifically to behavioural alterations of cancer cells in epithelial mesenchymal transition and in the acquisition of invasive faculty. Genes involved in cell cycle regulation and cell proliferation also form a part of the feature in the backdrop of cancer growth and dissemination. The discussion here revolves around how appropriate these genes are as molecular targets, how practicable are the constituents of their signal transduction systems as potential candidates and how accessible they are to targeted therapy. It is easy to expatiate on the advantages of being able to do this, but the perceived advantages might be fraught with difficulty, prominent being the influence of signalling cross talk with collateral gene activation or suppression. These themes and allied issues are addressed here as an extension of the deployment of growth factors and their receptors that I discussed in the previous work (Sherbet, 2011a).

I acknowledge with thanks the help and support I have received from Professors Satnam Dlay and Barrie Mecrow. I thank Dr M.S. Lakshmi for reading the manuscript and making helpful suggestions and also for providing the translation of the verse from Thirukkural. I recognise without reservation that Elsevier Insights personnel have been most helpful and supportive in the preparation of this book.

G.V. Sherbet
University of Newcastle upon Tyne
March 10, 2013

Abbreviations

ABC	ATP-binding cassette transporter
ADAM	A disintegrin and metalloprotease
AICAR	AMPK inhibitor 5-aminoimidazole-4-carboxamide 1-D-ribonucleoside
AIF	apoptosis inducing factor
APC	adenomatous polyposis coli
AMPK	AMP-activated protein kinase
AML	acute myeloid leukaemia
AREG	amphiregulin
ARF	alternative reading frame
ATM	Ataxia telangiectasia mutated
bFGF	basic fibroblast growth factor
bHLH	basic helix-loop-helix
Bmi-1	B lymphoma Mo-MLV insertion region 1 homologue
BMP	bone morphogenetic protein
CA	caffeic acid
CADM1	cell adhesion molecule 1
CaMKII	calcium/calmodulin-dependent protein kinase II
CAPE	caffeic acid phenethyl ester
Cdc4	cell division control protein
Chk2	checkpoint kinase 2
CLL	chronic lymphocytic leukaemia
CMG2	capillary morphogenesis gene 2
CML	chronic myeloid leukaemia
CSC	cancer stem cell
CtBP	C-terminal EIA binding protein
CTGF	connective tissue growth factor
CXCR4	C-X-C chemokine receptor type 4
Dachs	dachsous
DAG	diacylglycerol
DGCR8	DiGeorge syndrome critical region 8
Dkk	Dickkopf
DLC	deleted in liver cancer
DYNLL1	dynein light chain 1
DVL	dishevelled
EGF	epidermal growth factor

EGFR	EGF receptor
ELAM	endothelial leukocyte adhesion molecule
EMT	epithelial mesenchymal transition
EPB41L3	erythrocyte membrane protein band 4.1-like 3
ER	oestrogen receptor
ERK	extracellular signal-regulated kinase
ERM	Ezrin-Radixin-Moesin
ESA	epithelial cell-specific antigen
ESCC	oesophageal squamous cell carcinoma
EZH	zeste homologue protein
FADD	Fas-associated death domain
FAP	familial adenomatous polyposis
FGFR	fibroblast growth factor receptor
FOXO	forkhead box transcription factors
Fzd	frizzled Wnt receptor
GADD	growth arrest and DNA-damage inducible proteins
GAP	GTPase-activating protein
GLO1	glyoxalase I
GNAI2	guanine nucleotide binding protein alpha inhibiting activity polypeptide 2
GPCRs	G-protein coupled receptors
GRK2	GPCR kinase
GSK3	glycogen synthase kinase-3β
HAT	histone acetyl transferase
HCC	hepatocellular carcinoma
HDAC	histone deacetylase
HER2	human epidermal growth factor receptor 2
HGF	hepatocyte growth factor
Hh	hedgehog
HIC	hypermethylated in cancer
HIF-1	hypoxia-inducible factor 1
HMG	high-mobility group protein
HNPCC	hereditary non-polyposis colorectal cancer
HSGAG	heparan sulfate glycosaminoglycans
HUVEC	human vascular endothelial cells
ICAM	intercellular adhesion molecule
IDs	inhibitor of DNA binding proteins (bHLH transcription factors)
IGF	insulin-like growth factor
IGFR	IGF receptor
IKK	IκB kinase
ING	inhibitor of growth suppressor gene
IP3	inositol 1,3,5-trisphosphate
IP3R	IP3 receptor
IPF	idiopathic pulmonary fibrosis
JNK	c-Jun N-terminal kinase
LATS	large tumour suppressor gene

Lef	lymphoid enhancing factors
LeTx	lethal toxin
LKB1	liver kinase B1
LOH	loss of heterozygosity
LRP	LDL receptor-related Wnt co-receptor
MAPK	mitogen-activated protein kinase
MDR	multidrug resistance
MEK	MAPK/ERK Kinase
MDS	myelodysplastic syndrome
miRNA	microRNA
MO25	mouse protein 25
MPA	medroxyprogesterone acetate
MSI	microsatellite instability
mTOR	mammalian target of rapamycin
NDPK	nucleoside diphosphate kinase
NDRG1	N-myc downstream-regulated gene-1
NF2	neurofibromatosis gene 2
NO	nitric oxide
NOS	nitric oxide synthase
PAK	p21-activated protein kinase
PARP	poly (ADP-ribose) polymerase
PDCD4	programmed cell death 4 protein
PI3K	phosphoinositide 3 kinase
PAK	p21-activated protein kinase
PJS	Peutz–Jeghers syndrome
PKC	protein kinase C
PLC	phospholipase C
PLCD1	phospholipase Cδ1
PTCH	patched receptor
RAGE	Receptor for Advanced Glycation End products
RASSF	Ras-association domain family protein
Rb	retinoblastoma susceptibility gene
RECK	reversion inducing cysteine-rich protein with Kazal motifs
RES	resveratrol
RGS	regulator of G-protein signalling protein
RKIP	Raf kinase inhibitor protein
ROS	reactive oxygen species
RTK	receptor tyrosine kinase
Runx	Transcription Factor with Runt DNA-binding domain
SARAH	Salvador 1/RASSF1A/Hippo domain
Sav	Salvador component of Hippo signalling
SFRP	secreted frizzled-related proteins
SHh	sonic hedgehog
siRNA	small interference RNA
SMO	smoothened a downstream element in Hh signalling
SNP	single nucleotide polymorphisms

STRAD	STE20-related kinase adapter protein
STRAP	serine-threonine kinase receptor-associated protein; stress-responsive activator of p300)
TCF	T-cell factors
TEM8	tumour Endothelial Marker 8
TGF	Transforming growth factor
TIMP	Tissue inhibitors of matrix metalloproteinases
TNBC	triple negative breast cancer
TNF	tumour necrosis factor
TRADD	TNFR-associated death domain
TRAF2	TNF-associated factor 2
TrCP	transducin repeat containing protein
TSG101	Tumour susceptibility gene 101
TXNIP	thioredoxin interacting protein
uPA	urokinase-type plasminogen activator
VASH1	Vasohibin 1
VCAM	vascular cell adhesion molecule
VDR	vitamin D3 receptor
VEGF	vascular endothelial growth factor
VEGFR	VEGF receptor
VHL	von Hippel–Lindau syndrome
YAP	Yes-associated protein
YY1	Yin Yang 1 transcription factor

Introduction

The genome is inherently unstable and genetic instability generates genetic diversity. Genetic instability carries within itself the means of genetic imbalance and mutability. Many genetic changes occur in the development and progression of cancer. Genetic instability drives the processes of the generation of genetic diversity and the evolution of cellular phenotypes leading to tumour heterogeneity. Alterations in the genetic makeup would inevitably lead to alterations in cellular properties such as cell proliferation and population expansion, promotion and inhibition of apoptosis, induction of vascularisation, composition and characteristics of the extracellular matrix and intercellular adhesion and cell locomotion and the ability to evade immunological surveillance. The evolution of phenotypically distinctive subpopulations is a consequence of the accumulation of mutations that can be associated with distinctive behaviour and with distinct phases of disease progression. Genetic instability might be associated with a specific stage or stages of progression and with mutations, loss of heterozygosity or changes in the expression of genes whose function might coincide identifiably with the different phases of cancer progression.

The association of genetic instability with stage-specific gene function is stochastic in nature but would nevertheless provide a mechanistic mode of relationship between genetic instability and disease progression. A study of the pattern of the molecular changes occurring in these genetic determinants has allowed the construction of a model of progression of cancers with focus on colon and ovarian cancers, which might equally apply to tumour progression as a general rule. The molecular concept of cancer progression is the foundation on which the cancer stem cell hypothesis has been formulated of late and indeed it is implicit in the molecular concept of cancer progression. The restructured and restated postulate of the cancer stem cell has without any doubt sharply focused attention on targeting these subpopulations which are often resistant to chemotherapeutic agents. The development of chemoresistance by stem cells may be defined by the microenvironment, but a temporal acquisition of stem cell characteristics has also been advocated. However, this implies a distinction between cancer stem cells and other stem cells founded on plausible influences of the microenvironment and temporal acquisition of drug resistance that would be too artificial to contemplate.

The rationale and strength of the currency of the stem cell hypothesis are predictably easy to appreciate. By definition cancer stem cells would be akin to adult stem cells and would represent a clone that is amenable to the induction of differentiation, capable of self-renewal and possesses proliferative potential. Therefore they are potential targets for therapeutic intervention based on biological modifiers of differentiation, inducers of apoptosis and inhibitors of cell proliferation. So are many signalling pathways such as the Hedgehog, Wnt, miRNAs, and epithelial mesenchymal transition (EMT), among others. These are developmental pathways and are recognised as stem cell pathways. Inhibitors of EMT and promoters of the reverse process of mesenchyme to epithelial transition would be valuable aids in inducing differentiation of cancer stem cells. Accordingly here the emphasis is on the identification of strategic molecular targets (Table I.1) that might provide new and therapeutic approaches to patient management.

So we arrive at candidate gene identification in metastasis promotion or suppression. The candidate gene postulate revolves around the identification of an entity that is associated with and leads to pathogenesis. This has often invoked the genetic approach to equate a gene entity by linkage analyses employing various markers expressed by a family or siblings, the proximity of the markers to the genes of interest providing the linkage between the gene and the specified disease incidence and susceptibility to the disease state. This is often followed by studies to determine whether that gene is mutated in the process of pathogenesis, and if so the nature and frequency of mutations. Among approaches of note are whole-genome sequence analysis of families to determine the inheritance patterns; genotyping individuals for genetic variations and susceptibility to disease. Many genes such as the Rb (retinoblastoma susceptibility) gene and the breast cancer genes of BRCA family are indubitably linked with susceptibility to disease. Carriers of germline mutations of the latter show predisposition to develop breast and ovarian cancers. Many of these genes also interact and function in collusion with other genes. The Rb gene is a regulator of the cell cycle and interacts with p53 signalling cascade in that role, aside from leading in the mutated form to the development of retinoblastoma.

Function analysis of genes provides another means of approach to the problem of candidate gene identification. This would comprise the analysis of biological features and cellular properties associated with one or more traits of cells such as cell proliferation, invasion, angiogenesis and EMT. Stem cell maintenance and propagation and lineage specificity in general and in relation to cancer stem cells in particular also come into the consideration. The primary effector genes and proteins that initiate and produce the specified phenotype are the prime targets for identification. Several tumour and metastasis promoters and suppressors fall into this category. These have been quite justifiably classed as candidate genes on account of the broad spectrum of biological and cellular features that they appear to modulate. The downstream components and constituent links in signalling pathway which the primary effectors activate also prominently participate in and regulate successful transduction of the original signal. Numerous genes that can be included in this category have been called suppressor or promoter genes, whilst not overtly labelling them as candidate suppressors or promoters. The focus here is on metastasis promoter and suppressor genes that are widely perceived as candidate genes. It is needless to say that the commonly acknowledged tumour susceptibility genes, such as the Rb, BRCA genes, TSG101 and the cell cycle regulator p53 which is mutated in a majority of cancers, are cited and discussed in appropriate contexts. So also are the genes that modulate the signalling cascade referred in the relevant setting (Table I.1). This is the

Table I.1 Tumour Susceptibility, Metastasis Promoter and Suppressor Candidate Genes and Signalling Modulators

Genetic Syndrome/Neoplasm	Candidate Gene/Genetic Alteration
Tumour susceptibility genes	
Retinoblastoma susceptibility gene	Rb, cell cycle regulation, dysfunction is cancer associated
p53	Mutated p53 in human cancers
HNPCC (hereditary non-polyposis colorectal cancer)	MSH2, MLH1, PMS1, PMS2 germline mutations
FAP (familial adenomatous polyposis)	APC
Li–Fraumeni syndrome	Mutation of p53
Von Hippel–Lindau syndrome	VHLS; haemangioblastomas
Wilm's tumour	Mutation of one allele of Wt1
Breast and ovarian	BRCA1, BRCA2 germline mutations predispose carriers to cancers; participate in DNA repair
Neurofibromatosis	NF1 gene
TSG101	Cell growth, differentiation, negative regulator of growth, cell cycle regulation, mutations in breast cancer
Metastasis promoters	S100A4, osteopontin, MTA1 and miRNAs
Metastasis suppressors	nm23, KiSS1, CRSP3 and TXNIP in conjunction with KiSS1
	KAI1, NDRG1, BRMS1, 14-3-3 proteins, Maspin, EPB41L3, CADM1, PDCD4, PRDM2, miRNAs, Hippo and Lats, YAP/TAZ and Mats in tumour and metastasis (?) suppression, Merlin/NF2, MKK, RASSF
Signalling modulators	
Cell proliferation, apoptosis	Wnt, Notch and Hedgehog signalling, EGF, EGFR, TGF-β, TXNIP, HGF, PDGF, Hippo and its upstream and downstream regulators, TXNIP, MKK
Growth factor signalling, EMT	DLC, HIC1, CRSP3, LKB1, PLCD1, IDs, PDCD4 MTA3
	Transcription factors: twist, snail, slug, ZEB, Sip1 → down regulation E-cadherin
Invasion/metastasis promotion	CADM1 suppresses cell proliferation
	HGF/c-met in cell survival, motility
Cell–cell adhesion mediation	NF-κB in metastasis promotion by S100A4, osteopontin; targets MMPs, uPA, vimentin to promoting invasion; ICAM, VCAM, CD44, ELAM-1 to modulate adhesion properties; VEGF, NOS, TNF, interleukins in modulating angiogenesis
	RAGE interacts with activates NF-κB; and S100 proteins

Note: This table is based on information collated from Sherbet, 2001, 2003, 2011a; Sherbet and Lakshmi, 1997; Sherbet and Patil, 2006; and references cited in the text. Only a few prominent signalling modulators are listed here. The tumour/metastasis promoters and suppressors have been discussed at length in the text. Some of the genes listed as metastasis suppressors function as modulators of cell signalling and therefore discussed separately from the so-called conventional metastasis suppressors.

general scheme and design of discussion and debate here with the major objective that the identification suitable targets might enable one to manipulate their function, modulate, activate or totally inhibit their function as appropriate and to undo deregulated growth, invasion and metastasis of tumours in ways that might be beneficial in patient management.

The major motivation here is to identify and home in on items of therapeutic interest in the biological functioning of high-profile metastasis suppressor and promoter genes, uncover the links and nodes of their signalling pathways, so that the biological effects of these genes can be countermanded directly or indirectly. More often than not one encounters much interactive signalling leading to collateral gene activation or suppression and consequent contribution to the functioning of the intended target. This can aid the major objective, but can also be counterproductive by negating the main channels that are being tackled and generating new and subsidiary problems. In the same vein one can see branching of signalling systems into positive and negative end points. Finally the ubiquity of participation of signalling components makes specific focusing on and targeting the nodes and links in signalling an arduous task. These represent virtually limitless complexities that coerce one into a conservative rather than a comprehensive approach in the identification of molecular targets and to adopt a strategically and tactically pragmatic attitude to their therapeutic deployment.

Part 1

RNA Interference in Genetic Regulation

The regulation of genetic expression by RNA interference is a mechanism conserved in evolution. This involves a mode of suppression by double-stranded RNAs of target gene expression by post-transcriptional silencing by degradation of specific messenger RNAs (Caplen, 2004; Hannon, 2002). The deployment of RNA interference technology as a genetic tool has gained much importance in cell protection and in therapy. The overriding relevance of microRNAs (miRNAs), one of many classes of small RNAs, to cancer biology and pathogenesis is evident from their involvement in tumour suppression as well as promotion of tumour progression. Stem cells represent a clone of cells amenable to the induction of differentiation, capable of self-renewal and possess proliferative potential and believed to be potential targets for therapeutic intervention. Many signalling pathways including Hedgehog, Wnt, 14-3-3, RASSF/Hippo/TAZ, and not least miRNAs, are significant determinants of EMT and embryonic and cancer stem cell characteristics, namely development, self-renewal and maintenance of the pluripotent re-programmable state. So here in focus are the basics and the potential of RNA interference as a therapeutic tool, the biogenesis and association of miRNAs with human disease and the potential, suitability and practicality of miRNAs as targets in cancer management.

1 The Biogenesis and Functions of MicroRNAs

Much attention has been focused on RNA interference as mode of regulating gene expression. Three small RNAs have been identified, namely small interfering RNAs (siRNAs), microRNAs (miRNAs) and the repeat-associated small interfering RNA (rasiRNAs). The PIWI protein-interacting RNAs (piRNAs) are a distinct class of small RNAs differing greatly from miRNAs, but they are similar to rasiRNAs. RNA interference technology is being seriously considered for application in the clinical context.

MiRNAs are significant regulators of many biological phenomena, such as embryonic development and differentiation, regulation of the immune system and the pathogenesis of human disease; a varied role been now established. Unlike miRNAs, siRNAs are believed to regulate gene expression only in organisms which possess RNA-dependent RNA polymerase. So in mammals the biological functions subserved by siRNAs are still uncertain. But not unlike miRNAs, siRNAs have been found to be able to target mRNAs (messenger RNAs) possessing partially complementary binding sites in the 3′ UTR (Doench et al., 2003). Recently Watanabe et al. (2008) showed that endogenous siRNAs do participate in the regulation of gene expression.

MiRNAs may be expressed in a tissue-specific manner and have been implicated in development, differentiation, miRNAs more so than siRNAs; they have been linked with the regulation of the immune system; they participate in cell behaviour related tumour development and progression. Some are regarded as tumour suppressors, often down-regulated in tumour and therefore induced re-expression has been viewed as potential approach to therapy. MiRNAs are also key players in differentiation and pattern formation in early embryonic development. These systems together with neoplasia are characterised by phenotypic cellular changes such as epithelial mesenchymal transition (EMT) (see Sherbet, 2011a). These regulator RNAs influence the expression of oncogenes, suppressor genes, and growth and cell cycle regulator genes among others. These functions of the regulatory molecules have been highlighted and intensively investigated in the past few years.

The miRNAs and siRNAs are approximately 21–26 nucleotides long and possess similar function, but they differ in their modes of biogenesis (Carmell and Hannon, 2004; Kim, 2005). Importantly siRNAs are frequently derived from exons of genes and so match the corresponding mRNAs precisely, whilst miRNAs are derived from intronic sequences (Ambros et al., 2003; Lee et al., 2006; Duchaine et al., 2006; Chapman and Carrington, 2007).

Genes that encode miRNAs are first transcribed into primary miRNA which can form a stem-loop structure. These primary miRNA transcripts are processed by a complex called the microprocessor complex formed of an RNase III Drosha (the catalytic subunit) and the protein DGCR8 (DiGeorge syndrome critical region 8)

Therapeutic Strategies in Cancer Biology and Pathology. DOI: http://dx.doi.org/10.1016/B978-0-12-416570-0.00001-9

dsRNA **miRNA precursor**
↓ **Drosha** ↓ **Drosha**
(Dicer) **(Dicer)**
↓ ↓
siRNA **miRNA**
↓ ↓
Argonaute proteins
↓
↓
Effector Complex
RISC **RISC**
↓ ↓
DNA/histone methylation **mRNA degradation**
translation suppression

Figure 1.1 The biogenesis and functional routes of siRNAs and miRNA. Gene silencing is mediated by the formation of effector complex RISC in which the Argonaute protein is bound to siRNA or miRNA; the effector complex then suppresses the expression of target genes.

(the Pasha protein of Drosophila, the subunit that recognises the substrate) into pre-miRNAs, which are then translocated to the cytoplasm where they are further processed by Dicer (Bernstein et al., 2001; Lee et al., 2003; Denli et al., 2004; Gregory et al., 2004). The components of the microprocessor complex are said to mutually regulate one another, which might be a mode of miRNA biogenesis (Han et al., 2009a). The complete processing of miRNAs might be more complex than thought at one time and might involve steps specific to the maturation of individual miRNAs leading to much diversity in their function (Winter et al., 2009). Similarly siRNAs are also generated by Dicer-dependent processing complex of double-stranded RNA precursors.

The small RNAs form complexes with specific proteins to form the RNA silencing effector complexes, namely siRNA complexes called RISCs (RNA-induced silencing complexes) and miRNPs with miRNAs (Hammond et al., 2001; Meister and Tuschl, 2004). The mechanisms by which these small RNA modulate gene expression differ markedly. SiRNAs seem to be able to suppress gene expression by cleaving and degrading mRNAs that bear sequence identity with them and in this way inhibit protein synthesis (Valencia-Sanchez et al., 2006). They may also be able to suppress transcription of homologous DNA sequences (Grewal and Elgin, 2007; Zaratiegui et al., 2007). They can methylate promoters of genes and in this way suppress expression (Huettel et al., 2007). They can also influence heterochromatin modification. SiRNAs are implicated in heterochromatin assembly and associated chromatin condensation and re-organisation of nuclear domains which make it transcriptionally inaccessible and inactive (Volpe et al., 2002; Wassenegger, 2005; Grewal and Elgin, 2007; Zaratiegui et al., 2007). MiRNAs are non-protein-coding RNAs highly conserved in evolution and display a marked ability to negatively regulate gene expression. As stated earlier, miRNAs are approximately 22-nucleotide long and are double-stranded RNA molecules (Novina and Sharp, 2004; Meister and Tuschl, 2004). They repress translation of mRNAs of target genes. A further major difference between siRNAs and miRNA is that whereas siRNAs are not encoded by specific genes, miRNA are. Also siRNAs may have a viral origin, but miRNAs are totally endogenous (Figure 1.1).

2 Association of miRNAs with Pathogenesis

It is continually being recognised that non-coding RNAs including miRNAs might be associated with the pathogenesis of human diseases; among them are neurological and cardiovascular conditions, developmental abnormalities and tumour development and dissemination (Esteller, 2011). The participation of miRNAs in these processes has been anticipated by their involvement in cell proliferation, apoptosis, determination of cell lineage in haematopoiesis, neuronal patterning, among others, in various living systems (Table 2.1).

The Genesis of DiGeorge Syndrome

DiGeorge syndrome is a congenital condition resulting from defects in chromosome 22, more precisely a 22p11.2 deletion syndrome. The 22q11.2 microdeletion has been reported to occur with altered neurodevelopment and associated cognitive, behavioural and psychiatric disorders, cardiac abnormalities, deficiency of the immune system and proneness to infection, autoimmune conditions, abnormalities of the palate and parathyroid dysfunction (Philip and Bassett, 2011; Halder et al., 2010; Machado et al., 2010; Tison et al., 2011; Veerapandiyan et al., 2011). A vast majority of patients with DiGeorge syndrome show monoallelic deletion of 22q11.2 in 1/3000 live births (Shiohama et al., 2003), and further the deleted chromosomal region happens to contain the DGCR8 gene. But needless it would be to say a number of other genes related to developmental processes might be affected by the deletion. Of note in terms of elucidation of the modes of genesis of human disease is the perceived correlation between miRNAs and incidence of DiGeorge syndrome.

Association of the Glyoxalase Pathway with miRNA Function

Glyoxalase I (GLO1) has been attributed with anti-glycation mediated protection of cells. GLO1 together with glyoxalase II form the glyoxalase system which is an important route to break down of reactive free radicals and detoxification. GLO1 is highly expressed in many tumours, for example, colon, breast and prostate cancer (Ranganathan et al., 1993; Rulli et al., 2001; Davidson et al., 1999). In the past 5 years overexpression of glyoxalase 1 has been reported in melanoma (Bair et al., 2010) and pancreatic cancer (Wang et al., 2012d). Fonseca-Sanchez et al. (2012) found that GLO1 expression in breast cancer was associated with tumour stage.

Therapeutic Strategies in Cancer Biology and Pathology. DOI: http://dx.doi.org/10.1016/B978-0-12-416570-0.00002-0

Table 2.1 Pathogenetic Association of miRNAs in Human Disease

Phenotypic Feature	MiRNA Identifier	Reference
Cell lineage	miRNA-125/lin-4, Let-7	
CNS development, neuron migration	miRNA-103, 107	Mancini et al. (2011)
Bone remodelling	miRNA-29 family	Kapinas and Delany (2011)
Cellular senescence	miRNAs 23a, 26a, 30a	Lee et al. (2011)
Cardiovascular physiology, disease	miRNA-155, miRNA-21, miRNA-126	Urbich et al. (2008)
Tumours/Cell Lines		
Lymphoma (cells nasal)	EBV miRNAs	Ramakrishnan et al. (2011)
MCF-7 (breast cancer)	miRNAs 21, 182, let7-5a overexpressed	Roa et al. (2010)
Du-145 (prostate cancer), U118 (glioblastoma)	miRNAs-145, -155 reduced expression	
Synovial sarcomas and cell	Let-7e, miRNAs 99b, 125a-3p	Hisaoka et al. (2011)
Tumour Promotion		
EMT activation	miRNA-335	Zhang et al. (2012c)
Apoptosis suppression	miRNAs-17-92	Finoux and Chartrand (2008)
	miRNA-21	Carletti et al. (2010), Chan et al. (2005), Zhou et al. (2010b)
Promotion of cell motility	DNA damage/NF-κB → miRNA-21↑	Niu et al. (2012)
Breast cancer	miRNA-21 ↑	Si et al. (2007)
↑ Synovial sarcomas	Let-7e, miRNAs 99b, 125a-3p, among the 21 found to be regulated.	Hisaoka et al. (2011)
↑ Synovial sarcoma, rhabdomyosarcoma and colon cancer	miRNA-183	Sarver et al. (2010)
↑ Gastric cancer/cells	miRNA-223 among 16 miRNAs upregulated. miRNA-223 increased in vitro invasion	Li et al. (2011c)
↑ Gastric primary and metastatic tumours	miRNA-199a	Song et al. (2010a)
↑ oesophageal adenocarcinoma, high-grade dysplasia	miRNA-196a	Maru et al. (2009)
Inhibition of platelet-derived growth factor (PDGF)-induced downregulation of p27Kip1	miRNA-221	Davis et al. (2009)
Downregulation of PDGF	miRNA-219	

(Continued)

Table 2.1 (Continued)

Phenotypic Feature	MiRNA Identifier	Reference
Growth suppression targets MUC4	miRNA-150	Dugas et al. (2010) Srivastava et al. (2011)
Tumour Suppression		
Tumour suppression *in vivo* models	miRNA-34a Let-7f	Tivnan et al. (2011) Liang et al. (2011b)
+ ↓Suvivin and ↑ apoptosis	miRNA-708	Saini et al. (2011)
↓ Breast cancer	miRNA-125b miRNA-183	Zhang et al. (2011b)
↓ NSCLC	miRNA-200c	Ceppi et al. (2010)
↓ Liver cancer	miRNA-122	Tsai et al. (2009)
↓ Pancreatic cancer	miRNA-132	Zhang et al. (2011a)
↓ Astrocytomas grades III and IV	miRNA-124 miRNA-137	Silber et al. (2008)
↓ Oro/pharyngeal carcinoma	miRNA-9	Minor et al. (2012)
↓ Oral squamous cell cancers	miRNA-218, miRNA-585	Uesugi et al. (2011)
↓ Oesophageal cancer	miRNA-145, miRNA-133a and miRNA-133b	Kano et al. (2010)
↓ Colorectal cancer	miRNA-345	Tang et al. (2011a)
↓ Prostate cancer	miRNA-145	Suh et al. (2011)
↓ Endometrial cancer cells	miRNA-152	Tsuruta et al. (2011)
↓ Acute myeloid leukaemia, acute lymphoblastic, chronic lymphocytic, chronic myeloid, non-Hodgkin's lymphoma	miRNA-193a miRNA-124-1	Gao et al. (2011) Wong et al. (2011)
↓ Liposarcomas	Let-7	Finoux and Chartrand (2008)
↓ Gastric cancer	Let-7b, Let-7g Let-7a, Let-7f	Bianchini et al. (2011) Li et al. (2011c) Liang et al. (2011b)
EMT inhibition	Let-7d miRNA-34c miRNA-200s	Chang et al. (2011a) Yu et al. (2012a) Zhang et al. (2012f)
Via upregulation by p53	miRNA-200c	Chang et al. (2011b)
Via downregulation of BMI-1	miRNA-194	Dong et al. (2011)
Induction of apoptosis	miRNA-34a	Tarasov et al. (2007)
Cell cycle arrest in G1	miRNA-34a	
Upregulation by p53	miRNA-34a	
Induction of apoptosis via p21 inhibition	miRNA-101	Semaan et al. (2011)
Induction apoptosis via Akt targeting	miRNA-184	Foley et al. (2010)
↓ Survivin + ↑ apoptosis		

(*Continued*)

Table 2.1 (Continued)

Phenotypic Feature	MiRNA Identifier	Reference
↑Apoptosis, ↓HER2, EGFR, MAPK signalling	miRNA-708 ↑miRNA-451 → ↓14-3-3ζ	Saini et al. (2011) Bergamaschi and Katzenellenbogen (2012)
Angiogenesis		
Inhibition of angiogenesis	miRNA-221 and miRNA-222 miRNA-221 (?) and c-kit ↓ (?)	Kuehbacher et al. (2008) Davis et al. (2009)
Promotion of angiogenesis	miRNAs-17-92; Let-7f, miRNA-27b, miRNA-130a	Kuehbacher et al. (2008), Urbich et al. (2008)
Upregulated by twist, a promoter of angiogenesis	miRNA-223	Chamorro-Jorganes et al. (2011)
VEGF and bFGF regulation; VEGFR2 and FGFR1	miRNA-16, miRNA-424.	
Hypoxia/HIF signalling	miRNA-210	Camps et al. (2008)
↑P53→TSP-1↓	miRNA-194	Sundaram et al. (2011)
Targeting Metastasis-Associated Genes		
Suppressor genes		
PDCD4↓ (programmed cell death 4) protein, PTEN	miRNA-21	Asangani et al. (2008), Frankel et al. (2008), Yang et al. (2011a), Li et al. (2010e)
Maspin ↓	miRNA-21	Zhu S et al. (2008)
CADM1		Zhu S et al. (2008)
EPB41L3	miRNA-223	Li et al. (2011c)
RECK ↓ (reversion-inducing-cysteine-rich protein with Kazal motifs)	miRNA-21 ↑	Wu et al. (2011c), Reis et al. (2012)
Radiosensitivity/Drug Resistance		
	let-7 and miRNA-200 miRNA-214, 130a, miRNA-27a, 451, ↑ miRNAs, P-glycoprotein ↑	van Jaarsveld et al. (2010) Li et al. (2010g)
	Lin28-let7a	Oh et al. (2010)
Topotecan and irinotecan sensitivity	MiRNA-24 MiRNA-21	Gmeiner et al. (2010) Misawa et al. (2010)

Note: ↑indicates miRNAs upregulated expression; downregulation in tumours ↓ is given under tumour suppressors because loss or reduced expression suggests suppressor function.

In gastric cancers GLO1 overexpression correlated with invasion of the gastric wall and nodal metastasis. Significantly, overexpression was inversely related to patient survival (Cheng et al., 2012c). The enhanced GLO1 expression appears to be due to amplification of the GLO1 gene (Santarius et al., 2010). GLO1 has also been attributed with resistance to induction of apoptosis by anticancer agents (Taniguchi et al., 2012). Indeed overexpression has been linked with multidrug resistance.

Enhanced expression of GLO1 increases cell survival. It takes part in the cellular detoxification of reactive carbonyl compounds. The precise mode of its phenotypic effects is still unclear. De Hemptinne et al. (2007) have reported the involvement of GLO1. Indeed, De Hemptinne et al. (2009) showed that GLO1 is a substrate for CaMKII (calcium/calmodulin-dependent protein kinase II). GLO1 also undergoes nitric-oxide-induced post-translational modification. These changes seem to be able to suppress TNF/NF-κB inducible target genes. This could be one of the mechanisms adopted by GLO1 in promoting cell viability survival. Some of NF-κB responsive genes might have relevance to the formation of osteolytic metastasis. GLO1 is possibly a requirement for the generation of osteoclasts and appropriate inhibitors have been identified (Kawatani et al., 2008). For example, inhibition of the regulatory component IKK (IκB kinase) of NF-κB has been found to inhibit the osteoclast activity of NF-κB and inhibit osteolytic metastasis of breast cancer (Sherbet, 2011a). So inhibition of GLO1 could be helpful in preventing osteolytic metastasis. GLO1 is a downstream effector in the functional route of miRNAs and therefore can be targeted by inhibitors. Some miRNAs may counteract and suppress AGE (advanced glycation end product)-induced cell survival. Li et al. (2011b) have identified many miRNAs of rice (Oryza sativa indica) which have been projected to target mRNAs for important protein kinases, peroxidases and glyoxalases. They found that MiRNA-3981 is an exonic miRNA of the first exon of the putative glyoxalase gene and have proposed that its biogenesis pathway might be involved in the post-translational regulation of glyoxalase expression.

An indirect approach to targeted inhibition might be offered by the finding that miRNA-22 can regulate the expression of RGS2 (regulator of G-protein signalling protein) (Muinos-Gimeno et al., 2011), which itself can regulate the function of GLO1. RGS2 seems to regulate GLO1 by activating p38 MAPK and protein kinase C (PKC) signalling systems (Salim et al., 2011).

An exploration of potential inhibitors seems justified by findings that GLO1 expression is altered in many human neoplasms. However the status of expression seems uncertain at present. GLO1 is said to be downregulated in renal cell carcinoma (Cabello et al., 2010), but higher levels of GLO1 transcripts have been reported in primary prostate cancer (Romanuik et al., 2009). Bair et al. (2010) reported a marked upregulation of GLO1 expression in human melanoma (stages III and IV). Inhibition by siRNA of GLO1 expression in A375 and G361 melanoma cells led to inhibition of proliferation and induction of apoptosis. There are also other suggestions subject to the provision of further confirmation that GLO1 polymorphism is associated with breast cancer (Antognelli et al., 2009).

3 Are miRNAs Suitable Targets for Cancer Therapy?

The potential of miRNAs in targeted therapy against cancer was recognised with the finding that loss of certain miRNAs is associated with some forms of leukaemia and solid tumours, but equally miRNAs have been found to be overexpressed in other human neoplasia. Some miRNAs are differentially expressed in tumours and tumour derived cell lines. Possibly, miRNAs might be either suppressors or promoters of tumour development, which would be dependent upon the function of the target genes or proteins. Of this there are numerous examples, where miRNAs participate in the promotion or suppression of tumour by influencing the basic processes involved in tumour development and progression. Not infrequently, members of the same family can exert markedly different and diametrically opposite effects. MiRNAs are known to be able to target several genes that regulate biological processes highly relevant in the pathogenesis of human diseases. However, it has been recognised that even a single miRNA might target a multitude of genes (Bartel, 2009). According to Lal et al. (2011) miRNA-34a alone can regulate hundreds of genes. This equation makes the process of evaluating their relative significance a formidable task despite the advances in technology.

A Resumé of mTOR Signalling

The mTOR (mammalian target of rapamycin) signalling pathway has now pre-eminently associated with several cellular processes such as cell proliferation, growth, apoptosis, angiogenesis, cell motility and invasion. So its aberrant activation provides cancer cells with a huge proliferative and invasive advantage and in this way contribute significantly to the process of cancer metastasis. Recent identification of phosphoinositide 3 kinase (PI3K)/Akt pathway with mTOR signalling has brought growth factors into the arena of its activity. The mTOR pathway integrates oestrogen receptor (ER), epidermal growth factor receptor (EGFR), vascular endothelial growth factor (VEGF) and insulin-like growth factor receptor (IGFR) signalling and could facilitate cross talk between growth factor signalling pathways. The postulated regulation of mTOR by the versatile miRNAs by direct means or via PTEN, modulation of cytoskeletal dynamics, its perceived integration with the function of tumour- and metastasis-suppressor genes has contributed much to emphasise its potential as a therapeutic target. Inhibitors of mTOR signalling might offer potential new devices for the management of triple negative breast cancer (TNBC). With the coverage here concentrating on the modulation of biological response of cancer cells, especially

Therapeutic Strategies in Cancer Biology and Pathology. DOI: http://dx.doi.org/10.1016/B978-0-12-416570-0.00003-2

mediated by miRNAs, it might be appropriate to digress and provide here a resumé of mTOR signalling.

Activation of mTOR signalling involves the formation of two complexes, namely mTORC1 and mTORC2. mTORC1 is a complex of mTOR with Raptor (regulatory-associated protein of TOR) and other components GβL (MLST8), whereas mTORC2 is composed of mTOR, GβL MLST8 and Rictor. The PI3K/Akt activation leads to the phosphorylation of mTOR, which then phosphorylates the downstream targets p70S6K and 4EBP leading in turn to cell proliferation (Foster et al., 2010). Another component of the mTORC complexes is Deptor. It is an inhibitor of mTOR. In the absence of Deptor, active mTORC1 and mTORC2 kinases phosphorylate S6K1, 4EBP1 and SGK1 (a serine/threonine kinase) leading to the phosphorylation of downstream targets resulting in promotion of cell proliferation and survival and inhibition of autophagy (Efeyan and Sabatini, 2010). Compatibly, Deptor occurs at low levels in cancers (Figure 3.1).

miRNAs, Cell Proliferation and Apoptosis

The miRNA family Let-7 members have been extensively investigated for their biological function and the modes of regulation of their function. Let-7 is a tumour suppressor which is frequently downregulated in cancer (Finoux and Chartrand, 2008). But Let-7 family are not across the board tumour suppressors. Let-7e has been reported to be upregulated in synovial sarcomas. Furthermore, experimentally downregulating Let-7e (and also of miRNA-99b) has resulted in the suppression of cell proliferation. Arguably cell proliferation seems to be targeted by miRNAs. They show demonstrable effects on c-myc, but their influence on the expression of members of the apoptosis gene family is not clear. One would have liked to see if the effects on c-myc are translated phenotypically via cell proliferation regulation by gadd45, cyclins, cdc25A and other, or regulation of apoptosis through the function of anti-apoptosis Bcl-2 and Bcl-XL or pro-apoptosis genes Bax, Bad, Bak, etc. Nonetheless, it would be evident from the discussion below that the influences of miRNAs on some determinants of cell proliferation and apoptosis have been addressed.

In C5 molecular subtype of high-grade serous ovarian cancer, marked changes occur in the expression of N-myc, Lin-28B, Let-7 and HMGA2 (the high mobility group A2). Characteristic amplification and overexpression of N-myc, and overexpression of its targets Lin-28B together with loss of Let-7 expression and amplification and overexpression of HMGA2 protein frequently associated with tumour invasion and progression have been encountered (Helland et al., 2011). The expansion of cell population can also result from the inhibition of apoptosis. MiRNA-21 which is highly expressed in certain tumours has been reported to be anti-apoptotic (Carletti et al., 2010; Chan et al., 2005) as well as being able to actively induce cell proliferation (Asangani et al., 2008; Roldo et al., 2006; Si et al., 2007). It appears to be associated with promotion of invasion being able to suppress reversion-inducing-cysteine-rich protein with Kazal motifs (RECK) (Reis et al., 2012).

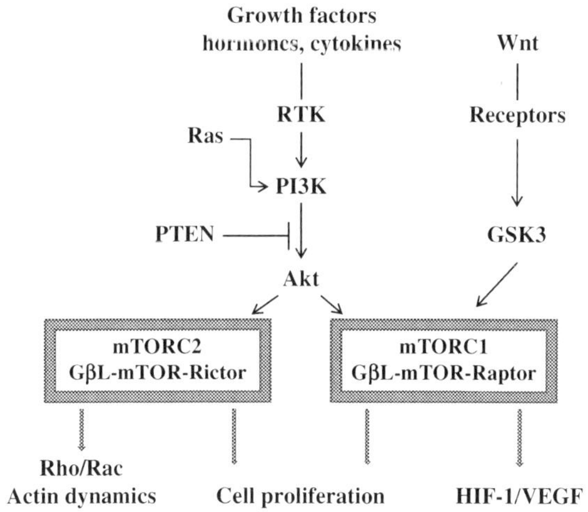

Figure 3.1 An abbreviated and simplified representation of the mTOR signalling pathway. The PI3K/Akt/mTOR pathway is the predominant pathway activated by growth factors and other biological behaviour modulators to bring about in cellular proliferation, growth, apoptosis and cell motility. Activation of mTOR signalling involves the formation of ternary complexes mTORC1 and mTORC2; mTORC1 in association with Raptor and GβL (G-protein beta-subunit-like protein) (MLST8 in HUGO [Human Genome Organisation] nomenclature), whereas the second complex mTORC2 contains GβL and Rictor (Rapamycin insensitive companion of mTOR). Both mTOR complexes negatively regulate Deptor. As shown in the figure, mTORC1 can activate VEGF signalling. This has been attributed with the ability to modulate actin dynamics through Rho/Rac and so might modulate cell invasion. The figure also shows how Wnt signalling might modulate cell behaviour by mTOR signalling. This representation is based on references that are cited in the text (Sarbassov et al., 2005; Efeyan and Sabatini, 2010; Sherbet, 2011a).

RECK is a negative regulator of MMPs. MiRNA-21 might influence MMPs by this means. Suppression of RECK would confer oncogenic properties on miRNA-21. Downregulation of its expression using anti-sense strategy has led to inhibition of glioma cell proliferation and to the induction of caspase-mediated apoptosis (Zhou et al., 2010b). MiRNA-21 can inhibit PTEN/Akt pathway and the pro-apoptosis Fas/FasL signalling. Also downregulation of miRNA-21 can upregulate FasL and PTEN and activation of Akt reverses this effect (Sayed et al., 2010). Foley et al. (2010) have demonstrated by some elegant experimentation that miRNA-184 is a pro-apoptosis miRNA that targets and inhibits Akt. They showed that N-myc inhibited miRNA-184 and increased Akt levels. They then co-transfected miRNA-184 with a vector carrying active Akt that lacked the miRNA-184 target site and found that this eliminated the pro-apoptosis effects of the miRNA. The recourse to PTEN signalling is also evident from the effects of miRNA-9 in oral/pharyngeal carcinomas. MiRNA-9 is suppressed by methylation in tumour tissue and 5-aza-deoxycytidine enhanced its expression. Furthermore, when it was transfected into tumour cell lines, PTEN expression was upregulated (Minor et al., 2012).

MiRNA-200 induces apoptosis but uses a different pathway possibly the CD95/Fas pathway. Zaman et al. (2010) found that transfection of miRNA-145 into PC-3 prostate carcinoma cells caused increased apoptosis. The miRNAs-17–92 are said to promote the development of myc-induced murine B cell lymphoma, by obliteration of mRNAs coding for proteins promoting apoptosis. The bcl-2 family genes which regulate apoptosis have been described as targets of certain miRNAs. The tumour-suppressor PTEN/Akt pathway has also been cited as an miRNA target. Suppression of PTEN and inhibition of apoptosis is one of the mechanisms (Bar and Dikstein, 2010, Zhang et al., 2010a, 2012a).

The regulation of anti-apoptosis gene bcl-2 might also be involved in some instances (Si et al., 2007). Cell proliferation might be influenced by other pathways that include altering the function of cell cycle regulatory proteins. MiRNAs have been reported to function as tumour suppressors by targeting c-myc, the cell cycle regulator cyclin A1 and the homeobox protein Six1. Six1 is targeted by miRNA-185 (Imam et al., 2010). Davis et al. (2009) reported upregulated expression of miRNA-221 with downregulation of p27kip1 in the induction of proliferation by PDGF. It is evident from recent work that miRNAs might function by regulating cyclin-dependent kinases (cdks). Zheng et al. (2011) found that miRNA-21 promotes proliferation and invasion by subduing p12 (CDK2-AP1). The latter suppresses cell proliferation by inhibiting cdk2 and G1/S transition. They have shown that miRNA-21 targets p12 (CDK2-AP1) by interaction with specific motifs of its 3′-UTR. MiRNAs might function via the cdc42 (cell division control) protein signalling, a Rho-type-like GTPase that is regulated by G1-cdk. Cdc42 is overexpressed in many tumours and activating mutations have been shown to lead to cell transformation, tumorigenesis and invasion (Stengel and Zheng, 2011). MiRNAs might use the cdc42 route to facilitate transduction of signals from RTKs, G-protein-coupled receptors (GPCRs) and integrins. Cdc42 activates downstream pathways regulated by PAKs (p21-activated protein kinases), namely MAPK/PAK3-ERK (extracellular signal-regulated kinase) promoting cell proliferation and also PAK1–JNK (c-Jun N-terminal kinase) pathways that influence apoptosis. With its ability to modulate cytoskeletal organisation cdc42 possibly could also modulate invasive behaviour. It is interesting therefore to note that MiRNA-137 is down-regulated in gastric carcinoma cells in association with upregulating of cdc42. When levels are restored miRNA-137 appears to target cdc42 and inactivates it leading to G1 arrest and apoptosis (Chen et al., 2011b).

It would be of much interest to note that activation of p53, a tumour suppressor and key regulator of cell cycle progression, was shown some while ago to upregulate the expression of some, notably of miRNA-34a and downregulate the expression of other miRNAs. Indications are that this miRNA might induce apoptosis and arrest cell cycle progression in G1. Also of interest is that p53 appeared to bind to a site proximal to the first non-coding exon of the miRNA (Tarasov et al., 2007). Transfection into the colorectal cancer HCT-116 cells and experimental upregulation of expression of miRNA-211 halved the expression of CHD5 (chromodomain helicase DNA-binding protein 5), a postulated tumour suppressor, and resulted in enhanced cell proliferation. Forced expression of miRNA-21 enhanced Bcl-2 and Bcl-xL by around 30% and reduced Bad gene expression by approximately 20% (Cai et al., 2012).

Another potential mode of miRNA function in promoting tumour progression has emerged from some recent studies, which show interaction between miRNAs and growth factor signalling. Several growth factor signalling pathways have been identified as being able to interact with miRNAs. Of considerable interest in this respect are the EGFR family. Of these EGFR and HER2 (human EGFR 2) signalling has received some attention. EGFR is highly expressed in a proportion of breast cancers, which also tend to be ER negative. Wang et al. (2012c) have addressed the question of the interaction of miRNAs in EGFR-mediated signalling via the PTEN/Akt signalling system in the pathogenesis of gliomas. Investigating the role of miRNA-143 in prostate cancer cells and tumour tissues, other studies have not only shown the involvement of PTEN/Akt system but also that PTEN mediation inactivates mTOR and its downstream effectors in response to the presence of certain miRNAs. For instance, Fornari et al. (2010) have suggested that miRNA-199a-3p might inhibit mTOR to generate its phenotypic effects. Indeed according to Uesugi et al. (2011) in oral squamous cell carcinomas, miRNA-218 and miRNA-585 are often silenced by methylation. But in experimental systems forced expression of miRNA-218 was found to inhibit mTOR.

Xu et al. (2011a) describe an inverse correlation of expression between miRNA-143 and K-Ras protein expression and further state that overexpression of the miRNA inhibited cell proliferation and migration by inactivating the EGFR/Ras/MAPK pathway. Downregulation of miRNA-181a is frequently encountered in oral squamous cell carcinoma cells. Forced expression of the miRNA inhibits cell proliferation and anchorage-independent growth. In this system miRNA-181a decreased the expression of K-Ras (Shin et al., 2011). MiRNAs are also known to modulate signalling by JAK/STAT/Akt and MEK/ERK signalling to effect cell proliferation. The membrane TRK c-Kit receptor for the cytokine stem cell factor (SCF), the kit receptor ligand, is involved with the regulation of cell survival and proliferation, mast cell biology and haematopoiesis, etc. and transduces signals via the PI3K and MAPK/ERK pathway to STAT transcription factors. MiRNA-193a was found to be downregulated by promoter hypermethylation in acute myeloid leukaemia (AML) cell lines and primary AML blasts; levels of expression of the miRNA showed a negative correlation with c-Kit receptor expression in leukaemia cell lines and primary AML (Gao et al., 2011). Liu et al. (2011b) found miRNA-125b frequently downregulated in osteosarcoma samples and human osteosarcoma cell lines. In the cell lines cell growth was suppressed and *in vitro* cell motility was inhibited when levels of the miRNA were restored. They also showed that STAT3 was a direct downstream target of miRNA-125b. On the other hand, in a feedback loop STAT3 itself seemed capable of regulating transcription and regulating the levels of miRNA-125b.

The GPCRs are an important family of transmembrane receptors which are known to co-ordinate and mediate several pathways of signalling, for example through the RTK/PLC/PKC and PKA systems leading to genetic transcription and promotion of cell proliferation. GPCR also regulates cell survival or apoptosis through PI3K, with the mediation of genes such as apoptosis-inducing Bax and inhibitory Bcl2 genes (Sherbet, 2011a). GPCRs are able to function with EGFR to promote cell proliferation and migration. G-proteins convey GPCR signals by dissociating into subunits. GNAI2 (guanine nucleotide-binding protein alpha inhibiting activity polypeptide 2) is said to be targeted by miRNA-30d. The enhanced expression of this miRNA in hepatocellular carcinoma (HCC) is said to correlate with greater cell migration *in vitro* and with incidence of liver metastases (Yao et al., 2010). Most of these findings are interesting *per se*, but they need to be followed up further to elucidate how and which signalling systems the miRNA modulates.

HER2 is highly expressed in a major proportion of human breast cancer and these receptors have been targeted by Herceptin (humanised antibodies against HER2) in breast cancer treatment. From this view point it is significant that miRNAs should interact with and influence HER2 signalling. Lin28 which inhibits the suppressor function of Let-7 miRNA seems able to upregulate the expression of HER2 and in this way promote cell proliferation and tumour growth. This is compatible with the fact that Lin28 correlated with poor prognosis of breast cancer (Feng et al., 2012). High expression of Lin28 is related to highly aggressive oesophageal cancer and also with poor prognosis (Hamano et al., 2012). Huang et al. (2009) reported that the expression of miRNA-21 correlated with HER2 upregulation. Furthermore, activation of HER2 and the MAPK (ERK1/2) pathway upregulated miRNA-21. In different mode, the miRNA downregulated PDCD4, the suppressor called the programmed cell death protein, in HER2+ cells. Thus miRNA-21 might promote tumour progression by HER2-mediated stimulation of invasion and by preventing induction of apoptosis by PDCD4 (also refer to discussion on pp. 209–212). The upregulation of HER2 should make tumours sensitive to Trastuzumab (Herceptin). On the contrary, upregulation of miRNA-21 has been shown to lead to the development of resistance to Trastuzumab therapy of breast cancer (Gong et al., 2011). Possibly, this is a consequence of lack of saturation binding of HER2 by Trastuzumab. High miRNA-21 could be counteracting the blockage of EGFR by Trastuzumab by activating EGFR/STAT signalling with apparent increase of resistance to Trastuzumab. In gastric cancer reduced expression of miRNA-125a-5p has been linked with larger tumours, enhanced invasion and metastasis to liver and poor prognosis. The miRNA appeared to suppress cell proliferation *in vitro* by directly targeting HER2. This inhibitory effect was compounded by Trastuzumab (Nishida et al., 2011).

MiRNA-21 seems to be able to control growth and apoptosis of glioblastoma cells. Ren et al. (2010b) tested the effects of miRNA-21 inhibition in glioblastoma U251 (PTEN-mutant) and LN229 (PTEN wild-type) cells. Inhibition of miRNA-21 rendered both cell types sensitive to Taxol. Also both cell types showed enhanced apoptosis. Therefore the authors have suggested that the miRNA inhibitor might be inhibiting EGFR/STAT3 signalling and diminishing the anti-apoptotic effect of the miRNA and apparently in this way render the cells sensitive to Taxol quite independently of PTEN function (Ren et al., 2010b). Also in the loop are ER and PR which not only regulate growth of tumours, but in breast cancer in the absence of ER a proportion of tumours seem to engage EGFR signalling. ERα is required for oestrogen-dependent growth and ERα positivity correlates with enhanced proliferation. Many miRNAs have been found to downregulate the expression of ERα, among them are miR-18a and b, miR-302, miR-193b, miR-22, and miR221/222 as well as miR-206 which have target sites in ERα 3′UTR and inhibit the promotion of proliferation by oestrogen (Leivonen et al., 2009; Pandey and Picard, 2009). On the other hand, the biogenesis of some miRNAs may be suppressed by ERα.

In this context it would be of interest to note miRNA-26a can regulate ERα and PR. Using HCC cells, Chen et al. (2011a) have shown that miRNA-26a can downregulate ERα and inhibit growth promotion by oestradiol. In the opposite mode of function in endometrial adenocarcinoma activated ER are said to upregulate the expression of tumour-suppressor Let-7 family miRNAs and miRNA-27a leading to the suppression of Bax and so increase bcl2/Bax ratio and promote cell proliferation and enhance cell survival. Seven Let-7 family members and miRNA-27a that were upregulated by ER were able to bind Bax transcripts but not Bcl2 transcripts (Zhang et al., 2012b). Although these authors have made a persuasive case, one cannot reconcile this with the tumour-suppressor properties of Let-7miRNAs. One could be losing sight here of the fact that ERα and ERβ exert opposite effects on proliferation; so it is essential to know which ER is involved in the given experimental conditions. Also Let-7miRNAs suppress stem cell pluripotency and oppose the effects of miRNA-290 cluster which promote pluripotency. This functional disparity is due to the ability of miRNA-290 to maintain the expression of Lin28 which inhibits the maturation of Let-7. The pluripotency transcription factors OCT4, SOX2, REX1, and NANOG induce miRNAs that promote pluripotency and cell survival, whilst Let-7s inhibit pluripotency (Li and He, 2012).

Bao et al. (2011b) found that CD44 binds to HER2 and inhibits the expression of mi-RNA139, which in turn downregulates the expression of CXCR4 (C-X-C chemokine receptor type 4). This chain of events has been attributed with loss of the invasive phenotype. These authors state that HER2 and CD44 are the initiators of invasive activity of cultured gastric cancer cells and loss of their expression suppresses tumour growth. In human gastric cancer, high levels of HER2, CD44, and CXCR4, and reduced levels of miRNA-139 correlated with metastasis to the lymph nodes.

Among other systems examined is the IGFR signalling pathway. MiRNA-223 reportedly targets IGF-1R and its downstream signalling pathway and inhibits cell proliferation (Jia et al., 2011). In a similar vein, miRNA-375 seems to target IGF-1R to achieve inhibition of growth and metastasis of ESCC (oesophageal squamous cell carcinoma). MiRNA-375 is downregulated by promoter methylation in primary ESCC, and furthermore the downregulation has significantly correlated with advanced stage, the presence of distant metastasis ($p<0.0001$) and poor prognosis (Kong et al., 2012). MiRNA-470, miRNA-669b and miRNA-681 also target and suppress IGF-1R (Liang et al., 2011a).

Hepatocyte growth factor (HGF) has been found to downregulate the expression of miRNA-200c and miRNA-27b. Non-small cell lung carcinoma (NSCLC) have been reported to show downregulated expression MiRNA-200c and the expression status was associated with poor differentiation and high lymph node metastases. The aggressive nature is compatible with the reduced expression of E-cadherin (Ceppi et al., 2010). Expression of the transcription factor ZEB1 (zinc-finger E-box 1), a target of miRNA-200c, was upregulated with further downstream downregulation of E-cadherin (Susuki et al., 2011). Earlier, Tryndyak et al. (2010) demonstrated that miRNA-200 transcriptionally upregulated E-cadherin by ZEB1 repression. This related closely with the inhibition of EMT and the acquisition of a less aggressive phenotype. EGF/EGFR signalling actively initiates EMT and this is associated with reduced expression of miRNA200s. Also, induced restoration of this miRNA is able to block EGF-mediated induction of EMT and lead to reversal of the EMT phenotype (Zhang et al., 2012f).

One should include a comment here that the c-met receptor of HGF is also known to be targeted by miRNAs. Upregulating the expression of miRNA-199a-3p has been found to target c-met and reduce the invasive ability of HCC cells (Fornari et al., 2010). Tan et al. (2011) have shown that miRNA-198 directly targets c-met and binds to a single cognate binding site of c-met 3′UTR and regulates its expression. It inhibits HGF/c-met signalling of HCC cell migration in Matrigel *in vitro* assays.

The angiogenic growth factors VEGF and basic fibroblast growth factor (bFGF) both seem to upregulate the expression of miRNA-16 and miRNA-424. The expression of primary transcripts primary-miRNA-16-1 and primary-miRNA-16-2, but not of primary-miRNA-424, was upregulated. Thus both growth factors seem to regulate transcription as well as post-transcription maturation process of these miRNAs. Activated ERα has been reported to inhibit the maturation of primary-miRNA into pre-miRNA (Yamagata et al., 2009) possibly by impinging upon the function of the Drosha/Pasha complex. On the other hand, this may not apply as a general rule, for there are instances where ER do upregulate miRNAs.

It would appear that overexpression of miRNA-16 or miRNA-424 might regulate VEGF and bFGF function in terms of angiogenic signalling by targeting VEGFR2 and FGFR1 (Chamorro-Jorganes et al., 2011). The operation of this upregulatory loop can lead to the amplification of a primary angiogenic signal with the aid of miRNAs. MiRNA-194 seems to influence angiogenesis by a p53/TSP-1 pathway. Sundaram et al. (2011) have shown that p53 upregulates miRNA-194 expression and this then suppresses TSP-1 levels resulting in the promotion of angiogenesis *in vitro*. The suppression of TSP-1 is mediated by a single complementary site in the THBS1 gene which encodes TSP-1.

In a preliminary report, Porter et al. (2011) have implicated transforming growth factor beta (TGF-β) signalling in the regulation of miRNA-21 in glia and neurons. Hao et al. (2011) have found that miRNA-483-3p is highly expressed in pancreatic cancer. They have also provided convincing evidence that forced enhancement of miRNA-483-3p in pancreatic cancer cell lines is able to downregulate the expression of Smad4 and also promote cell proliferation. MiRNA-128a has been shown to bind to the 3′UTR region of the gene encoding the TGF-βRI receptor protein and negatively regulate its expression in MCF breast cancer cells (Masri et al., 2010).

miRNAs in EMT, and Cell Motility and Invasion

The establishment of the possible involvement of miRNA in metastatic spread has touched many steps of metastatic cascade. The approach to this has included investigating if the biological faculty of invasion and cell motility, induction of angiogenesis and metastasis associated genes are targeted by miRNAs. Invasion is a prerequisite for the secondary spread of cancer. Some of the tumour promoter miRNAs have been found to enhance cell motility. Li et al. (2011c) have claimed that cell lines derived from metastatic gastric cancers overexpressed miRNA-223 and indeed stimulated non-metastatic gastric cancer cells to migrate and invade. This is highly significant given that miRNAs can accentuate and augment angiogenic signalling, as discussed earlier.

MiRNAs actively influence EMT. Indirect support is also adduced by the fact that miRNA-223 is inducible by the basic helix-loop-helix (bHLH) transcription factor Twist which is associated with EMT and embryonic differentiation and which has been deeply correlated with regulation of genes associated with metastasis. The transcription factors Twist, Snail, ZEBs, etc. are negative regulators of E-cadherin expression. Loss of E-cadherin expression initiates EMT. Downregulation of Twist has led to the inhibition of EMT (Yu et al., 2011). The miRNA Let-7d expression inversely correlated with the expression of Twist and Snail in oral squamous cell carcinoma cell lines and inhibited EMT (Chang et al., 2011a). MiRNA-21 seems to be involved in the induction of EMT by TGF-β. It is often upregulated in cancers and promotes TGF-β-mediated activation of EMT (Zavadil et al., 2001). Forced downregulation of miRNA-21 has been shown to promote apoptosis and inhibit cell motility of ovarian carcinoma cells *in vitro* (Lou et al., 2011). This could quite possibly involve p53, although not demonstrated to be the case here. Some miRNAs are regulated by p53; p53 can and does upregulate miRNA-200c and inhibit EMT (Chang et al., 2011b). P53 also upregulates miRNA-34a. Upon being induced to re-express in human pancreatic cancer stem cells (CSCs) and in human pancreatic cancer cell lines miRNA-34a inhibited cell proliferation, cell cycle progression, EMT and invasion (Nalls et al., 2011). Enforced expression of miRNA-34c in breast cancer cell lines has been shown to inhibit the activation of EMT (Yu et al., 2012a). Others such as miRNA-200 and miRNA-205 are said to be downregulated in EMT stimulated by TGF-β. MiRNA-200 can transcriptionally upregulate E-cadherin by ZEB1 repression and apparently in this way inhibit the EMT phenotype (Tryndyak et al., 2010). The Snail-mediated regulation of EMT activation seems to involve miRNA-661. Vetter et al. (2010) have suggested that this process involves downregulation of the immunoglobulin-like adhesion molecule Nectin-1 through the upregulation of miRNA-661. The expression of Let-7f is reduced in gastric cancers, and upon transfection into cell lines significantly suppressed cell invasion and the transfected cells also showed reduced ability to form tumours when xenografted into immune-compromised mice (Liang et al., 2011b). These authors state that Let-7f directly targets and binds to the 3′UTR of MYH9 (myosin-9 also known as cellular myosin heavy chain, type A) gene. It should be noted here many functions that could impinge upon cell motility and invasion have been attributed to MYH9 such as the reorganisation of the actin cytoskeleton. So the targeting of MYH9 could be important. MiRNA-335 has been found to be able to upregulate BRCA1 expression (Heyn et al., 2011). This is in agreement with the view that miRNA-335 is an inhibitor of metastasis. MiRNA-335 is said to suppress cell migration and lung colonisation in the less than ideal assay for metastasis upon tail vein injection. This in fact seems to occur by inhibiting SOX4, an activator of EMT (Zhang et al., 2012c).

Some years ago, Yang et al. (2004) demonstrated the involvement of Twist in EMT as indicated by the expression of EMT markers together with alterations in E-cadherin and acquisition of cell motility. Further definitive evidence is now available implicating Twist in EMT. Twist overexpression is accompanied in parallel by increased cell motility and alteration in the expression of E-cadherin, N-cadherin and MMP-9 expression (Yu et al., 2012b). Jouppila-Matto et al. (2011) examined a series of pharyngeal carcinomas and noted that Twist and Snail expression was associated with stromal component of higher grade tumours, and these recurred more frequently. They also noted that Twist and Snail negative tumours reflected favourable prognosis. Similar relationship was reflected in gastric and renal cancers also (Ru et al., 2011; Mikami et al., 2011).

We now know that besides Twist, the transcription factors such as Snail (SNAI1 Snail 1 Drosophila homologue), Slug, ZEB and Sip1 are also involved in EMT. MiRNA-205 has been reported to suppress EMT by inhibiting transcription factors ZEB1 and ZEB2, which are known to promote EMT. MiRNA-138, which is capable of inhibiting EMT, has been shown to target ZEB2 and regulate E-cadherin expression (Liu et al., 2011e). The miRNA-200s and miRNA-205 block the repression of E-cadherin expression (Burk et al., 2008; Gregory et al., 2008; Korpal et al., 2008; Park et al., 2008). MiRNA-200 family can transcriptionally upregulate E-cadherin accompanied by ZEB1 repression and inhibit EMT generating a less aggressive phenotype (Tryndyak et al., 2010). Not only this, Twist induces the expression of Snail and this is essential for Twist-mediated activation of EMT (Casas et al., 2011). PDGF overexpression in PC3 cells has been found to activate EMT. This seems to result from the suppression of miRNA-200s. This leads to the upregulation of ZEB1, ZEB2 and Snail2, with consequent repression of E-cadherin transcription (Kong et al., 2008, 2009). ZEB1 recruits the co-repressor BRG1 (Sánchez-Tilló et al., 2010). Blocking this interaction induces the expression of E-cadherin and inhibits EMT. Logistically this membrane associated disposition of the ZEB1/BRG1 complex and E-cadherin could be a mechanism linking EMT with tumour invasion. Whist in many cell types miRNA-200 suppresses EMT by upregulation of E-cadherin expression, it is the overexpression of miRNA-375 that inhibits migration in gastric cancer cells. This seems to occur by a different route via inhibition of JAK2 (Ding et al., 2010). MiRNA-194 has been shown to inhibit EMT by upregulating the expression of E-cadherin and downregulating that of Vimentin. MiRNA-194 suppresses the transcriptional repressor BMI-1, a Polycomb group (PcG) gene by directly binding to its 3′-UTR (Dong et al., 2011). Thus miRNAs might achieve the same phenotypic effect by targeting different genes, possibly through the function of different transcription factors. It should also be borne in mind in this context that many miRNAs are downregulated and others upregulated (Angeles et al., 2011), but whether they induce the expression of specific transcription factors needs to be fully understood. Twist expression shows marked correlation with tumour associated microvascular density in HCCs (Che et al., 2011), which indeed adds another twist, to use a well-worn cliché, to the argument. Snail expression has been linked with choroidal neovascularisation (Hirasawa et al., 2011). One should include a comment here that the c-met receptor of HGF/scatter factor is also known to be targeted by miRNAs. Upregulating the expression of miRNA-199a-3p has been found to target c-met and reduce invasion ability of HCC cells (Fornari et al., 2010).

Modulations in cell mobility and the acquisition of invasive potential are an inevitable consequence of alterations in intercellular adhesion and the transduction of signals imparted by cell membrane associated adhesion molecules to the cytoskeletal system. As referred to above, miRNAs have been shown to alter the expression of cadherins. E-cadherin is an established suppressor of cell motility and invasion. It is downregulated in cancers and this correlates with acquisition of invasive ability and metastatic spread. Equally, miRNAs can target molecules such as Ezrin, the product of the Vil2 gene. Ezrin links up with the actin cytoskeleton and by virtue of this participates in cell motility, morphogenesis and intercellular adhesion. When overexpressed it promotes anchorage-independent growth *in vitro*, and motility and invasion. Downstream it regulates signalling pathways of PI3K/Akt/MAPK and ERK1/2, and so its function has clear implications for apoptosis and cell proliferation. Therefore the targeting of Ezrin by miRNAs is of some considerable significance to the life of the cell and its biological behaviour. Lowery et al. (2010) found miRNA-183, which is downregulated in breast cancer, targets Ezrin. *In vitro* also overexpression of miRNA-183 has led to the downregulation of Ezrin.

MiRNAs do affect other adhesion molecules such as CADM1 and EPB41L3 (see pp. 134–135), which interact downstream with actin elements to regulate cell motility. MiRNAs are known to modulate the expression of ECM components and regulate adhesion mediated processes including EMT, and cell motility and invasion. As stated on many previous occasions the transcription factor Snail 1 promotes the activation of EMT. Vetter et al. (2010) have suggested that this process involves downregulation of another immunoglobulin-like adhesion molecule Nectin-1 through the upregulation of miRNA-661 (Vetter et al., 2010). VCAM-1 mediates leucocyte adhesion and is possibly involved in this way in diapedesis across endothelia and also known to influence the processes of metastasis. This is another adhesion molecule which is also regulated by miRNAs. VCAM-1 expression in endothelial cells is regulated by miRNA-126. Overexpression of miRNA-126 has been correlated with increased expression of VCAM-1, whilst TNF-α-stimulated VCAM-1 expression is inhibited in endothelial cells transfected with anti-sense miRNA-126 (Harris et al., 2008).

Finally, the hyaluronate receptor glycoprotein CD44 which participates in cell adhesion and of which the splice variant CD44v6 participates also in metastasis (Lakshmi and Sherbet, 1997) is directly targeted by miRNA-34a (Liu et al., 2011a).

miRNAs and Tumour Angiogenesis

Angiogenesis is another essential requirement for tumour growth and spread. MiRNA-221 and 222 have been implicated in the inhibition of angiogenesis on the basis that they block endothelial cell migration, proliferation and angiogenesis *in vitro*, either by targeting the SCF receptor c-Kit or by altering the expression of endothelial nitric oxide synthase; whilst in contrast, miRNA-17–92 promotes tumour angiogenesis *in vivo*. Let-7-f, miRNA-27b and miRNA-130a are believed to be pro-angiogenic *in vitro* (Kuehbacher et al., 2008, Urbich et al., 2008). The role of miRNA-27b requires further study. Fang et al. (2011) found that this miRNA inhibited angiogenesis. Their work has indicated that miRNA-29b directly targeted MMP-2. Fang et al. (2011) have reported that experimental tumours derived from HCC expressing miRNA-29b showed reduced tumour associated microvessel density. Davis et al. (2009) noticed that miRNA-221 occurred in parallel with downregulation of ç-Kit, which is normally linked with induction of angiogenesis. Certain miRNAs, for example miRNAs-17–92 and miRNA-378, might show specific association with tumour angiogenesis. Of note would be the association of miRNA-223 which is inducible by the bHLH transcription factor Twist, itself linked with EMT and embryonic differentiation and which has been deeply correlated with angiogenesis.

The mTOR signalling pathway is closely involved in growth factor receptors function (Figure 3.2). The PI3K/Akt activation leads to the phosphorylation of mTOR, which then phosphorylate the downstream targets p70S6K and 4EBP leading in turn to cell proliferation. Angiogenesis might be regulated by mTOR signalling as Grundmann et al. (2011) have shown. MiRNA-100 regulated mTOR expression *in vitro* and *in vivo*. It targeted mTOR in vascular cells and downregulated its expression and inhibited endothelial cell proliferation and so attributed with anti-angiogenic properties (Figure 3.3).

Hypoxia is another major regulator of angiogenesis. HIF-1α (hypoxia-inducible factor-1α) is a transcription regulator in hypoxia. Signals transduced via RTK/PI3K/Akt into PDK1/PKC or mTOR and downstream to the phosphorylation of S6K1/EBP1 can lead to HIF-1α protein expression. HIF-1α is recognised and degraded by VHL (von Hippel–Lindau tumour-suppressor protein), which is a component of the E3 ubiquitin ligase complex. In hypoxic situations VHL is inactivated and HIF-1α is stabilised and accumulates. This then can activate VEGF to induce angiogenesis and TGF-α and promotes cell survival (Pantuck et al., 2003). Hypoxia involves miRNAs

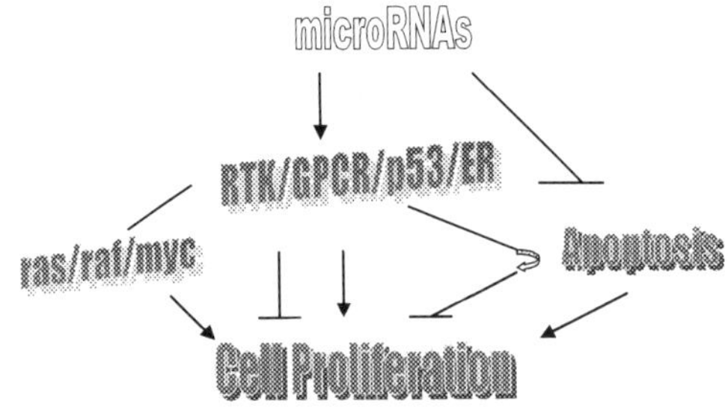

Figure 3.2 A general notion of the instigation by miRNAs of the regulation of cell proliferation via RTKs, GPCRs and the nuclear ER. Also shown is the involvement of p53 in controlling cell proliferation by activating the apoptotic pathway, which is also modulated by miRNAs. It ought to be borne in mind here that some miRNAs may promote apoptosis. This is not shown in this figure. Further details of the signalling systems are provided in Figures 3.3–3.7 and described in the text.

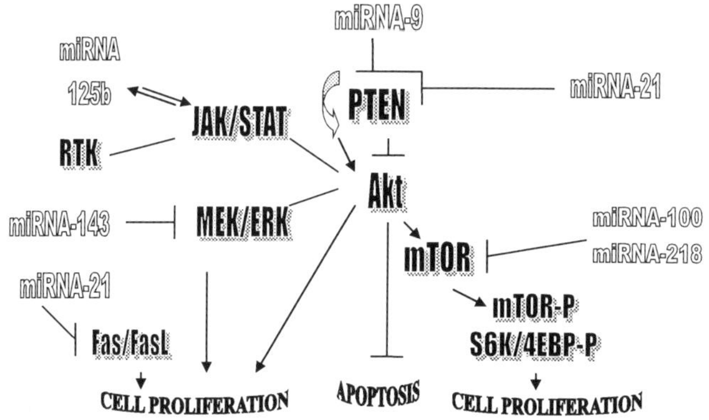

Figure 3.3 A chart of PTEN/Akt pathway in which PTEN inhibits Akt and promotes apoptosis. Interacting with PTEN/Akt are JAK/STAT, MEK/ERK and mTOR pathways. Fas receptor of the TNF receptor family upon being activated by binding of FasL induces apoptosis. MiRNA-21 inhibits Fas/FasL and can also inhibit PTEN promoting cell proliferation. Survivin, an inhibitor of apoptosis, is believed to be suppressed by miRNA-708 leading to the promotion of apoptosis. The figure also shows how these signalling systems are inhibited or mutually modulated by certain miRNA. References are provided in the text.

in two ways. Hypoxia induces many miRNAs, for example miRNAs-23, -24, -26, -27, -103, -107, -181, -210 and -213; at least some of them via an HIF-1-dependent mechanism (Kulshreshtha et al., 2007). HIF1-α activity is involved in the promotion of angiogenesis by certain miRNA, for example miRNA-210. In some experimental systems such as MCF-7 breast cancer cells hypoxia induced an overexpression of miRNA-210 and this was mediated by HIF-1α/VHL (Camps et al., 2008). The authors also reported a negative correlation of disease-free and overall survival with miRNA-210 levels. Hypoxia induces miRNA-424 in an HIF-dependent manner and promotes angiogenesis (Ghosh et al., 2010). On the other hand, hypoxia has been shown to be able to reduce the expression of miRNA-34a with promotion of EMT by targeting Notch signalling (Du et al., 2012).

The ADAM (a disintegrin and metalloprotease) proteins are Zn-dependent metalloproteinases which release the extracellular domains of membrane-bound growth factors, cytokines and their receptors. MMPs and ADAMs have similar domain structure; they both possess a N-terminal propepetide region and a Zn catalytic domain. MMPs have a C-terminal hemoplexin-like module. The ADAM proteins possess a disintegrin domain C-terminal to the catalytic domain, then a cysteine-rich domain and an EGF-repeat domain. ADAMTS has a TSP-1 (thrombospondin-like) domain interposed between the disintegrin and cysteine-rich domains (Wee-Yong et al., 2001; Paulissen et al., 2009).

ADAMs have been ascribed a potential role in angiogenesis being present in endothelial cells. Inhibition of ADAMs inhibits endothelial cell migration and adhesion *in vitro*. ADAM-17 has been shown to possess angiogenic property. It is capable of enhancing *in vitro* invasion. Furthermore, inhibition of ADAM-17 leads to reduced expression of VEGF and MMP-2 activation. Several ADAM proteins have been identified with chronic inflammation and cancer. The ADAM proteins, ADAM-9, ADAM-12, ADAM-15 and ADAM-17, have been implicated in tumour development. ADAM-17 is overexpressed in primary and metastatic colonic cancers compared with normal colonic mucosa. Inhibition of EGFR using RTK inhibitors and inhibition of ADAM-17 and effectively blocking EGFR activation and downstream signalling, suppresses cell proliferation, invasion and angiogenesis (Sherbet, 2011a). According to the findings of Tsai et al. (2009) miRNA-122 is downregulated in liver cancers with intrahepatic metastasis, but when present at normal levels miRNA-122 suppresses cell migration and invasion and *in vivo* suppresses angiogenesis. The import of the significance of ADAM-17 is accentuated as it turns out to be a target of miRNA-122. Nonetheless, these are presented as parallel findings and no causal link has emerged. Further study into the biological outcome of targeting of ADAM-17 is warranted.

miRNAs, Tumour Growth, Invasion and Metastasis

The Influence of miRNAs on Metastasis Suppressors

With the identification of many metastasis suppressor and promoter genes, efforts have been directed towards determining whether any of these genes might be targeted by miRNAs. Several miRNAs that generate their phenotypic effects by influencing suppressor gene function have been identified. There are many reports relating to the effects of miRNAs on putative tumour-suppressor genes, notably BRCA1 and BRCA2, BRMS1, PDCD4 (programmed cell death 4), Maspin, EPB41L3, CADM1 among others. These exert their suppressor effects by different mechanisms and the association of miRNAs with suppressor with diverse modes of function reiterates the multiplicity of targeting by miRNAs. The suppressor function of BRCA1 is induced and modulated by miRNAs through the mediation of other genes and signalling elements such as ID4, ERα, IGF-1R, and Sp1 that can exert exemplary effects cell growth and behaviour. BRMS1 is another example of a suppressor gene that is regulated by miRNAs. PDCD4 is a suppressor gene that is targeted and regulated by miRNA. So are Maspin, EPB41L3 and CADM1 modulated by miRNAs. The specifics of suppressor gene regulation by miRNAs and vice versa are discussed in detail in their respective locations. Conversely suppressor genes such as the ING can manipulate the expression of miRNAs. The contention here simply is that there is overwhelming evidence that tumour- and metastasis-suppressor genes and signalling systems are greatly influenced and impelled into regulating their biological functions. Some important and classical suppressor genes do not appear to have been investigated for potential regulation by miRNAs; possibly they were and no miRNAs might have been involved. There is a prevalent view, erroneous one, that negative results may not be worthy or publications.

miRNAs and Tumour/Metastasis Promoter Genes

Far less attention has been received by tumour promoter genes with regard to possible association with miRNAs and their function. Li et al. (2010f) found that the expression of miRNA-146a in pancreatic cancer cells was lower than in corresponding normal epithelium. *In vitro*, induced overexpression of the miRNA inhibited the invasive capacity. These changes occurred in parallel with the downregulation of EGFR, NF-κB and the tumour promoter MTA (metastasis associated antigen)-2. One would recall here the MTA family genes are a well-studied family of metastasis associated proteins. They are repressors of transcription by virtue of their function in histone deacetylation and are a component of the NuRD (nucleosome remodelling and deacetylating) complex.

MTA2 is believed to interact with p53 and interfere with its cell cycle regulatory function and apoptosis pathway. At the level of cell behaviour, MTAs are involved in cancer cell migration by regulation of cell adhesion proteins in consort with cytoskeletal proteins such as ezrin. MTA1 can repress ERα activation by chromatin deacetylation of the ER-response element of responsive genes (Toh and Nicolson, 2009). However, given that ERα promotes and ERβ inhibits cell proliferation, it may be that the effects of MTA are partly attributable to the suppression of ERβ (see pp. 77–79.).

S100 family proteins exert marked negative or positive effects on tumour progression. S100A4, which actively promotes metastasis and counters the suppressive effects of nm23, appears to produce a >fourfold increase in expression in EMT-mediated aggressive behaviour engendered by miRNAs. Besides, the expression of several other genes possibly associated with EMT and promotion of metastasis was also influenced by miRNA (Lopez-Lago et al., 2010). No specific miRNAs have been ascribed with these changes. However, Fassina et al. (2012) recently demonstrated that miRNA-205 is downregulated with EMT in malignant mesotheliomas and in parallel there was increased expression of S100A4 and the conventional EMT markers. Further they showed that forced expression of miRNA-205 inhibited EMT as indicted by the suppression of ZEB1 and ZEB2 and enhanced E-cadherin expression. In contrast, higher levels of miRNA-196a have been reported in oesophageal carcinoma and Barrett's oesophagus (BE) and dysplastic lesions than normal squamous mucosa, and in high-grade dysplasia than in BE and low-grade dysplasia. Also miRNA-196a expression showed an inverse with its molecular targets among them S100A9. Experimentally enhancing the levels of miRNA-196a in carcinoma cell lines resulted in the suppression of transcription of S100A9 (Maru et al., 2009). These findings are compatible with the well-established relationship between S100A9 and tumour progression. S100A9 is highly expressed in many tumours and higher S100A9 levels have correlated with enhanced cell proliferation and greater invasive potential.

Idiopathic pulmonary fibrosis (IPF) is an age-related condition wherein activated alveolar epithelial cells induce mesenchymal cell proliferation and stimulate EMT (see King et al., 2011). Several miRNAs are downregulated in IPF, among them is let-7d, which is targeted and downregulated by TGF-β via the downstream effector Smad3 binding to the let-7d promoter. This results in the expression of some EMT markers. *In vivo* let-7d inhibition was associated with increased S100A4 expression (Pandit et al., 2010). Reading between the lines, the change of cell morphology could be linked with let7-d-mediated effects of S100A4.

miRNAs and Chemo/Radiosensitivity of Tumours

The pronounced effects on cell proliferation and growth of tumours has inevitably led to examining the influence of miRNAs on the modulation of chemo/radiosensitivity of tumours since active proliferation makes cells susceptible to the action of cytotoxic agents and to radiation. Several miRNAs have been identified to be able to modulate tumour sensitivity.

A large number of miRNAs have been linked with chemo- and radiosensitivity (Table 3.1). Some miRNAs have been reported to directly correlate with drug sensitivity, whilst others might be inversely related. No specific associations have appeared, although some have received more attention than others. This would be due to accident than by design in the sense that some miRNAs have been focused upon because of their occurrence in and relationship to cancer and the biological behaviour of cancer cells. In the light of this more attention is devoted here to the possible mechanisms and modes of perceived link with drug resistance and sensitivity to radiation.

The question for determining the state of therapeutic resistance has prompted investigations into the mechanisms by which miRNAs might confer drug resistance or radio/drug sensitivity to tumours. Among the lines of investigation are (a) the modes of regulation of miRNA expression by natural and integral inhibitors, (b) effects of miRNAs on links in signalling systems and (3) the intrinsic modes of cell defence against stress and environmental insults, including drugs (see Table 3.1).

Many miRNAs can determine the state of drug resistance or sensitivity of tumours. Instances are not many where the effects exerted by miRNA effects on tumours is controlled or regulated by endogenous or natural inhibitors. The expression of the composite Lin28-let7 does offer an illustration of this. Overexpression of let-7a and inhibition of its negative regulator Lin28 reduce K-ras expression and radiosensitise A549 cells (Oh et al., 2010). The mode of action of miRNAs in chemoresistance is yet to be elucidated.

MiRNAs are known to markedly modulate the expression of a family of proteins called regulators of G-protein signalling (RGS) proteins, which modulate the function of heterotrimeric G-proteins. RGS proteins activate GTPase function leading to the hydrolysis of G-protein alpha subunits, which inactivates the G-protein and G-protein-coupled signalling pathways (De Vries et al., 2000) and so regarded as negative regulators of G-protein-dependent signalling. GPCRs use G-protein as intermediary in their signalling function of activating several important systems such as Src, Ras, MAPK and JAK/STAT pathways. GPCRs co-operate with EGF and PDGF in mitogenic function and possibly also with VEGF. Their mediation may be seen in the functions of some cytokines, including interleukins and TNF. It may be relevant also to recall here that certain miRNAs are known to regulate the activation by TNF-α of apoptosis signalling (Sherbet, 2011a).

Table 3.1 miRNAs in Drug Resistance and Radiosensitivity

miRNA	Drug Resistance, Radiosensitivity	Reference
Let-7 Lin28-let7a		van Jaarsveld et al. (2010) Oh et al. (2010)
miRNA-200		
miRNA-214		
miRNA-130a		
miRNA-27, miRNA-451	P-glycoprotein ↑	Zhu et al. (2008a)
miRNA-326↓	MCP-1↑	Liang et al. (2010b)
miRNA-451↑	Cisplatin sensitivity↑	Bian et al. (2011)
miRNA-224↓	Methotrexate cisplatin sensitivity↓	Mencia et al. (2011)
miRNA-106a↑	5-FU sensitivity↑	Hummel et al. (2011)
miRNA-148a↑	Cisplatin, 5-FU sensitivity↑	Hummel et al. (2011)
miRNA-195↑	5-FU sensitivity↑	Yang et al. (2012)
MiRNA-141	Cisplatin	Imanaka et al. (2011)
miRNA-27a	miRNAs↑ P-glycoprotein↑ Paclitaxel P-glycoprotein Doxorubicin sensitivity↑	Li et al. (2010g) Zhang et al. (2010b)
miRNA-199a-3p miRNA-326	Doxorubicin, VP-16 sensitivity↑	Fornari et al. (2010) Liang et al. (2010b)
miRNA-331-5p; miRNA-27a	P-glycoprotein↓ in doxorubicin-resistant cells; sensitivity↑	Feng et al. (2011a)
miRNA-122↑	Adriamycin vincristine sensitivity↑MDR-1	Xu et al. (2011b)
MiRNA-24	Topotecan, Irinotecan sensitivity↓ Topotecan IFN-α/5-FU Trastuzumab (Herceptin)	Gmeiner et al. (2010)
miRNA-21	Doxorubicin	Misawa et al. (2010) Tomimaru et al. (2010) Gong et al. (2011)
miRNA-21	Taxol sensitivity↑ with inhibitor	Tao et al. (2011)
miRNA-21↓ inhibition	Ara-C (arabinosylcytosine) sensitivity↑	Ren et al. (2010b)
miRNA-21↓ (anti-sense oligonucleotide)	Docetaxel	Li et al. (2010e)
miRNA-21↓ (inhibited) miRNA-100, miRNA-130a, miRNA-197↓ miRNA-101, miRNA-181b, miRNA-181d, miRNA-195↑	Docetaxel sensitivity↑	Dai et al. (2011b)
miRNA-143↑	Paclitaxel	Xu et al. (2011a,b)
miRNA-125b	Paclitaxel	Zhou et al. (2010a)
miRNA-135a		Holleman et al. (2011)

Note: ↓indicates downregulation of expression or reduction of drug sensitivity; ↑ indicates upregulation of expression or increased resistance; where no symbols are shown and a direct correlation between expression and drug resistance is implied.

MiRNA-22 is able to regulate the expression of RGS2 (Muinos-Gimeno et al., 2011). Several RGSs are expressed at lower levels in cells resistant to chemotherapy. Cytotoxic agents downregulate RGS10 and RGS17 expression. Inhibition of these RGSs by siRNA reduces cytotoxicity of the agents suggesting the possibility that cytotoxic agents could be inhibiting specific miRNAs leading to the development of chemoresistance. Both RGSs inhibited the Akt pathway (Hooks et al., 2010). However, miRNA-22 suppresses PTEN and would lead to activation of Akt signalling and in a self-regulatory loop Akt seems able to upregulate miRNA-22. So valuable it would be to consider that several miRNAs can and do influence Akt signalling in either direction to modulate cell migration and apoptosis.

It would be worthwhile examining the possible effects in such a system of the expression of miRNAs. However, it would also be worthy of note that RGS protein are not expressed at levels consistent with neoplastic status. Jiang et al. (2010) reported not only that RGS2 expression was lower in recurrent colorectal cancer as well as in cell lines derived from metastatic tumour but also the lower levels correlated with poor prognosis. Hurst et al. (2009) noted that whilst RGS2 occurred at lower levels in ovarian cancer cells as compared with immortalised epithelial cells and RGS19 showed similar levels of expression in both. In contrast, RGS4 and RGS6 were occurred at a magnitude higher in epithelial cells as compared with cancer cells. Perhaps levels of RGS are not relevant to chemoresistance, but it begs the question whether RGSs differ in respect of their chemoresistance potential. This is not to be construed as a total denial of their significance. But the possibility that miRNAs might function independently of the RGSs has to be envisaged.

Genetic changes have been detected in RGSs. Dai et al. (2011a) found single nucleotide polymorphisms (SNPs) in NSCLC that significantly correlated with poor prognosis. The situation is complicated further by claims that RGS2 might indeed promote angiogenesis albeit by an unrelated mechanism (Boelte et al., 2011). So this topic has to rest at present.

The MDR (multidrug resistance) genes (mdr1 and mdr2) and the mdr1 encoded P-glycoprotein (p170) form an intrinsic cellular system for dealing with efflux of cytotoxic substances from the cell. P-glycoprotein mediates efflux of chemotherapeutic agents from the intracellular sites thus produce drug resistance. It seems that miRNAs might be involved in P-glycoprotein-mediated drug resistance or sensitivity. Zhang et al. (2010b) found that downregulation of miRNA-27a might decrease the expression of P-glycoprotein. In parallel they noticed downregulation of the apoptosis family Bcl-2 and upregulation of Bax. These genes also function downstream of p53 and since miRNAs can and do modulate p53 expression, it may be that these effects have resulted from parallel signalling by the p53 pathway (Figure 3.7). Indeed, miRNAs-15b and miRNA-16 target Bcl-2 in human gastric cancer cells (Xia et al., 2008), whilst others, for example miRNA-27a and miRNA-451, have been shown to regulate mdr1/P-glycoprotein expression (Zhu et al., 2008a). Also the PTEN/Akt pathway might be operating here. These various interactive signalling systems have to be taken into account while assessing the role of miRNAs in drug resistance. Li et al. (2010g) found upregulation of expression of miRNA-27a and P-glycoprotein in paclitaxel-resistant ovarian cancer cells. According to

Feng et al. (2011a) miRNA-331-5p and miRNA-27a expressions inversely correlated with P-glycoprotein expression in doxorubicin-resistant cells, and besides when transfected into doxorubicin-resistant K562 and HL60 cells, these miRNAs increased the sensitivity of the cells to the drug. Quite clearly the P-glycoprotein pathway of drug resistance is mediated by miRNAs. Another protein associated with multidrug resistant, namely MCP (monocyte chemoattractant protein)-1 is also influenced by miRNA-326 as Liang et al. (2010b) have reported. Downregulation of miRNA-326 was inversely related to MCP-1 expression in advanced breast cancers. When expressed at high levels miRNA-326 not only downregulated MCP-1 expression but also sensitised these cells to VP-16 and doxorubicin.

Potential operation of mTOR signalling and its attenuation by miRNAs have been adduced as another means by which drug resistance might be modulated. When miRNA-100 occurs in a downregulated state, mTOR shows parallel upregulation. In agreement with this, when overexpressed miRNA-100 inhibits mTOR signalling (Nagaraja et al., 2010), a clear indication that the miRNA-100 targets mTOR. The work of Fornari et al. (2010) has suggested the possibility that miRNA-199a-3p might also be inhibiting mTOR signalling and increase doxorubicin sensitivity. Inhibitors of mTOR are known to be able to reverse doxorubicin resistance (Grünwald et al., 2002).

ABC (ATP-binding cassette) transporters are another family of proteins implicated in drug resistance in tumour cells. There is evidence that miRNAs are also involved in the regulation of these. Borel et al. (2011) reported the upregulation of five ABC genes by 13 cellular miRNAs in human HCC samples. ABC2 was downregulated by miRNA-379 (Haenisch et al., 2011). These may offer themselves as suitable therapeutic targets in the light of their acknowledged drug resistance function and having been linked with targeting of the relevant miRNAs. Equally, one should not lose sight of the possibility that drugs can induce the expression of miRNAs, some of which might be able to activate signalling systems for cell survival and or proliferation and growth.

Massive amounts of data have accumulated concerning the role of miRNAs in conferring drug resistance or sensitivity; nonetheless no specific patterns of response, specific attributions of induction or alleviation of drug resistance, or any tumour or tissue specificity have emerged from these data.

The Therapeutic Potential of miRNAs

The potentialities of miRNAs as therapeutic tools have been largely appreciated in the light of their ability to inhibit, influence or modulate a number of features of cell behaviour that are relevant in the contexts of cancer development and spread. It is patently obvious that they affect cell proliferation, apoptosis, invasion and angiogenesis and in this way might be expected to have a major bearing on metastatic spread of cancer. As depicted in Figures 3.4–3.7, several miRNAs have been identified to be able to affect these phenotypic features (see also Table 2.1). Also in many cases their molecular targets have been identified and their functional involvement with

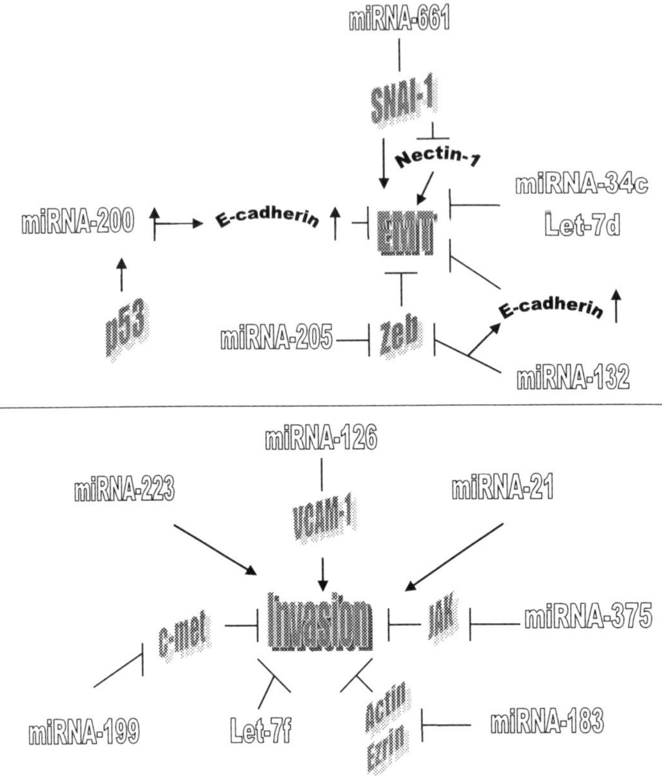

Figures 3.4 and 3.5 The modulation of EMT and invasion/motility by miRNAs and the signalling ligands and factors involved therein both upstream and downstream of miRNA function. These two figures should read together. It should be borne in mind that EMT does lead to the acquisition of invasive capacity. Here the depiction of c-met and invasion in two separate illustrations is dictated merely by expediency. However, the routes of signalling differ markedly for the different miRNAs. MiRNA-128a blocks the canonical TGF-β signalling cascade and could effectively block EMT. MiRNA-194 has been reported to inhibit EMT by the downregulation of the transcriptional repressor BMI-1, a PcG (Polycomb group) protein. Suppression of BMI-1 upregulates E-cadherin and downregulates Vimentin (this is not shown here).

these targets has been elucidated to a certain extent. There is nothing uncommon with the three signalling pathways that are apparently involved in signalling by miRNAs. These are shown in Figures 3.4–3.7 together with the miRNAs that are known to activate and transduce the signals to their phenotypic conclusion. The numerical abundance of miRNAs has inevitably led the investigations towards the well-established signalling systems with precious little attention to cross talk between lines of signal transduction and the function of the identified targets. These again represent well-established molecular targets of which some are known to be able to transmit signals downstream using different pathways and occasionally generating a diversity of phenotypic outcome.

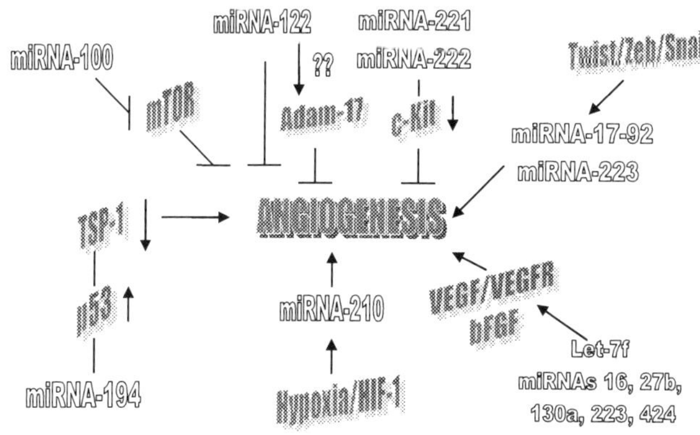

Figure 3.6 The routes of signalling adopted by some miRNA to promote angiogenesis; some activate angiogenic growth factors such as VEGF and bFGF suppress angiogenesis inhibitors. Some transcription factors are also able to activate angiogenic miRNAs. References are provided in the text. Hypoxia involves miRNAs in two ways. Hypoxia induces many miRNAs, for example miRNAs-23, -24, -26, -27, -103, -107, -181, -210 and -213, some via an HIF-dependent mechanism. HIF1-α activity is involved in some such as miRNA-210 to promote angiogenesis. Hypoxia induces miRNA-424 in an HIF-dependent manner and promotes angiogenesis. It is said to reduce the expression of miRNA-34a with promotion of EMT by targeting Notch signalling (Du et al., 2012). In contrast, miRNAs-221 and -222 seem to downregulate c-Kit RTK and inhibit angiogenesis. There are some indications that mTOR signalling might be inhibited by miRNA-100 leading to the inhibition of endothelial cell proliferation and angiogenesis.

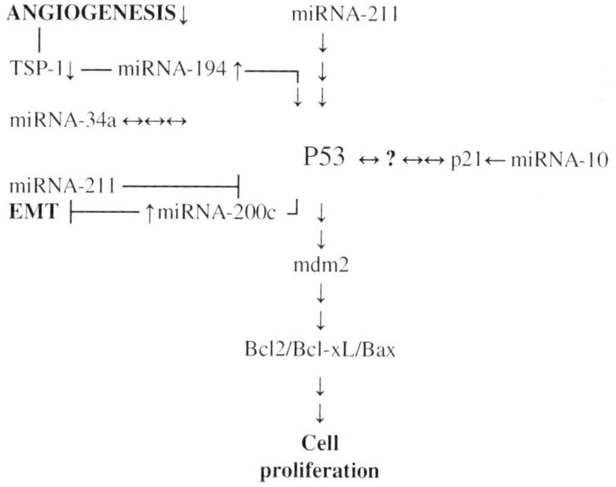

Figure 3.7 p53/mdm2/bcl2-bax regulation of cell proliferation and apoptosis and the modulation of signalling by miRNAs. The effects might occur as a consequence of altering the balance of Bcl family pro- and anti-apoptosis genes. ER is also suggested as being able to do this, but some reservations and caveats may be expressed in relation to that. References are given in the text.

Part 2

EMT Associated Gene Targeting

The conversion of epithelial cells into mesenchymal cells (EMT) is a developmental programme, which is accompanied by characteristic alterations in cell morphology, reduced intercellular adhesion and acquisition of cell motility. The Wnt, Notch and Hedgehog (HH) pathways are developmental pathways that are commonly activated in many types of cancer and directly or indirectly linked with EMT. Hence, EMT has been ascribed an integral role in cancer development and secondary spread. Significantly, it is induced by the presence of EGFR family of growth factor receptors and TGF-β, among others, with clear implications of different modes of signalling to EMT activation. E-cadherin plays a prominent part in this process leading to the manifestation of EMT. The major changes involve the loss of E-cadherin, acquisition of N-cadherin and vimentin and also activation of Wnt/β-catenin, which are the highlights of the modes of signalling. Whereas TGF-β might promote EMT other TGF-β family ligands such as the BMPs affect EMT in an opposite direction. Furthermore, both canonical and non-canonical systems of signalling are functional in the action of these growth factors. The tumour promoter S100A4 has been regarded by some as a potential marker of EMT and this aspect has been discussed in detail in a separate location. Therefore, EMT as a target for therapeutic assault has several open and adoptable provisions (see Sherbet, 2011a).

The targeting of TGF-β signalling may be regarded as a potentially rewarding avenue to approach. In human HCC cell lines, the inhibition of TGF-β signalling by using the TGF-β receptor kinase inhibitor LY2109761 blocks the activation of the canonical signalling pathway. It upregulates E-cadherin mRNA and protein levels and inhibits the migration of cells in vitro (Fransvea et al., 2008). Inhibition of TGF-β receptors could block not only the Smad pathway but also inhibit the Wnt/β-catenin pathway, thus providing a dual route to the inhibition of EMT. Examining intraductal breast carcinomas for the operation of Wnt signalling, Prasad et al. (2009) found Wnt/β-catenin signalling via Slug down regulated the expression of E-cadherin.

As discussed at length below, also involved with the regulation of EMT is Hedgehog (Hh) signalling, which possesses the potential to interact with Wnt signalling. With the demonstration that TGF-β1 can upregulate the expression of β1-integrin (Yeh et al., 2010), one can visualise the possibility of altering the function of more than one membrane component. Inhibitors that can block the Smad as well as the non-Smad pathways are currently available.

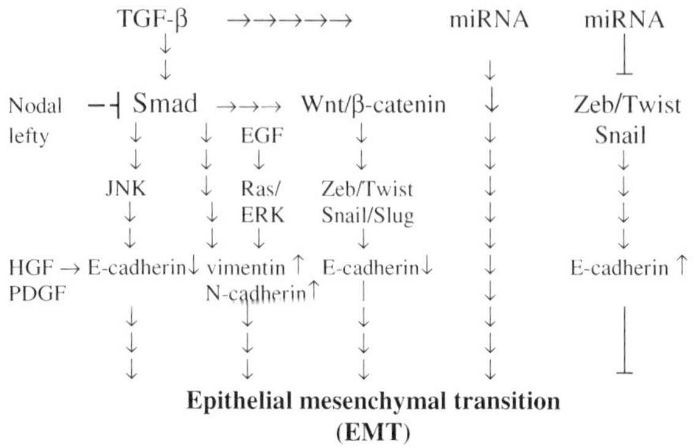

Figure P2.1 The possible modes of signalling by some growth factors and miRNAs in the promotion of EMT. The signalling pathways do interact to collude in or negate the phenotypic effect. EMT activation by TGF-β is counteracted by certain members of the TGF-β family, e.g. Nodal and certain BMPs. HGF and PDGF are believed to function by the mediation of E-cadherin; EGF reportedly activates the Ras/ERK pathway and upregulates vimentin and N-cadherin to promote EMT. This pathway is effectively blocked by Resveratrol. It has been suggested that ERK activation can block Smad-mediated activation of EMT. This is not shown in the figure. MiRNA-128a is said to inhibit the canonical TGF-β signalling cascade by directly targeting the gene encoding the TGF-β receptor I protein and block Smad activation. The intervention of E-box zinc-finger transcription factors in the regulation of E-cadherin expression in the Wnt/β-catenin and miRNA driven signalling is also shown. The varied effects of miRNAs are discussed and illustrated in Figures 3.4 and 3.5 (also see Sherbet, 2011a for further discussion and references).

Resveratrol (RES), a naturally occurring phytoalexin, is a potent promoter of apoptosis. The efficacy of this agent can be attributed to its being able to suppress proliferation as well as induce apoptosis by inhibiting the Notch pathway besides influencing the operation of the interacting apoptosis pathways mediated by p53 and PI3K/Akt (Cecchinato et al., 2007). Furthermore, in addition to modulating three different pathways that regulate cell survival and cell death, RES markedly inhibits EMT. It can downregulate TGF-β2 expression and concurrently increase the expression of E-cadherin and result in the suppression of invasion. Resveratrol achieves this effect by inhibiting Smad and also independently of Smad (Kim et al., 2011). It is also able to inhibit the EMT-promoting effects of EGF. The latter has been found to upregulate the transcription factors Slug, ZEB1, ZEB2, and vimentin and N-cadherin. This does not happen when Resveratrol is present with EGF (Vergara et al., 2011). These authors also suggest that EGF activates ERK1/2 signalling to activate EMT and that this pathway is inhibited by Resveratrol (Figure P2.1). With this prelude, it would be judicious to delve into discussions of the intricacies of EMT signalling.

4 Hedgehog Signalling in EMT

Hedgehog (Hh) signalling represents a well-established developmental pathway of embryogenesis. With the widespread recognition of EMT also as a developmental programme, it is little wonder that Hh signalling can markedly influence EMT in embryonic development, in the formation and maintenance of CSCs and acquisition of EMT characteristics. Hh expression has correlated with aggressive cancers with enhanced tumour vascularisation, marked extension of the tumour to regional lymph nodes and poor prognosis. Kelleher et al. (2011) have recently highlighted the fact that as many as a dozen signalling pathways are deregulated and that in a large majority of cases Hh signalling is affected.

It would be appropriate to open this analysis with a discussion of stem cells and how their expansion and perpetuation might be shaped and affected by Hh signalling. Albeit subject to much debate and discussion, CD44+/CD24+/ESA+ (epithelial cell specific antigen) are regarded as putative stem cell markers and a clone of cells carrying these markers might be expanded and tumour enriched in respect of this subpopulation. This subpopulation is seen as representing a stem cell clone capable of initiating tumour development. The CD44+/CD24+/ESA+ cells form a tiny proportion of the total tumour cell population in pancreatic cancer. Pancreatic cancer cell lines do not show a consistent pattern of expression of these different antigens and also the diversity of expression depends upon the microenvironment, namely whether grown as cell cultures or cell spheres (Wei et al., 2011). In xenograft experiments, the percentage successful takes using a small number of stem cells is virtually the same as xenografting a total cancer cell population. CD44+/CD24+/ESA+ clone is found in breast cancers (Fillmore and Kuperwasser, 2008). But CD133 positivity seems to be associated with glioma initiation and resistance to therapy (Woodward and Sulman, 2008).

Here of considerable interest is that not only does this clone of cells showed expansion in pancreatic cancer but also this stem cell subpopulation showed 46-fold enhancement of Sonic hedgehog (SHh) expression as compared with normal epithelial cells. Furthermore, the stem cell subclone of pancreatic cancer is believed to be resistant to ionising radiation and to conventional chemotherapy. It has been suggested that this chemoresistant subpopulation is accountable for the high recurrence rate of breast cancer (Fillmore and Kuperwasser, 2008). The development of chemoresistance by stem cells may be defined by the microenvironment, but a temporal acquisition of stem cell characteristics has also been advocated. However, the distinction between CSCs and other stem cells founded on plausible influences of the microenvironment and temporal acquisition of drug resistance would be too artificial to contemplate. An extended definition of microenvironment would include

Therapeutic Strategies in Cancer Biology and Pathology. DOI: http://dx.doi.org/10.1016/B978-0-12-416570-0.00004-4

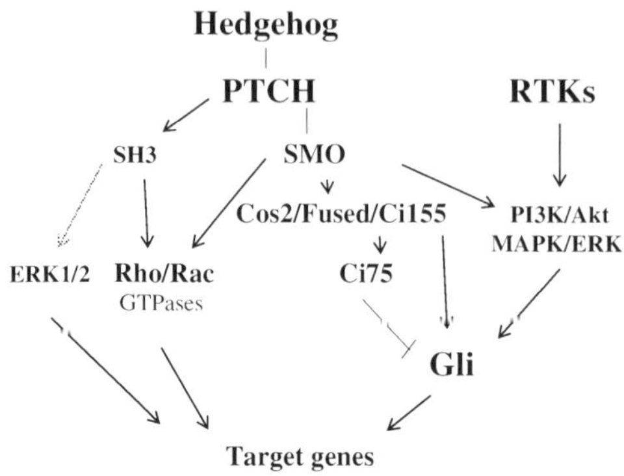

Figure 4.1 The canonical Hh signalling pathway involving Gli activation and the non-canonical SMO-dependent and independent routes. In the canonical pathway, in the absence of Hh, SMO targets the Cos2/Fused/Ci155 protein complex leading to the truncation of Ci155 to the repressor Ci75 and as a consequence Gli is not activated. When Hh ligand binds PTCH, this truncation does not take place and Ci155 can activate Gli and target downstream genes to generate the phenotypic effect (see Figure 4.2). Also shown here is the integration of RTK signalling with SMO leading to positive transduction of signals via Gli. Overall the phenotypic result is cell transformation and activation of EMT.

the infiltration of inflammatory cells and fibroblast into the tumour microenvironment, which can quite conceivably modulate the signalling functions of regulatory ligands. From a reading of the literature on CSCs, it is difficult to escape from a general feeling that there is much teleological thinking in relation to the concept. Notwithstanding the debate about the role of stem cells, there are overwhelming arguments why modulation of Hh signalling can be justified and viewed as a potential path to therapy.

The major elements of the Hh signalling cascade are PTCH (PATCHED), the 12-pass trans-membrane receptor, 7-pass GPCR SMO (smoothened) of the Frizzled family of receptors and Gli1. PTCH is the regulator of SMO activity. Hh binds to PTCH and in this bound state SMO is able to activate the Kruppel family of zinc finger transcription factor Gli1. When Hh is not present PTCH suppresses SMO and Gli1 is not activated. SMO functions by signalling to an intracellular protein complex composed of Cos2, the protein kinase Fused and the transcription factor the full-length Ci155 (Cubitus interruptus) (Robbins et al., 1997). In the absence of Hh binding PTCH is able to inhibit SMO and this complex triggers the cleavage of the full-length Ci155 to a truncated Ci75 repressor form, in a process involving Ci155 phosphorylation by PKA (Price and Kalderon, 2002). When PTCH is bound by Hh, SMO is active and reduces the cleavage of Ci155 to the repressor form, thus allowing the transduction of downstream Hh signalling (Figures 4.1 and 4.2).

As a general rule, Hh signalling can be distinguished into the canonical and non-canonical systems which both seem eventually target genes whose activity or suppression is required in cell transformation and activation of EMT. The canonical form of Hh signalling pathway involves Gli activation and the non-canonical

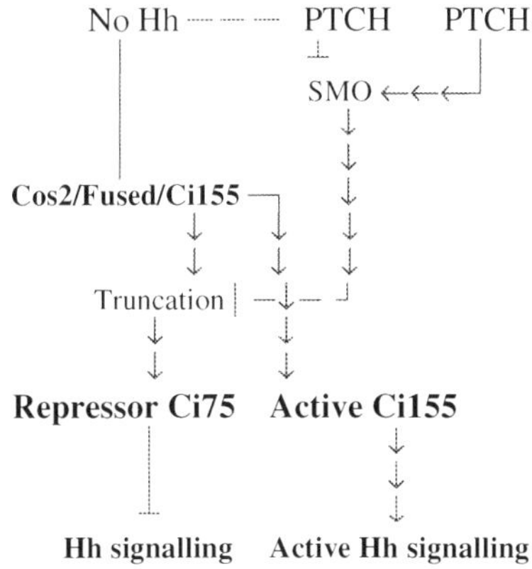

Figure 4.2 A flowchart of the Hh signalling cascade. SMO targets the protein complex made up of Cos2/protein kinase Fused and the full-length transcription factor C155. In the absence of Hh, the complex leads to the cleavage Ci155 to the truncated Ci75 which is a genetic repressor and inhibits Hh signalling. PTCH regulates SMO function. When PTCH is bound to Hh, it cannot inhibit SMO and this inhibits the truncation of Ci155 which allows the transduction of active Hh signalling downstream to generate the phenotypic effects.

SMO-dependent and -independent routes. To reiterate, in the canonical pathway in the absence of Hh SMO targets, the Cos2/Fused/Ci155 protein complex leading to the truncation of Ci155 to the repressor Ci75 and as a consequence Gli is not activated. When Hh ligand binds PTCH this truncation does not take place and Ci155 can activate Gli and target downstream genes to generate the phenotypic effect.

The SH3 (src homology) domain are small domains occurring in signalling proteins and said to function by binding to proline-rich motifs of proteins. SH3 domains have been implicated many biological functions including signalling systems. In the present functional view point, SH3 domains can provide links with cytoskeletal proteins thus influencing cell motility. The organisation of the actin cytoskeleton involves many SH3 domain proteins. The involvement of Rho/Rac GTPases in membrane activity, for example promotion of focal clustering of integrins and cell adhesion processes, is well established. It would follow from this that they would be involved in cell motility. These functions are also dependent upon SH3 domains. The C-terminal region of PTCH has SH3 domains which could be affecting the cytoskeletal organisation. PTCH can function independently of SMO to activate ERK1/2 (Chang et al., 2010). Activation of MEK/ERK is a requisite for integrin and cadherin-induced adhesion (Avizienyte et al., 2004). Equally, it has been advocated that cytoskeletal changes mediated by Rho/Rac can also occur via the agency of SMO (Polizio et al., 2011). These can be viewed as the non-canonical SMO-independent and -dependent means of Hh signalling.

Sonic hedgehog (SHh) activates Gli1 as shown in Figure 4.1. The transcription of Gli1 is upregulated in many cancers. In gliomas levels of its expression have correlated with tumour grade. Suppression of SHh/Gli1 signalling suppressed *in vitro*

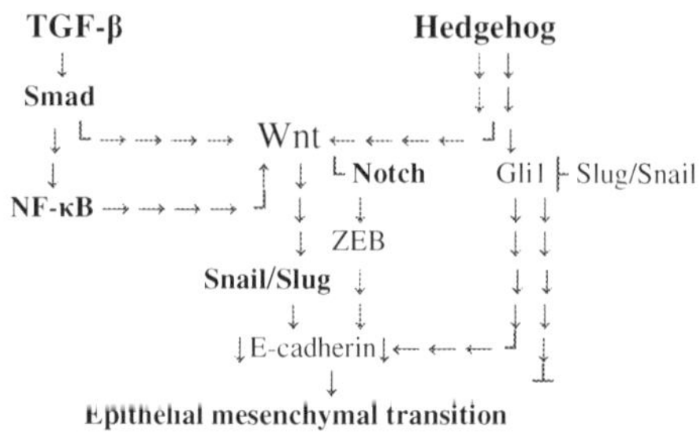

Figure 4.3 Pathways activated by Hh and TGF-β towards EMT. Here emphasised is the interaction with Wnt and Notch in the downregulation of E-cadherin expression leading to the manifestation of EMT. Certain miRNAs modulate ZEB expression and influence EMT. Indeed, there is evidence that Notch, Wnt and TGF-β pathways might be integrated by Notch ligands.

cell migration. In parallel with Gli1 downregulation cells showed downregulation of expression of Snail, MMP9 and enhancement E-cadherin expression (Wang et al., 2010b). The expression of Gli1 has been found to correlate negatively with the expression of E-cadherin and positively with the tumour promoter S100A4 in primary hepatocellular carcinoma tissues (Zheng et al., 2010). Liao et al. (2009) observed that overexpression of PTCH/Gli1correlated with poor survival of ovarian cancer patients. Also they found high levels of SHh mRNA in carcinomas as compared with normal tissue and benign conditions. But Joost et al. (2012) reported that inhibition of Gli1 promoted EMT and further that decreased Gli1 levels corresponded with more aggressive and metastatic phenotype in pancreatic ductal carcinoma. The latter progresses from intraepithelial neoplasm through several steps with associated cellular changes; of these mucin5AC expression is one. Gli1 increases MUC5AC expression in pancreatic carcinoma cells and suppression of Gli1 and Gli2 negates this. Gli1-mediated upregulation of MUC5AC also leads to localisation of E-cadherin to the membrane and consequent decrease in E-cadherin-dependent cell adhesion and migration (Inaguma et al., 2011). So it may be that pancreatic carcinoma represents a unique model. There is also the possibility of a switch of signalling pathways.

These findings provide the background and serve also a prelude to the topic of deployment of the amalgamation of Hh signalling with TGF-β and Wnt to target them for inhibition. TGF-β upregulates the transcription of Gli1. The transcription factors Slug/Snail which suppress E-cadherin also can downregulate Gli1 (Katoh and Katoh, 2008, 2009). It would seem therefore that whereas in collusion Hh and TGF-β could be functioning via the Smad/Wnt pathway to activate EMT, when TGF-β is absent Gli1 activation might be suppressed by Slug/Snail. It ought to be stated here that SHh can directly induce Snail (Wanshura et al., 2011). The role of Hh signalling in EMT induction is also supported by reports that inhibitors of the pathway do inhibit EMT activation (Gadgeel et al., 2011; Li et al., 2012d). Thomas et al. (2011) found high levels of Gli1 expression in the triple-negative, highly invasive breast cancer cell line, SUM149. Suppression of GLi1 by siRNA or Gli1 inhibitor decreased cell proliferation and increased apoptosis and cell migration (Figure 4.3).

Hh signalling is known to interact with the Wnt pathway and so expected to modulate EMT (Figure 4.1). Wnt5A occurs as two isoforms with alternative exons 1A and 1B. Smad binding elements and bHLH binding sties occur in the promoter A close 5′ to exon 1A, whilst NF-κB and two Smad binding sites occur within the promoter B region, 5′-adjacent to exon I B. In the absence of Gli binding site, it seems likely that the TGF-β can directly upregulate the expression of Wnt5a possibly through the agency of Smad and indirectly by the Smad/NF-κB route (Katoh and Katoh, 2008, 2009). TGF-β is capable of upregulating SHh in NSCLC cells accompanied by the acquisition of EMT features. Inhibition of Hh signalling leads to reduction in the expression of ZEB-1 and fibronectin and upregulation of E-cadherin indicating the reversal of EMT. Inhibition of SHh also reduced EMT inducing effects of TGF-β (Maitah et al., 2011). Besides TGF-β, Hh can also function in conjunction with EGF, FGF, HGF, together with the transcription factors Snail, Slug, Twist and ZEB and often implicated in this way in many facets of cell proliferation, embryonic development and pattern formation and possibly also in stem cell pluripotency and perpetuation (Sherbet, 2011a).

Besides Hh and TGF-β, Wnt can interact with Notch signalling in differentiating systems as well as in the activation of EMT. Indeed, Notch, Wnt and TGF-β and other members of the TGF-β family are linked together and possibly also functionally by the Notch ligand Delta-like 1 and lef and Smad components of signalling downstream (Bordonaro et al., 2011). There is a strong possibility that activation of Notch signalling might downregulate E-cadherin expression. Overexpression of Notch-1 activates EMT results in the expression of ZEB-1 which suppresses E-cadherin and other markers such as CD44 and EPCAM, the epithelial cell adhesion molecule which is overexpressed in many tumours (see also Figure 4.3). The expression of a number of miRNAs is also altered; the expression of miRNA-21 is upregulated, and that of miRNA-200b, miRNA-200c, let-7a, let-7b and let-7c is downregulated. Forced re-expression of miRNA-200b decreased the expression of ZEB-1 and vimentin, and increased expression of E-cadherin (Bao et al., 2011a). The intervention of miRNA in Notch-mediated induction of EMT is supported by other studies. Brabletz et al. (2011) showed that downregulation of miRNA-200 leads to enhanced expression of ZEB-1, a suppressor of E-cadherin and activator of EMT, and Jagged 1, which is a downstream effector of Notch. A general reading of these data would be that Notch suppressed the expression of E-cadherin and activated EMT.

On the other hand, it has been argued that enhanced E-cadherin inhibits the Notch pathway and conversely that Notch activity is enhanced in E-cadherin deficient cells (Ferreira et al., 2012). Inhibition of Notch as indicated by using DAPT ((*N*-[*N*-(3,5-difluorophenacetyl)-L-alanyl]-*S*-phenylglycine *t*-butyl ester) also suppressed growth of E-cadherin deficient cells. Since DAPT functions independently of SHh, the possibility of influences of SHh can be ruled out. However, the status of E-cadherin expression upon drug treatment is uncertain, but this is somewhat important. For it is possible that Notch indeed could be inhibiting E-cadherin expression. A further possibility is that in certain differentiation systems the canonical Wnt/β-catenin signalling enters into a reciprocal interaction and regulation with Notch (Li et al., 2012a).

5 Targeted Inhibition of Hh, Wnt, TGF-β Signalling Complex

The perceived integration of Hh, Wnt, TGF-β signalling pathways, cross talk and mutual regulation has made this signalling complex a significant target for therapeutic intervention. Several inhibitors of Hh have been designed, synthesised and clinically tested with more than a modicum of success.

SMO Is a GPCR Component of Hh Signalling

GPCRs have been closely linked with the regulation of cell proliferation and possibly also with cell motility and invasion. With the recognition that the major element of the Hh signalling cascade the 7-pass SMO (Smoothened) might belong to the GPCR Frizzled family, there is general agreement that drugs targeting GPCR might provide a practical approach to the inhibition of deregulated Hh signalling in cancer. The precise links of SMO with the G-protein-transduced signalling are continually being elucidated. Recently a G-protein has been shown to be a direct downstream effector of SMO signal transduction (Ogden et al., 2008). GPRK2 (G-protein-coupled receptor kinase 2) participates in several cellular events. GPRK2 often manifests divergent effects, for example blocking the regulation of cell proliferation and apoptosis by TGF-β, but it promotes cell proliferation signalling by EGF or by SMO activation (Ho et al., 2005; Meloni et al., 2006). In Drosophila, GPRK2 is required for the regulation of SMO activity. Following upon its activation SMO is internalised. Phosphorylation of SMO by GPRK2 is suggested to be involved here. SMO is phosphorylated at several sites in the C-terminus; phosphorylation is induced or blocked by SMO agonists and the natural antagonist cyclopamine respectively. Whilst CK (casein kinase) 1α initiates phosphorylation, further phosphorylation is promoted by CK1α and GPRK2 in a positive regulatory loop. GPRK2 is also said to be able to bind to the C-terminal tail of SMO and stabilise its active conformation quite independently of GPRK2 activity (Chen et al., 2010, 2011d). PTCH is said to regulate SMO by a different mechanism (Cheng et al., 2010). Other factors such as GAS8 (the growth arrest specific 8) could also be involved in this intracellular trafficking and stimulation of SMO (Evron et al., 2011). GAS8 is regarded as a putative tumour suppressor. It is a microtubule associated protein and so involved in cell motility and invasion. GAS proteins inhibit cell proliferation and motility. On the other hand, GAS8 is described as a positive regulator of Hh signalling being able to promote SMO function. Possibly the function of GAS as a cytoskeletal component

Therapeutic Strategies in Cancer Biology and Pathology. DOI: http://dx.doi.org/10.1016/B978-0-12-416570-0.00005-6

occurs quite independently of SMO. This conclusion is justified by the finding that GAS3 does not inhibit EMT (Roux et al., 2005).

Small Molecule Inhibitors of SMO

GDC-0449 (Vismodegib, 2-chloro-*N*-[4-chloro-3-pyridin-2-ylphenyl]-4-methylsulfonyl benzamide), an Hh pathway inhibitor is in phase II multicentre clinical trial to assess tumour response and drug toxicity in patients with metastatic and locally advanced BCC (basal cell carcinoma). Clinical trials are also being conducted to evaluate GDC-0449 treatment for other forms of cancer such as pancreatic, small cell lung cancer, oesophageal, stomach, breast and prostate cancers, among others (Curis Inc. and Genentech in collaboration with NCI). A partial response was reported in a patient with a refractory metastatic medulloblastoma. Investigations of the tumour revealed that the Hh pathway was active and the activation seemed to be due to loss of heterozygosity (LOH) at the PTCH locus and mutation of the PTCH-1. Skulic et al. (2012) noted that objective response rate was 33% (30 patients in metastatic BCC cohort) and the response rate was 43% (63 of the locally advanced BCC cohort) and 13 of the 63 patients showed complete response. The duration of objective response in both cohorts was 7.6 months and progression-free survival was 9.5 months. Low response rates were recorded in patients with other types of cancer (11.8% in pancreatic and <5% in others) (Rudin et al., 2009; LoRusso et al., 2011). GDC-0449 has displayed a range of oral bioavailability (Wong et al., 2009). Mean bioavailability in female subjects was 35% (Graham et al., 2010). GDC-0449 (up to 95%) binds with high affinity to plasma alpha-1-acid glycoprotein (AAG) and often levels of GDC-0449 have corresponded with those of AAG. *In vitro* this binding is reversible. This binding might be increasing single dose half-life of the drug by approximately 200-fold (Giannetti et al., 2011). GDC-0449 (at 15 μM) did not significantly inhibit efflux transporter P-glycoprotein in MDR1-MDCK cells carrying the MDR1 gene that codes for P-glycoprotein (ABCB1) (Wong et al., 2009). But Zhang et al. (2009c) have reported inhibition of both P-glycoprotein (IC_{50}=3 μM) and ABCG2 (IC_{50}=1.4 μM). Overall, there could be important implications for treatment from the GDC-0449/AAG binding influencing both drug clearance and distribution.

Several new putative candidates for the inhibition of Hh signalling are being studies *in vitro* and in animal models, a prominent one being LDE225 (*N*-(6-((2S,6R)-2,6-dimethyl morpholino) pyridin-3-yl)-2methy1-4′-(trifluoromethoxy) biphenyl-3-carboxamide), which has been described as a potent SMO inhibitor (Pan et al., 2010). Using the transgenic Rip1Tag2 mice, which develop islet cell neoplasms, Fendrich et al. (2011) were able to show that treatment with LDE225 inhibited tumour growth substantially and there was demonstrable downregulation of SMO in the tumours. Preliminary studies with BCC, LDE225 inhibited basaloid tumour nest formation and induced regression of preformed tumours in organ cultures of PTCH1 heterozygous murine skin. Also in a provisional clinical study involving nevoid BCC syndrome patients with BCCs, of 13 LDE225-treated BCCs three showed complete and nine partial response, and in one case no response was recorded (Skvara et al.,

2011). A limited phase I study of LDE225 of medulloblastoma, BCC, lung adenocarcinoma, spindle cell sarcoma and breast cancer has been reported. Some tumour inhibition has been found (Rodon et al., 2011, Tawbi et al., 2011). The safety, pharmacokinetics and pharmacodynamics of LDE225 administered orally on a daily dosing schedule to children with recurrent or refractory medulloblastoma is under way at present. Many other forms of human tumours are also included in this study (Novartis, NCT01125800). Several other trials are also in progress. Another SMO inhibitor is IPI-926 (Saridegib) that is currently in clinical trial in patients with chondrosarcoma (NCT01310816). The effects of combination of IPI-926 with Gemcitabine in pancreatic cancer have been reported by Infinity Pharmaceuticals (ASCO 2012 Gastrointestinal Cancers Symposium). Of 16 patients, five showed partial responses and median progression-free survival was 7.6 months and median overall survival was 10.2 months. Three further SMO inhibitors undergoing clinical trials are BMS-833923 (XL139), LEQ506 and TAK-441. The paucity of information in the public domain concerning the biological effects of these compounds is somewhat surprising.

A putative SMO inhibitor was described some time ago. CUR61414 (*N*-((3S,5S)-1-(benzo[d][1,3]dioxol-5-ylmethyl)-5-(piperazine-1-carbonyl)pyrrolidin-3-yl)-*N*-(3-methoxybenzyl)-3,3-dimethylbutanamide) was highly active in BCC carrying PTCH1 mutations, achieving inhibition of signalling with IC_{50} in the 100–200 nM range, causing regression by inhibiting proliferation and inducing apoptosis (Williams et al., 2003). According to Frank-Kamenetsky et al. (2002), CUR61414 directly binds to SMO. When tested *in vivo* on murine BCCs in Ptch1(+/−) K14-CreER2 p53 fl/fl mice CUR61414 seemed able to induce regression of BCCs. However, it was ineffective when applied topically (at 0.09%, 0.35%, 1.1% and 3.1%) for up to 28 days in human superficial or nodular BCCs (Tang et al., 2011c).

Combination of Hh inhibitors with other agents, for example inhibitors of PARP (poly (ADP-ribose) polymerase) and HDAC (histone deacetylase) inhibitors, has been attempted with some success *in vitro* and xenograft models. GDC-0449 has been tested in combination with other inhibitors. Bahra et al. (2012) tested the effects of the steroidal alkaloid cyclopamine and Gemcitabine allied with GDC-0449. Cyclopamine and GDC-0449 inhibited tumour cell growth *in vitro* and *in vivo*. When combined with Gemcitabine, cyclopamine showed synergistic growth inhibition of pancreatic adenocarcinoma xenografts. Cyclopanine binds SMO (Chen et al., 2002) and has been previously shown to be able to inhibit the proliferation of medulloblastoma cells in culture and the growth of murine allografts and xenografts (Berman et al., 2002). However, although GDC-0449 showed no signs of toxicity when administered to mice it produced permanent skeletal abnormalities (Kimura et al., 2008).

A cyclopamine based derivative called IPI-926 (Infinity Pharmaceuticals) has shown complete tumour regression in an allograft medulloblastoma model with active Hh (Tremblay et al., 2008, 2009). IPI-926 is under several clinical trials for the assessment of safety and activity in BCC and Chondrosarcoma (NCT01609179), metastatic pancreatic cancer in combination with Gemcitabine (NCT01130142l) and with FOLFIRINOX (leucovorin) (NCT01383538) and recurrent head and neck cancer in combination with Cetuximab (EGFR inhibitor).

Strand et al. (2011) have identified another small molecule benzamide derivative inhibitor MS-0022 (2-bromo-*N*-(4-(8-methylimidazo[1,2-*a*]pyridin-2-yl)phenyl) benzamide) of SMO. MS-0022 at nM levels effectively inhibited SMO. Its effect at lower μM levels extended downstream to Suppressor of Fused (see Figures 4.1 and 4.2). MS-0022 inhibited cell proliferation *in vitro*. In tumour xenografts, delayed growth correlated with reduced levels of Gli1. Many other SMO inhibitors are known. The Smoothened antagonists SANT74 and -75 and SAG (SMO agonist) are chlorobenzothiophene derivatives with small structural differences yet functioning markedly antagonistically. SANTs and SAG regulate SMO by modulating its molecular conformation, possibly involving alltosteric alteration (Yang et al., 2009a; Rominger et al., 2009).

Combination therapy has much ground to make. In pancreatic cancers for instance neither cyclopamine nor rapamycin on its own had any effect *in vitro* or *in vivo*. Only when both Hh and mTOR pathways were blocked along with standard chemotherapy, there was any discernible reduction in the stem cell pool of pancreatic cancer (Mueller et al., 2009). The possibility that acquired resistance to the inhibitors might be a factor needs to be taken into account. Buonamici (2010) have reported that the SMO antagonist LDE225 produced tumour regression in animal models of medulloblastoma but resistance to the antagonist developed. Upon detailed analysis they noticed that Gli1 gene had amplified and also PI3K signalling system had been activated. The appearance of resistance could be delayed by employing the dual PI3K/mTOR NVP-BEZ235 at the start of the experiment.

At the risk of reiteration, the problem basically is that currently combinations are virtually random selections. The combinations have to be designed because more often than not more than one signalling systems might be operating in a tumour model and so many combinations turn out ineffective. This is where one has to look at inhibitors TGF-β and family, Wnt and miRNA as potential candidates for combination between them and with conventional chemotherapeutic agents. Deregulation of Hh signalling has been associated with aggressive cancers coupled with enhanced tumour vascularisation. So it is rather surprising that *de novo* clinical trials combining Hh and inhibitors of neovascularisation such as avastin have not received the degree of attention they merit.

HDAC Inhibitors Combination with Hh Inhibition

Histone acetyl transferases (HATs) acetylate histones and lead to increased transcription of genes. In contrast, histone deacetylases (HDACs) remove acetyl groups of lysine residues of histones and lead to chromatin condensation and silenced gene transcription. The HDAC inhibitor suberoylanilide hydroxamic acid (SAHA) arrests growth and induces apoptosis and so inhibits tumour growth. Combination of SAHA and SANT-1 has led to growth suppression of Gemcitabine-resistant pancreatic adenocarcinoma cell lines Panc-1 and BxPC-3. The additive inhibitory effects are attributed to upregulation of human hedgehog-interacting protein (HHIP), a negative

Table 5.1 Clinical Trials of SMO Inhibitors

SMO Inhibitor	Tumour Type
GDC-0449	Phase II BCC, pancreatic, ovarian, lung, gastric colorectal, medulloblastoma
	Phase I breast
LDE225	BCC
	Medulloblastoma
	Lung adenocarcinoma, spindle cell sarcoma, breast cancer
IPI-926	BCC, chondrosarcoma
IPI-926+ Gemcitabine	Metastatic pancreatic cancer
IPI-926+ leucovorin	Metastatic pancreatic cancer
IPI-926+ Cetuximab	Recurrent head and neck cancer
BMS-833923 (XL139)	BCC, advanced or metastatic solid tumours refractory to standard therapy
BMS+ lenalidomide + dexamethasone, bortezomib	Refractory multiple myeloma
BMS+ Dasatinib	Leukaemia
BMS+ Cisplatin; Capecitabine	
BMS+ Carboplatin+ Etoposide	Metastatic gastric, gastroesophageal, osophageal adenocarcinomas
LEQ506	Small cell lung carcinoma
TAK-441	Advanced solid tumours, recurrent or refractory medulloblastoma, locally advanced or metastatic BCC
	Non-haematologic malignancies; BCC

Source: Data collated from www.clinicaltrials.gov and from references given in the text.

regulator of hedgehog signaling, by SAHA and accentuated downregulation repression of PTCH-1 mRNA expression (Chun et al., 2009).

Numerous clinical trials are being carried out with HDAC inhibitors on their own or in combination with other agents), which are not enumerated in Table 5.1 or discussed here for the focus is on HDAC inhibitors that potentiate the effects of inhibitors of SMO function. The current or completed clinical trials are listed in Table 5.1, which is based on the information collated from www.clinicaltrials.gov.

Targeting Gli1 in Deregulated Hh Signalling

As discussed earlier, SMO is an important element to target in negating deregulated Hh signalling encountered in cancer. Downstream of SMO is the transcription factor Gli1 whose activity is regulated by SMO. Of course other G-proteins are known to activate it quite independently of SMO. Indeed, several G-proteins possess the ability to activate Gli transcription factors (Douglas et al., 2011). This information is highly valuable since SMO inhibitors can function in synergy with Gli inhibition. Furthermore, PTCH, SMO and Gli transcription factors may be differentially expressed; this can affect the response of the target cells to inhibitors (Decker et al., 2012). Also Hh signalling can adopt the non-canonical systems and inter alia function via the Smad, Notch and PI3K/Akt and MAPK/ERK pathways. Thus besides the considerable focus on SMO, Gli has been recognised and targeted to inhibit aberrant Hh signal transduction.

As noted earlier, Hh signalling activates the embryonic developmental programme EMT and induces the formation and maintenance of CSCs. Hh expression has correlated with aggressive cancers with enhanced tumour vascularisation, marked extension of the tumour to regional lymph nodes and poor disease prognosis. Therefore it is noteworthy that Gli transcription factors could be directly related to the process of angiogenesis. Overexpression of Gli3 seems to induce endothelial cell migration *in vitro* and this was dependent upon activation of Akt and ERK1/2 pathway (Renault et al., 2009). These findings provide additional support to the deployment of Gli inhibitors in a therapeutic mode. Growth factors such as EGF and VEGF activate PI3/Akt and ERK signalling systems. EGFR mediated MEK/ERK activation appears to co-operate with Gli1 and Gli2 to induce cellular transformation. Blocking EGFR and Gli1 inhibits the proliferation of murine BCC cells with activated Gli signalling (Schnidar et al., 2009). Since MAPK signalling promotes Gli1 activity, inhibition of MAPK/MEK pathway is a target worthy of pursuit. So it seems that not only is Gli a valuable target, but it can be inhibited directly or by inhibiting the upstream signalling links. Of the other signalling pathways integrating with Hh/Gli1 is Notch signalling. Notch possibly directly targets Gli. Schreck et al. (2010) have found that the downstream effector of Notch, namely Hes1 binds to the first intron of Gli1 and might be inhibiting its expression.

6 Encountering Aberrant Wnt Signalling

The Wnt system is a major developmental signalling pathway whose activation is an important requisite for many biological phenomena such as stem cell maintenance, maintenance of potential for cell replication, and in the aberrantly activated state it participates in the induction of angiogenesis and invasion. The wide spectrum of pleiotropic function has led to the formulation of and lends significant support to the concept that Wnt signalling might integrate and interact with other developmental pathways such as Notch, Hedgehog and co-ordinate the activation of EMT, cell proliferation and tumour progression. This concept of integrated signalling has been developed and discussed in detail in an earlier location. EMT forms an integral part of cancer development and represents the phenotypic outcome of a closely knitted network of growth factor, Wnt and Hh signal transduction. Indeed EMT is activated by interactive signalling by TGF-β and Hh, both of which possesses the potential to interact with Wnt signalling leading eventually to the downregulation of E-cadherin and upregulation of N-cadherin and Vimentin and to the activation of EMT. Other growth factors such as HGF, PDGF and EGF do promote EMT albeit via different routes. The perceived integration of Hh, Wnt, TGF-β signalling pathways, their cross talk and mutual regulation has made this signalling complex a significant target for therapeutic intervention. In essence then, unsurprising it is that Wnt signal transduction has intuitively formed the focus of attention in the identification, designing and the development of putative small-molecule inhibitors that can block the flow of signal by the interrelating and colluding courses.

The Canonical and Non-canonical Wnt Routes of Signalling

The formation of Wnt ligand-Fizzled receptor complex activates several signalling systems, the canonical/β-catenin and the non-canonical and calcium pathways. The basic feature of Wnt signalling is that in the absence of ligand engagement to the Fzd/LRP5/6 (low-density lipoprotein receptor-related proteins 5/6), a multiprotein complex composed of Axin, GSK-3β, CK1 and APC is formed with β-catenin; Axin promotes the phosphorylation of β-catenin with the mediation of GSK-3β leading to the degradation of β-catenin. The turnover of β-catenin is interceded by β-TrCP (β-transducin repeat containing protein), which is responsible for its ubiquitination and degradation. The degradation of β-catenin enables repressor factors to bind to the TCF (T-cell factors)/Lef (lymphoid enhancing factors) transcription factors to block the transcription of β-catenin responsive genes. On the other hand, when Wnt

Therapeutic Strategies in Cancer Biology and Pathology. DOI: http://dx.doi.org/10.1016/B978-0-12-416570-0.00006-8

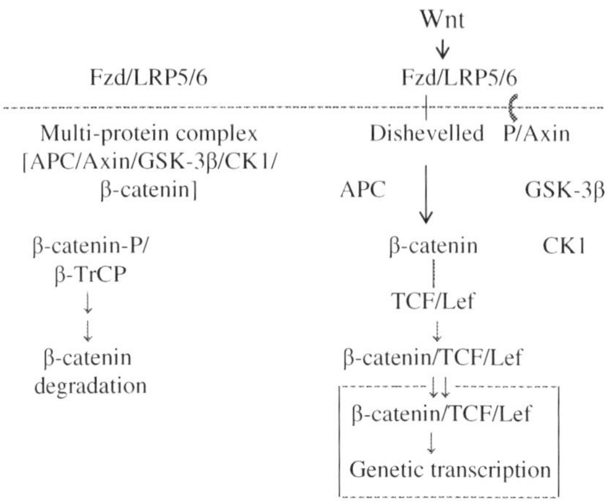

Figure 6.1 A schematic portrayal of the canonical Wnt/ β-catenin signalling pathway. DVL contains DIX, the protein interacting PDZ (PSD-95 a synaptic signalling protein, DLG and the zonula occludens ZO-1 protein) module and the DEP and functions directly downstream of Fzd, and transduces the Wnt signal into the several pathways listed in the text. These are not shown here.

ligand bind the phosphorylated LRP5/6 co-receptor engages Axin. This process leads to the breakup of the multiprotein complex and accumulation of β-catenin in the nucleus. In the canonical pathway, the binding of Wnt ligand to the receptor inhibits GSK-3β and stabilises β-catenin. β-Catenin now forms a complex with the transcription factor TCF/Lef and the complex is translocated into the nucleus leading to the transcription of responsive genes.

The phosphoprotein DVL (Dishevelled) is involved in the disintegration of the multiprotein complex. DVL is multidomain proteins containing DIX (DVL and axin domain), the protein interacting PDZ (PSD-95 a synaptic signalling protein, DLG and the zonula occludens ZO-1 protein) module and the DEP (DVL, Egl-10 G-protein signalling regulator, Pleckstrin) domain. By virtue of the presence of these domains, the DVL effector functions directly downstream of Fzd and transduces the Wnt signal into the several pathways. Among those identified are (a) the CamKII (calcium/calmodulin-dependent kinase II) and PKC; (b) the phospholipase C (PLC) and phosphodiesterase (PDE) activated by the recruitment of heterotrimeric GTP binding proteins; (c) the non-canonical or planar cell polarity pathway) with Rho/Rac and Ra/JNK pathway with implications for cell motility; (d) the Wnt/mTOR pathway and (e) the conventionally recognised one with β-catenin as the main mediator leading to genetic activation. These pathways are integrated into specific developmental patterns and cell fate decisions and markedly implicated in tumorigenic conduits (Figure 6.1).

Several elements can be identified in the canonical and non-canonical Wnt signalling pathways that could be targeted for impeding or preventing the flow of signals along these pathways. The salient ones are (1) the Wnt receptor Fzd, (2) the co-receptors for Wnt ligands, namely the LDL receptor-related proteins LRP5/6, (3) the secreted proteins Dickkopf (Dkk), inhibitor of Wnt signalling, which also bind LPR5/6, (4) SFRPs (secreted Fzd-related proteins), (5) the downstream components β-catenin/TCF, (6) Axin which negatively correlates with β-catenin/TCF and (7) the non-canonical signalling via mitogen activated protein kinase.

Fzd and LRP Inhibition, Dkk and SFRPs

Several natural inhibitors of Wnt signalling have been identified. Wnt ligands bind to the Fzd receptor and the LRP5/6 co-receptors. The ligands bind to the cysteine-rich domain of Fzd. The Dkks and SFRPs are established inhibitors of the pathway. The Dkk are a small family of four secreted proteins which function by blocking the interaction between Wnt ligand and the LRP5/6 co-receptors. Dkks act as ligands for and bind with high affinity to the transmembrane proteins Kremen 1 and 2. The LRP5/6 co-receptors may be endocytosed upon formation of a complex with the Kremen proteins. In this way, Kremens can inhibit Wnt signalling. There is a view that Kremens can indeed enhance Wnt signalling when Dkks are absent, which leaves the LRPs unaffected at the membrane, and further that Dkks might indeed be the determinants of the functioning of the Wnt pathway (see Niehrs, 2006; Cselenyi and Lee, 2008). Five SFRPs have been identified. SFRPs contain around 300 amino acid residues with a signal sequence, an Fzd-like cysteine-rich domain suggestive therefore of Wnt binding, plus a short hydrophobic C-terminal domain (Rattner et al., 1997). SFRPs have generally been regarded as antagonists of canonical Wnt signalling on account of this sequence homology with the Wnt-binding domain of Fzd. Their potential role in pathogenesis is being scrutinised with vigour.

The SFRPs have been suggested to possess tumour suppressor ability possibly by inhibiting Wnt signalling. The expression of SFRPs is reduced in medulloblastomas by methylation and a majority of tumours have shown loss of SFRP1. Induced expression of SFRPs reduces the expression of phosphorylated DVL2 (Kongkham et al., 2010). Epigenetic silencing might involve both promoter methylation and histone acetylation (Meng et al., 2011; Shin et al., 2012). All SFRPs 1, 2, 4 and 5 were almost uniformly methylated in head and neck cancers (Marsit et al., 2006). In contrast, SFRP1 was hypermethylated and silenced in a majority of colorectal cancers (Caldwell et al., 2004). The degree of silencing of SFRP2 by methylation in these tumours might be related with progression. Takeda et al. (2011) identified two methylation sensitive regions in the SFRP2 promoter. In adenomatous polyposis, the methylation SFRP2 was only partial, that is it occurred in either of the two sensitive regions, as compared with rather extensive methylation of both regions in carcinomas (Takeda et al., 2011). These authors also found that mutation of Ki-Ras appeared to be associated with SFRP methylation. Ki-Ras mutations are found frequently in the later stages of adenomas before carcinomatous foci are noticed. Epigenetic alterations seem to begin early in the progression of colonic neoplasia (Belshaw et al., 2008). So by and large there is a distinct possibility that methylation and silencing of SFRP2 occurs before the carcinomatous foci appear. Rawson et al. (2011) have confirmed SFRP silencing by methylation in colon carcinomas. They also noted Dkk1 suppression in a much smaller proportion of patients. The SFRP1 methylation status inversely related to MSI (microsatellite instability), but Dkk1 methylation was found in tumours with MSI. Although conceptually there is a causal relationship between methylation and MSI statuses, the significance of this difference needs further elucidation, since less than 20% of colon cancers show MSI. At the phenotypic levels, SFRPs have been implicated in angiogenesis. SFRP2 is expressed in the microvasculature associated with breast cancers and has demonstrable angiogenic properties *in vitro* (Bhati et al., 2008; Courtwright et al., 2009). SFRP2 induces endothelial cell migration and tube formation *in vitro*. The Ca^{2+}-regulated transcription factor NFAT (nuclear factor of activated T cells) seems to control gene expression required for SFRP2 stimulated angiogenesis (Courtwright et al., 2009). In contrast, SFRP4 has been found to inhibit endothelial cell migration and the development of endothelial sprouts *in vitro*. It is also able to block the effects of VEGF. There was also inhibition of tumour growth of xenografted SKOV-3 cells implanted subcutaneously into BALB/c nude mice (Muley et al., 2010). Unfortunately vascular density associated with tumours was not been determined. Previously He et al. (2005) reported silencing of SFRP4 gene in mesothelioma cells. Indeed some tumours show uniform methylation of SFRPs. Some native molecules possessing Fzd motifs actively inhibit Wnt signalling. Lavergne et al. (2011) have described such a molecule which they call V3Nter. The latter is an SFRP-like molecule that seems to inhibit canonical Wnt/β-catenin signalling and the β-catenin responsive genes. V3Nter inhibits cell proliferation *in vitro* as well as *in vivo* tumour growth.

A secreted protein called the Tsukushi (TSK) protein binds to the cysteine-rich domain and serves as a competitive inhibitor of Wnt ligands (Ohta et al., 2011). While most of these studied focus on Wnt signalling, other pathways of signalling might be involved in some of the phenotypic effects such as migration. SFRP1 seems able to activate ERK mediated increase of MMPs (Foronjy et al., 2010) that could be conducive to significant changes in cell behaviour. Equally downregulation of SFRP1 may promote Wnt mediated activation of EMT or by TGF-β mediation as suggested by Gauger et al. (2011).

Fzd and DVL Interaction

Another site of intervention is the Fzd–DVL binding. This interaction occurs via the PDZ domains of DVL. PDZ domains are protein–protein recognition and interaction modules. The nomenclature PDZ is derived from three protein (PSD-95 a synaptic signalling protein, DLG the Drosophila melanogaster Discs Large protein, and the zonula occludens 1 protein ZO-1) (Harris and Lim, 2001). PDZ domains occur in many signalling proteins and participate in the assembly of signalling complexes. They can recognise specific C-terminal as well as internal sequences that resemble a terminal sequence.

Fujii et al. (2007) designed a small-molecule inhibitor called FJ9. This disrupts the interaction between the Fzd7 and the PDZ domain of DVL, so diminishing canonical Wnt signalling. They also demonstrated that FJ9 suppressed tumour cell growth *in vitro* with induction of apoptosis. Also it markedly suppressed growth of xenografted human cancer cells. Small peptides that appear to inhibit Fzd7 activity by disrupting the Fzd–DVL have also been described by Nambotin et al. (2011). These induced apoptosis of hepatocellular carcinoma cells *in vitro*. Using a transgenic murine tumour with many features corresponding with human hepatocellular carcinoma, Nambotin et al. (2011) have been able to show intratumoral injection of these peptides retarded tumour growth.

The importance of this approach of targeting aberrant Wnt signalling is underscored by its latent value in the management of TNBC. Transcriptional suppression of LRP6 or FZD7 in TNBC cells inhibits tumour growth *in vivo*. The TNBC are most refractory to treatment by their very nature of being ER−/PR−/HER2− and therefore show no response to tamoxifen, aromatase inhibitors or to Herceptin. Around a tenth of breast cancers turning out to be triple negative, TNBCs are highly aggressive and associated with enhanced recurrence and metastasis and poor prognosis. Thus the problem of managing these patients assumes a serious dimension. Several therapeutic avenues have been investigated, such as targeted DNA homologous recombination, inhibition of EGFR and VEGFR signalling, inhibitors of mTOR, src kinase signalling and cell cycle checkpoint inhibitors, among others (Sherbet, 2011a). Fzd7 and LRP6 are overexpressed in the triple negative breast cancer cell lines MDA-MB-231 and BT-20 cells. Suppression of Fzd7 has led to inhibition of cell proliferation and colony formation and invasion *in vitro*. *In vivo* formation of tumours from xenografting of cells with suppressed Fzd7 was also significantly inhibited (Yang et al., 2011d).

In the context of TNBC, it would be worthwhile stating that Salinomycin has turned out to be a useful inhibitor of Wnt signalling with associated consequent effects on cell proliferation, apoptosis and migration. It is suggested to be a putative cancer stem cell inhibitor. Salinomycin has been found to inhibit proliferation of breast cancer cells and prostate cancer cells. In breast cancer cells, its inhibitory effect is over a 100-fold greater than paclitaxel. The susceptibility of breast cancer cells to salinomycin correlated with the proportion of $CD44^{high}/CD24^{low}$ CSC cells. The antibiotic also inhibited tumour growth in murine models with downregulation of genes associated with the CSC phenotype (Gupta et al., 2009). In prostate cancer cells besides reducing the numbers of cells, Salinomycin reduced the expression of myc, AR and ERG, induced oxidative stress and inhibited NF-kB activity and cell migration (Ketola et al., 2012). In osteosarcomas too, Salinomycin could be targeting stem cells by indication of its effects on stem cell marker expression (Tang et al., 2011b). These authors have also implicated interference with Wnt signalling. Interestingly, in Wnt-transfected HEK293 cells Salinomycin prevents phosphorylation and promotes the degradation of LRP6. In CLL cells with constitutively activated Wnt, the antibiotic downregulated the expression of the transcription factor LEF1 and cyclin D1 (Lu et al., 2011a). With this background, it might be worthwhile pursuing Salinomycin in the treatment of TNBC. Quite clearly this route has to traverse a long way, but most certainly it is worth the effort.

The non-steroidal anti-inflammatory agent sulindac has been found to interfere with Wnt signalling. The β-catenin/TCF transcription complex targets many genes. The metastasis promoter S100A4 is regulated by this complex. Stein et al. (2006) found TCF binding sites in the S100A4 promoter and the β-catenin/TCF complex directly bound S100A4 promoter and enhanced its activity. Stein et al. (2011) intra-splenically xenografted colon cancer cells over expressing S100A4 and treated the host animals with sulindac. This treatment seems to have interfered with Wnt signalling, reduced β-catenin expression and its nuclear accumulation together with a decrease in the formation of β-catenin/TCF transcription complex, and importantly here in the downregulation of S100A4 promoter activity and expression. The tumour that formed in the spleen and its metastases in the liver both had reduced levels of β-catenin and S100A4. Sulindac does operate by other means and more effective agents such as the phospho-sulindac (OXT-328) have been described. Phospho-sulindac is believed to be more effective than sulindac. Zhu et al. (2012) have claimed that it targets breast CSCs *in vitro* and in breast cancer xenografts, inhibits cell proliferation and induces apoptosis of breast CSCs. The demonstration of the inhibitory effects on Wnt signalling via suppression of S100A4 provides a reasonable basis for further investigations into the mode of action and anticancer properties of sulindac and its derivatives.

7 Therapeutic Targeting of TGF-β Signalling

Morphogenesis, neoplastic transformation and EMT are a phenotypic outcome of a closely integrated network of growth factor, Wnt and Hh signal transduction. We have seen how Hh signalling interacts with and interrelates to the Wnt pathway. TGF-β also can activate Wnt signalling. Arguably TGF-β can do this by means of the canonical Smad system or indirectly by the Smad/NF-κB route. Hh can function in concert with growth factors such TGF-β, FGF and together with the trail of transcription factors that can lead to the activation of Hh signalling and activation of the EMT programme. TGF-β can augment the expression of some miRNAs and perform a co-operative induction of EMT. Growth factor signalling by the activation of RTKs can activate Gli and aid transcription of Hh responsive genes. Among key characteristics of EMT are the loss of E-cadherin, acquisition of N-cadherin and vimentin, together with the activation of Wnt/β-catenin, which highlight a major mode of signalling. Finally, certain tumour promoters have been seen as potential activators of EMT. This network of signalling with possible cross talk and synergistic as well additive functioning provides a firm foundation on which to build a multifaceted therapeutic edifice.

The targeting of TGF-β signalling may be regarded as a potentially rewarding avenue to approach. Inhibition of canonical Smad signalling also inhibits the Wnt/β-catenin pathway and in consequence modulates E-cadherin expression, thus providing a dual route to the inhibition of EMT. It is well known that TGF-β can act as a tumour suppressor or switch to a promoter mode of function. Therefore detailed information about the factors involved in the switch over and if this switch operates at specific time points in tumour progression is crucially important in developing strategies for therapeutic deployment of TGF-β inhibitors. One can see a spectrum of effects following upon TGF-β inhibition. Modulation of the microenvironment including the infiltration of inflammatory cells and fibroblasts into the tumour microenvironment can quite conceivably modulate the signalling functions of regulatory ligands such as TGF-β. The derepression of TGF-β induced suppression of immunological responses is another salient phenomenon. Neutralising TGF-β with antibodies or RTK (receptor tyrosine kinase) inhibitors has led to a remarkable inhibition of progression of triple negative mouse tumours to the metastatic state (Tan et al., 2009). In this study by Tan et al. (2009), no effects were seen on cell proliferation and apoptosis, which might suggest a stage dependent effect of inhibition possibly reflecting the well-known switching of TGF-β function with disease progression. The potential involvement of miRNA-128a in the suppression of the Smad signalling pathway would deserve to be mentioned here. Masri et al. (2010) reported that miRNA-128a bound to the 3′UTR region of the gene encoding the TGF-β RTK

Therapeutic Strategies in Cancer Biology and Pathology. DOI: http://dx.doi.org/10.1016/B978-0-12-416570-0.00007-X

protein and negatively regulated canonical TGF-β signalling. This could inhibit EMT activation by the ligand. So the obvious means of approach to the inhibition of TGF-β signalling are monoclonal antibodies raised TGF-β family receptors, targeting Smad and related signalling components and using antisense RNAs to inhibit TGE-β ligands and their receptors at the mRNA level, which was proposed many years ago.

Antisense Oligonucleotide Trabedersen (AP 12009)

Trabedersen (AP 12009) is an antisense oligonucleotide inhibitor which targets TGF-β2 mRNA. Many studies have been reported on the effectiveness of Trabedersen on several forms of cancer. A randomized controlled international phase III clinical trial of recurrent or refractory anaplastic astrocytoma and glioblastoma began in early 2009, wherein Trabedersen is administered by convection-enhanced delivery (CED), an intratumoral drug delivery method, which is highly effective in patients with malignant brain tumours.

Schlingensiepen et al. (2011) studied the antitumour activity of Trabedersen in pancreatic cancer cells and *in vivo* xenograft of metastatic pancreatic cancer in muirine hosts. *In vitro* Trabedersen at μM concentration effectively reduced TGF-β2 secretion by pancreatic cell lines and inhibited cell proliferation, and migration of pancreatic cancer cells. *In vivo* xenografts of metastatic pancreatic cancer treated intraperitoneally significantly inhibited tumour growth, lymph node metastasis and angiogenesis (Schlingensiepen et al., 2011). A phase I clinical trial is under way in advanced pancreatic carcinoma, metastatic melanoma and metastatic colorectal carcinoma. Trabedersen is administered intravenously to patients with advanced tumours over producing TGF-β2 (NCT00844064).

Anti-TGF-β Monoclonal Antibodies

Lerdelimumab, Metelimumab and Fresolimumab

Two monoclonal antibodies against TGF-β have received some attention. Lerdelimumab (CAT-152) might be useful in treating some ophthalmological conditions. It reduces TGF-β fibrotic effects in Graves' ophthalmopathy (Pohlers et al., 2009). It was shown some time ago to be able to inhibit TGF-β induced effects and suppress posterior capsule opacification, which occurs after extracapsular cataract surgery and intraocular lens implantation (Wormstone et al., 2002). Lerdelimumab prevents, not invariably, corneal scarring after trabeculectomy. Metelimumab (CAT-192) is a human monoclonal TGF-β1 antibody which is being touted as potential treatment of diffuse cutaneous systemic sclerosis (Sorbera, 2004). The human anti-TGF-β antibody Fresolimumab is said to neutralise all isoforms of TGF-β and has immunomodulatory ability. It has been the subject of phase I trial in patients with refractory primary focal segmental glomerulosclerosis (FSGS) to assess the safety

TGF-β family ligand

|

Type IIR

|

Heterotetrameric complex
I/II-II/I

----------------------↓----------------------

RI-P

↓

R-Smad-P

|

Smad4

----------------------↓----------------------

R-Smad/Smad4

+

DNA binding co-factors

+

Co-activator/co-repressors

↓

Responsive target genes

Figure 7.1 A flowchart of the TGF-β signalling cascade leading to the formation of Smad transcription factors (Sherbet, 2011a).

and pharmacokinetics of single-dose infusions. The antibody is well tolerated but preliminary findings have revealed some side effects (Trachtman et al., 2011). A study of systemic sclerosis is under way at present to assess the effects of the antibody on the skin of patients (NCT01284322). Relevant in the present context are two studies that began in 2011. One of them relates to the safety and effects on possible tumour regression of combining radiotherapy and Fresolimumab administration to patients with metastatic breast cancer (NCT01401062). The second one is a phase II study of malignant glioma (NCT01472731).

Small Molecule Inhibitors of the TGF-β Receptor Family

TGF-β family of ligands transduce their signals via the type I and type II (RI and RII) receptors and type III accessory or co-receptors. These transmembrane receptors consist of an extracellular ligand-binding domain, a transmembrane domain and a cytoplasmic serine/threonine kinase domain. Seven type I ALK and five type II receptors have been identified. Upon ligand binding, the type II receptor initially engages type I to form a heterotetramer receptor complex, which phosphorylates and activates type I (I–P*). The activated type I receptor in turn phosphorylates cytoplasmic R-Smads. These bind to Smad4. The R-Smad–P*/Smad4 complex now within the nucleus integrates with co-factors leading to the recruitment of transcription co-activators or co-repressors (Figure 7.1) (Sherbet, 2011a).

Inhibition of TGF-β RTKs has been tested in dermal sclerosis. Blocking TGF-β RI ALK-5 using LY36494 (*SD208*) has been shown to suppress expression of type I collagen and integrin β1 in dermal fibroblasts derived from systemic sclerosis and in parallel reduce the adhesive and contractile abilities of the fibroblasts (Chen et al., 2006). Amplified expression of ALK-5 receptor gene has been encountered in neuroblastomas. The occurrence of somatic mutations of the kinase domain and germ line mutations in two neuroblastoma families has led to the suggestion that the ALK gene might be associated with the development of neuroblastomas. ShRNA (short hairpin RNA)-mediated knockdown of the gene has resulted in the suppression of cell proliferation (Janoueix-Lerosey et al., 2008).

LY2109761 is a specific receptor kinase inhibitor commonly employed *in vitro* and *in vivo* studies into TGF-β signalling. In human HCC cell lines, inhibition of TGF-β signalling using the TGF-β LY2109761 blocks the canonical signalling pathway and upregulates E-cadherin mRNA and protein levels and inhibits the migration of cells *in vitro* (Fransvea et al., 2008). The finding that pre-treatment of breast cancer cells with the inhibitor enhances the radiosensitivity of the cells (Bouquet et al., 2011) augments the potential value and prospective therapeutic deployment of the TGF-β RTK inhibitor. Type RI ALK4/5 function is inhibited by SB431542 and SB525334. These compounds inhibit Smad phosphorylation. Ogunjimi et al. (2012) have found that SB431542 is a selective inhibitor of ALK kinases with the exception of ALK2, which indeed suggests that it might not inhibit signalling by BMP (bone morphogenetic protein). Further they show that Ser280-Thr mutation in RI makes it resistant to SB431542 and mutation of the corresponding Thr283-Ser in ALK2 makes it sensitive to the inhibitor. Thus Ser280 seems to determine the selectivity for SB431542. Petersen et al. (2008) have described another type IR kinase inhibitor GW788388 which inhibits both type I and type IIR kinase activities. As in SB525334, no inhibition occurs of BMP type IIR kinase. At the phenotypic level, GW788388 is able to suppress EMT. One has to consider these findings in the context of the fact that the Src kinase inhibitors PP1 and PP2 negate the growth inhibitory effects of TGF-β1 and also suppressed TGF-β1 induced EMT. Tatton et al. (2003) found that PP1 can inhibit c-Kit and Bcr-Abl RTKs. Imatinib inhibits the c-Abl kinase and suppresses the proliferative influence of TGF-β (Wang et al., 2005). The imatinib related inhibitors dasatinib and nilotinib effectively counteract TGF-β and PDGF signalling (Distler and Distaler, 2010). *In vitro* kinase assay indicated that inhibitory effects were comparable to that of SB431542 (Ungefroren et al., 2011). So the question of selectivity of inhibition needs further study. With Src kinase inhibitors being promiscuous in their function, perhaps there is a case to be made for combining Src inhibitors with TGF-β receptor inhibitors.

SKI-606 (Bosutinib) seems capable of inhibiting migration of breast cancer cells *in vitro*, which is accompanied by peripheral localisation of β-catenin and inhibition of focal adhesion kinase (FAK) signalling (Vultur et al., 2008). It inhibits the phosphorylation of many key signalling elements, for example, Akt, MAPK besides FAK and alters the expression of N-cadherin, fibronectin and MMPs. Also notably, SKI-606 inhibits the expression of TGF-β and BMP in prostate cancer cells (Rabbani et al., 2010). It is needless to say there are many such inhibitors that might be tested for possible synergistic or additive effects. SB431542 and SB525334 have been found to be particularly effective for increasing the cytotoxicity of Gemcitabine on pancreatic cancer cells that are resistant to the drug. SB525334 significantly increased apoptosis of cell death in the resistant cells and this appears to be due to the inactivation of Akt signalling (Kim et al., 2012b). Finally, the ubiquitous miRNAs should not go unheeded from this scenario. According to Mizuno et al. (2009), miRNA-210 inhibits the TGF-β R1b receptor.

Inhibition of Type RIII Function

Among the regulatory components of the canonical Smad signalling system are betaglycan (herein referred to as RIII) and endoglin (CD105), which share some sequence homology and which are often described as accessory type III receptors of the TGF-β family. RIII mediates both canonical and non-canonical signalling by TGF-β family members and also participate in ligand independent signalling.

The co-receptor RIII is widely regarded as a tumour suppressor. It has been implicated in the regulation of cell proliferation, and angiogenesis and its loss is believed to promote tumour progression. RIII expression is reduced or lost in many neoplasms. Restoration of its expression in multiple myeloma cells inhibits cell proliferation and motility (Lambert et al., 2011). A similar situation has been reported in prostate cancer where loss of expression of RIII has been linked with LOH of the RIII encoding gene. With LOH found only in around 58% of specimens, loss of expression seems to be compounded by changes in epigenetic regulation. Although RIII expression reduced cell invasion, but it had no effect on cell proliferation. Underscoring an obvious link with invasive capacity, the loss of RIII correlated with the expression of PSA (Prostate-specific antigen) in prostate cancer suggesting a potential relationship with progression (Turley et al., 2007). LOH at the RIII locus has also been reported in breast cancer, where loss of RIII expression correlated with disease progression and poorer prognosis (Dong et al., 2007a).

Compatible with these findings, loss of RIII, but not of RI and RII, appears to be conducive to EMT. The activation of EMT accompanying the loss of RIII possibly occurs without the mediation of E-cadherin or its regulator transcription factors. Although this occurs subsequent to the loss of RIII, maintaining RIII levels suppresses EMT without affecting the expression levels of E-cadherin. Also suppression of RIII with shRNA promotes EMT without affecting the expression of E-cadherin or Snail (Gordon et al., 2008). It seems possible that signalling might operate via other pathways and independently of both canonical and non-canonical systems. MicroRNAs are known to activate EMT independently of both. Also TGF-β can promote EMT by the mediation of microRNAs. Another example of RIII functioning independently of both canonical and non-canonical pathways is the induction of apoptosis by experimentally upregulation of RIII *in vitro* and *in vivo* murine models of clear cell renal carcinoma. Here apoptosis was induced by p38 MAPK signalling (Margulis et al., 2008). Once again one can invoke the intervention by microRNAs. For instance, miRNA-137 can target cdc42 and inactivate it and lead cells into G1 arrest and into the apoptosis path.

A phase I study (NCT00656825) designed to evaluate tolerability and bioavailability of P144 peptide, which blocks the interaction between ligand and type RIII in healthy subjects, has been reported. The P144 peptide covers amino acid residues 730–743 (TSLDASIIWAMMQN) of the ligand binding external domain of the RIII receptor. This peptide inhibitor is also being investigated for its safety and efficacy by topical application in patients with skin fibrosis related to systemic sclerosis (NCT00574613). There are many experimental studies on the effects of this peptide on fibrosis, but these have not been discussed here on criteria of relevance.

8 EGFR Signalling in EMT

The EGFR family have been known for a while to be actively related to activation of EMT and the acquisition of biological features associated with cellular transformation to the EMT phenotype and drug resistance. EGF is a proven promoter of cell proliferation and is known to induce EMT. It decreases cell adhesion and enhances cell motility. Investigations into EGFR family in relation to EMT have acquired much impetus due to the problem of drug resistance encountered in patient management. The mechanisms of acquired resistance to Trastuzumab in patients expressing HER2 at high levels are being seriously addressed at present. It has been suggested that upregulation of EGFR might be one of the mechanism involved in the induction of the EMT phenotype and this acquired resistance (Figure 8.1).

E-cadherin in EGFR Signalling

The induction of the EMT phenotype by EGFR activation may occur by several routes. The involvement of E-cadherin with its expression being suppressed in EGFR-mediated induction of EMT is now a distinct possibility, indeed recognised many years ago. Jones et al. (1996) reported that the loss of E-cadherin markedly correlated with levels of EGFR expression. Activation of EGFR and transactivation of HER2 initiates signalling cascades that lead to increased motility and invasion. EGF treatment and EGFR activation have the effects of downregulating E-cadherin expression, whereas inhibition of EGFR enhances E-cadherin expression (Zhang et al., 2012f; Cheng et al., 2012a). On the other hand, EMT and the expression of EMT markers are suppressed when EGFR activity is inhibited. Also E-cadherin expression is upregulated with attendant reduction in the Snail and Slug transcription factors that are negative regulators of E-cadherin expression (Chang et al., 2012b). Similarly serous borderline ovarian tumours EGF treatment upregulated Snail, Slug and ZEB1 leading to the suppression of E-cadherin and induction of EMT.

EGFR kinase inhibitors geftinib and erlotinib constitute effective treatment of lung cancers where activating mutation of EGFR are present (Engelman et al., 2007). Using geftinib-sensitive PC9 cells and a resistant clone (PC9/gef), Chang et al. (2011d) were able to demonstrate an enhanced expression of Slug in the resistant PC9 cells as compared with the sensitive ones. However, the mechanism involved in acquired geftinib resistance can involve secondary EGFR mutation as well as overexpression of HGF and amplification of its receptor c-met (Engelman et al., 2007; Kosaka et al., 2011). HGF-mediated induction of cell invasion and EMT does indeed involve upregulation of Snail. This seems to occur via the MAPK/Egr (early growth response)-1 pathway which leads to Snail upregulation and repression of E-cadherin (Grotegut et al., 2006).

Therapeutic Strategies in Cancer Biology and Pathology. DOI: http://dx.doi.org/10.1016/B978-0-12-416570-0.00008-1

EGFR Signalling Path to EMT

Much effort has gone into determining which signalling systems are operational in EMT initiation by EGF/EGFR. Cheng et al. (2012b) besides confirming the downregulation of E-cadherin expression and the corresponding upregulation of the negative regulatory transcription factors have stated that EGF/EGFR effects were mediated by MAPK family kinases ERK1/2 and p38 and Akt and the effects were blocked upon inhibition of MAPK and Akt signalling. Conversely, inhibiting E cadherin expression has resulted in the upregulation of EGFR (Wang et al., 2011a). Interestingly EGF treatment also activated PI3K/Akt and ERK1/2 signalling (Cheng et al., 2012a). However, it ought be stated here that expression of many genes that straddle stage I/II and early stage III epithelial ovarian cancer is modulated, including E-cadherin and also notably CD44, MMP and uPA, fibronectin, laminin, all closely related to the EMT phenotype (Sherbet and Patil, 2006). In particular, CD44 and MMPs have been closely associated with EGFR signalling and EMT. Jensen et al. (2012) while unable to confirm this relationship between EGFR and PI3K activation do suggest that PI3K mutations and activation of signalling might be more resistant to anti-HER2 therapy. Loss of PTEN has been held to predict Trastuzumab resistance, but Jensen et al. (2012) reported no such correlation. Activation of PI3K pathway following from PI3K mutation or loss of PTEN has been implicated in the development of resistance to lapatinib and Trastuzumab (Wang et al., 2011a).

Earlier, Gan et al. (2010) propounded a differential function of PI3K/Akt and ERK signalling. EGFR activation suppressed E-cadherin mediated by the upregulation of Snail and EMT activation, but this depended upon Akt activation and occurred independently of ERK signalling. The latter seemed to negatively regulate EGFR activation. The significance of the activation of Akt pathway has to await further clarification since others have shown that the activation of this pathway or the loss of PTEN is unrelated to Trastuzumab resistance in HER2 expressing breast carcinoma (Barbareschi et al., 2012).

Whilst there is general agreement that the downregulation of E-cadherin is due to the negative regulatory ZEB, Snail and Slug transcription factors, Zuo et al. (2011) have reported that EGFR activation not only initiated PI3K/ERK signalling but also led to the production of MMP9. The latter might be promoting the cleavage and release of E-cadherin. These changes were accompanied by the acquisition of the EMT phenotype. They have reiterated further the postulate that the soluble E-cadherin might be activating ERK1/2 and PI3K/Akt pathways.

ECM and Cell Membrane Components in EGFR Signalling

EGFR activation by ligand binding involves lateral mobility, dimerisation and internalisation of the receptor. It is to be expected therefore that cell membrane and ECM components would influence EGFR signal transduction. As noted elsewhere CD44, MMPs and scaffolding proteins that form links with the cytoskeletal machinery are closely involved with EGFR signal mediation to generate phenotypic effects of motility and proliferation.

The hyaluran receptor CD44 expression has been closely associated with EGFR signalling and EMT. CD44, another marker of EMT, has been described as being able to activate EGFR. Overexpression of CD44 correlated closely with EMT and interestingly also with increased expression of EGFR together with the activation of PI3K/Akt and enhanced GSK-3β expression. Also in parallel the expression of E-cadherin was reduced but that of N-cadherin and vimentin increased (Cho et al., 2012).

The ERM (Ezrin–Radixin–Moesin) proteins are a family of proteins that link the plasma membrane with the actin cytoskeleton. The ERM binding protein EBP50 forms this link. EBP50 contains the PDZ (PSD-95 a synaptic signalling protein, DLG Discs Large protein and the zonula occludens ZO-1 protein) domain and by virtue of this it is able to recognise and interact with several target proteins involved in biological processes including cell adhesion, motility and signal transduction. With the ability to interact with E-cadherin, EBP50 has been attributed with a function in EMT. EBP50 interacts with and forms complexes with EGFR and FAK and in this way promote focal adhesion and cell migration (Song et al., 2012b). This interaction between EBP50 and EGFR is an essential requirement in the activation of EMT by EGFR (Claperon et al., 2012).

CA125 (cancer antigen 125, MUC16) is a membrane glycoprotein involved in cell adhesion, cell proliferation and apoptosis and linked with tumour progression. It is a marker of malignancy in many tumours, prominently ovarian cancer. High serum levels of the marker correlate with peritoneal dissemination of ovarian cancer and poor prognosis (Emoto et al., 2012). Tumour tissues also express the antigen in a variable pattern, but high tissue expression too correlates with poor prognosis (Strappel et al., 2012). Ogmundsdottir et al. (1996) measured serum and tissue CA125 in breast cancers and noted difference in the surface and cytoplasmic distribution between normal breast tissue and carcinomas; in a majority of carcinomas CA125 was associated, albeit not exclusively, with cytoplasmic granules. The expression of both may be relevant in predicting recurrence and prognosis. The possibility that the loss of membrane associated CA125 might contribute to EMT has to be entertained. EMT is induced under certain circumstances such as downregulation of CA125. It could be facilitating the formation of peritoneal metastases by aiding the binding of the tumour cells to mesothelial cells. At first sight these findings look intrinsically contradictory. However, CA125 which is believed to interact with E-cadherin and β-catenin and it could be stabilising the membrane location of E-cadherin. Downregulation of CA125 not only activates EMT but also activates EGFR signalling via ERK and PI3K/Akt (Comamala et al., 2011).

An important aspect of EGFR activation is internalisation of the ligand-receptor complex. Receptor dimerisation, activation of RTK and autophosphorylation of the receptor are important requirements for the internalisation process. E-cadherin interacts through its extracellular domain with EGFR and other receptor tyrosine kinases and this leads to regulation of EGFR by decreasing receptor mobility and affinity of ligand to the receptor (Qian et al., 2004). So it seems possible that the interactions between E-cadherin and EGFR or between CA125 and EGFR might prevent internalisation of EGFR upon ligand binding. In other words, inhibition of E-cadherin or its stabilisation at the membrane might have similar outcomes in terms of EMT activation and the associated alterations in marker expression upon EGFR signalling. Indeed Zhao et al. (2010) showed that E-cadherin expression is reduced via EGFR/ERK signalling.

EGFR and TGF-β Signalling Pathways Interact in EMT

EGFR might function with several other cellular factors. Not least among them is TGF-β1, which itself is capable of initiating EMT. EGFR has also been shown to engage in cross talk with TGF-β1/Smad canonical signalling in activating EMT. TGF-β1 could not induce EMT when EGFR was inhibited (Kang et al., 2012a), possibly indicating a regulatory role (see below). Cell subpopulations overexpressing EGFR and expressing the ZEB negative regulators of E-cadherin appear to be more prone to undergo EMT in response to TGF-β stimulation; also this competence for EMT was augmented by the presence of p53 mutations (Ohashi et al., 2010). Further, Kang et al. (2012a) showed that TGF-β signals by the canonical route involving Smad to activate EGFR function. They found that activation of EGFR is required for TGF-β to induce EMT.

Supplementary evidence of TGF-β/EGFR cross talk comes from the demonstration that the transmembrane 4 L6 family member 5 (TM4SF5) might be implicated in the induction of EMT. TM4SF5 is highly expressed in HCC with attendant activation of EMT. Inhibition of TM4SF5 seems to suppress EMT. The involvement of TGF-β and EGF signalling in this process has been emphasised by their ability to induce TM4SF5 (Kang et al., 2012b). As noted before, TGF-β can indeed do that in different ways, canonical and non-canonical pathways together with other modes of signalling. Nonetheless, in the hepatocellular carcinoma model, it is clear that EGFR is an en route requirement for induction of EMT by TGF-β. This view differs markedly from the one expressed earlier by Murillo et al. (2005) who found that whilst TGF-β was capable of activating EGFR and Akt signalling, TGF-β mediation of EGFR activation may involve the proteinase ADAM 17 inhibition of which suppresses Akt activation. EGFR involvement did not appear to be essential component in the induction of EMT by TGF-β. However, this dissension does not detract from the convincing demonstration of the role of EGFR in activating EMT.

Finally, it would be relevant to mention here the considerable role that miRNAs play in the regulation of cell proliferation and apoptosis and growth factor signalling, discussed in an earlier location (see pp. 5–18 and Figures 3.4–3.7). MiRNAs target and influence the function of several genes such as p53, Bcl-2 family members, and myc and so regulate cell cycle progression and cell proliferation and apoptosis. MiRNAs also influence signalling by growth factors and growth factor receptors, for example the EGFR signalling to EMT. They have been seen to function by interacting with and modulating the signalling pathways that are activated by growth factors.

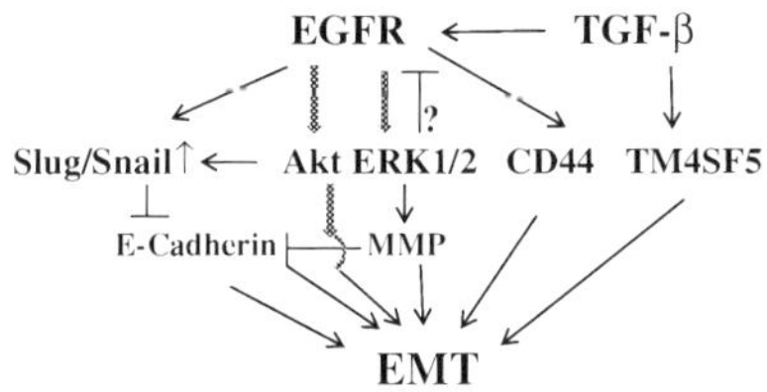

Figure 8.1 The postulated pathways of the induction of EMT by EGFR activation by its own ligand and TGF-β. TGF-β-mediated activation of EGFR may involve the proteinase ADAM 17, inhibition of which suppresses Akt activation (not shown here), but the precise mode of action needs elucidation. MiRNAs play a substantial role in the regulation of signalling by EGFR family. This is not shown here nor discussed in this section since this topic has been reviewed in detail in another location (see pp. 5–24 and Figures 3.4–3.7).

Overall there is overwhelming evidence that EGF/EGFR activates EMT, although the modes of initiation of EMT are yet to be fully elucidated. Nevertheless there are very clear leads that can be followed up, especially in relation to the interactions of EGFR with cell membrane determinants of invasion and adhesive function and the signalling systems that might be relevant in the context of EGF/EGFR-mediated changes in cell behaviour (Figure 8.1).

Part 3

Therapeutic Deployment of Metastasis-Associated Gene Function

Abnormalities of cell proliferation, motility and intercellular and cell–substratum adhesion are distinguishing features of cancer, which are phenotypically revealed in the initiation, growth and distant dissemination of tumours. These arise from deregulated signalling derived from the overexpression or inappropriate expression of genetic determinants that are identifiable with these cellular attributes. Diverse determinants can co-operate in their distinctive modes and contribute to the totality of the phenotype; the signalling systems can function interactively and influence each other. Equally, one can identify genetic determinants that transduce signals using diverse routes. It is essential that the complexities of these pathways and how they interact and regulate the flow of information are elucidated in order that specific chemotherapeutic targets can be identified and new treatment strategies can be evolved.

The most obvious targets of therapy are (a) physiological/differentiation-related targets: cell signalling pathways as for instance by modulation of hormonal and growth factor signalling, (b) physical/morphology related: motility factors; remodelling of the cell membrane and the extracellular matrix; cytoskeletal targets and (c) molecular targets differentially expressed and related to tumour growth and cell proliferation and apoptosis (Sherbet, 2009). Many tumour promoters have been identified which fulfil one or more of these essential attributes.

9 S100A4 as a Potential Target

S100A4 is a member of the S100 family of calcium binding proteins. The 1q21 locus harbours the epidermal differentiation complex (EDC) encompassing a 2.05 Mbp of human genomic DNA. The S100 family genes except the S100beta are arranged in the following order: 1 cen-S100A10-S100A11-THH (trychohyalin)-FLG (profilaggrin)-IVL (involucrin)-LOR (loricrin)-S100A9-S100A12-S100A8-S100A7-S100A6-S100A5-S100A4-S100A3-S100A2-S100A13-S100A1-1qtel.S100beta is located on 21q22.3 (Schäfer et al., 1995; Marenholz et al., 1996; Mischke et al., 1996).

The Spectrum of Biological Function of S100A4

S100A4 signalling can focus on factors associated with normal and aberrant proliferation, apoptosis and growth, and differentiation. It is able to activate signalling pathways leading to the remodelling of the cell membrane and the extracellular matrix, modulation of cytoskeletal dynamics, acquisition of invasiveness and induction of angiogenesis. Thus S100A4 protein can promote metastasis. This has been confirmed by gene transfer studies. Its expression strongly correlates with tumour spread to lymph nodes and with prognosis. Significantly, S100A4 appears to act as a counter point to metastasis suppressor nm23. The routes of signalling by S100A4 are listed in Table 9.1. S100A4 is arguably a molecular target of considerable potential possessing a wide ranging biological activity that can alter and regulate the major phenotypic features of cancer (Sherbet, 2009, 2011b).

The ideal therapeutic strategy would involve the identification of targets most likely to be effective in the disease process with minimum side effects. The ability of S100A4 to activate interacting and multi-functional signalling systems and induce several biological activities highly relevant in the context of cancer would make it an eminently valuable chemotherapeutic target. Some downstream effectors of the S100A4 pathways might also lend themselves as targets of interest. It is now an indisputable fact that S100A4 signalling is deregulated under certain circumstances and that this leads to abnormal cell behaviour (Table 9.1).

Influence of Wnt Signalling on S100A4 Expression

Despite the recognition of the wide ranging function of S100S4, there has been little notable effort towards comprehending, let alone accepting, the potential value of targeting S100A4 for therapy. Admittedly hitherto efforts have been concentrated on

Therapeutic Strategies in Cancer Biology and Pathology. DOI: http://dx.doi.org/10.1016/B978-0-12-416570-0.00009-3

Table 9.1 S100A4 Signalling Pathways in Cancer

Phenotypic Feature	Signalling Pathway
Cell proliferation, tumour growth	Binds to and forms complexes with P53, now confirmed as target of S100A4; stabilises; binds to C-terminal regulatory region of p53; p53 binding sites occur in S100A4 promoter; Rb/stathmin/p53 downstream effectors, for example p21^{waf}, p16, etc.; bind to N-terminal domain of mdm2; P53/stathmin/ microtubule dynamics/cell division; p53, apoptosis family genes; caspases; calpain/Fas (?)
	EGFR expression correlates with S100A4 expression; postulated interaction with HER2
	S100A4 stimulates EGFR/HER2
	Signalling; an HER2 response element occurs on human S100A4 promoter
Invasion, intercellular adhesion, EMT	Interacts with the TGF-β system via Smad; postulated S100A4 binding to the N-terminal region of Smad3; postulated to lead to EMT; influenced by Wnt signalling; targeted by β-catenin
	Modulation of cytoskeletal dynamics; Cadherin/catenin complex cytoskeletal linkage; CD44/cytoskeletal linkage
	ECM-associated proteolytic enzyme system/ECM remodelling
Tumour vascularisation	Activation of MMP/TIMP
	Activation of angiogenic factors
	VEGF/endothelial cell proliferation
	MetAP2/p53-mediated inhibition of endothelial cell proliferation

Source: Based on Sherbet (2009, 2011b).

the biological manifestations and molecular mechanism of S100A4 function. The patently obvious involvement of S100A4 in EMT has spurred much activity towards elucidating the influence of Wnt signalling on its expression together with its relationship to the expression of downstream effectors of Wnt signalling and to that of standard EMT markers.

It has emerged from early work that S100A4 expression inversely correlated with E-cadherin in NSCLC specimens. S100A4 was expressed in a majority of specimens and its expression correlated with metastasis to the lymph nodes and with poor patient survival (Kimura et al., 2000). Transfection of E-cadherin into gallbladder carcinoma G-415 cells increased cell adhesion and inhibited motility and cell proliferation and *in vitro*. These changes were paralleled by markedly reduced the expression of S100A4, although comparatively minor increases of c-myc and membrane type 1-MMP also occurred (Kohya et al., 2003). In oral squamous cell carcinoma cell lines with different invasive abilities, forced overexpression of S100A4 suppressed the expression of E-cadherin (Moriyama-Kita et al., 2005). High levels of expression of E-cadherin coupled with S100A4 negative state correlated with enhanced disease-free survival in melanoma patients (Andersen et al., 2004).

The status of β-catenin was uncertain in early work. For instance Kimura et al. (2000) and Moriyama-Kita et al. (2005) reported a downregulation of β-catenin in a background of reduced S100A4 expression. However this would not be compatible with the operation of active Wnt signalling. With the avowed heterogeneity of tumours, it seems possible that this might have occurred in a subpopulation not directly linked with metastatic ability. Indeed, Stein et al. (2006) identified TCF binding sites in the S100A4 promoter and noted that β-catenin/TCF complex directly bound to the promoter and enhanced its activity. This β-catenin/TCF-mediated regulation of S100A4 has been amply confirmed recently. Stein et al. (2011) intrasplenically xenografted colon cancer cells overexpressing S100A4 and treated the host animals with sulindac. This treatment seems to have interfered with Wnt signalling, reduced β-catenin expression, the formation of β-catenin/TCF transcription complex and importantly here in the downregulation of S100A4 promoter activity and expression. The tumour that formed in the spleen and its metastases in the liver both had reduced levels of β-catenin and S100A4. High S100A4 expression together with reduced β-catenin expression has been linked significantly with tumour growth, invasion and clinical stage of gastric cancers (Yoon et al., 2008).

The inhibitory effects of the Ca2+ ionophore calcimycin (A23187) on the expression of S100A4 were studied in human and murine hematopoietic cell lines some years ago by Grigorian et al. (1994) who showed that it markedly reduced the expression of S100A4 mRNA in the treated cells. This they suggested to be a result of the degradation of mRNA since it was preventable by cycloheximide treatment, possibly by blocking nucleases. Sack et al. (2011a) have reiterated and considerably amplified the earlier findings with the demonstration that calcimycin inhibits the transcription of S100A4. Treatment of colon (HCT116, SW620, LS174T and SW480) cancer cells with calcimycin reduced S100A4 mRNA and protein levels. Calcimycin reduced S100A4 promoter activity in a concentration-dependent manner. Typically, it inhibited cell proliferation and motility. The treatment also reduced β-catenin expression suggesting that the effects of the ionophore might have been brought about by inhibition of Wnt signalling. Among other Wnt effectors affected were Dkk-1 and cell cycle related proteins. Sack et al. (2011a) also showed that tumours developing from calcimycin treated HCT116 cells xenografted into intrasplenic location showed markedly reduced metastatic spread to liver.

Sack et al. (2011b) have used a similar experimental regime to show that the antihelminth drug Niclosamide also downregulated S100A4 expression with parallel inhibition of cell proliferation, migration and metastatic spread of colon cancer cells in xenograft tumour models. Of much interest in relation to the S100A4 linkup with Wnt signalling is the finding that Niclosamide promoted the degradation of Wnt co-receptor LRP6. It also induced apoptosis of prostate and breast cancer cells at IC_{50} of 1 μM. This dosage was comparable with that which inhibited Wnt signalling (Lu et al., 2011b). Osada et al. (2011) also reported the inhibition of cell proliferation of colon cancer cell lines and that oral administration of the drug controlled the growth of xenografts of tumours. Inhibition of Wnt signalling here seemed to be due to the downregulation of DVL2. Other suggested routes of functioning include NF-κB mediation in AML cells, where Niclosamide seems to inhibit the transcription and DNA binding of NF-κB. It blocked TNF-induced IκB phosphorylation, translocation of p65, and expression of NF-κB responsive genes (Jin et al., 2010). Highly powerful cytotoxicity of Niclosamide on multiple myeloma cells has been attributed to both NF-κB inhibition and STAT3 signalling (Khanim et al., 2011).

A novel approach has been advocated by the Rudland/Barraclough group which has shown specific mutations that inhibit self-association of S100A4 markedly reduce its metastasis promoting effects measured by *in vivo* metastasis assays and *in vitro* motility assays. The mutations reduce self-association and reduce the affinity of S100A4 to two target proteins, namely p53 and non-muscle myosin heavy chain isoform A (Ismail et al., 2010). An apparent contradiction might be construed in that Malashkevich et al. (2010) found that trifluoperazine (TFP) promoted S100A4 oligomerisation and in this way interfered with the interaction of S100A4 with myosin-IIA interaction. Inhibition of S100A4 function occurs at concentrations that promoted its oligomerisation. But here the disruption of interaction between S100A4 and myosin-IIA seems to be due to the packaging of S100A4/TFP dimers into the formation of pentameric rings leading to the sequestration of the S100A4 dimers.

Osteopontin an Intermediary Target of S100A4

Many strands of evidence have reflected strongly the possibility of taking advantage of osteopontin being an intermediary in a S100A4 signalling pathway. Osteopontin was identified as a metastasis-associated protein some time ago. Oates et al. (1996) demonstrated quite convincingly that progression of a murine cell line Rama37 to the metastatic line Ca2-5-LT1 was accompanied by marked expression of osteopontin mRNA. Rama37 with forced expression of osteopontin produced tumours that were capable of metastasis. An early study of breast cancer has shown expression of osteopontin generally correlated with poor patient survival. The median survival time of osteopontin negative patients was >228 months and 68 months of the osteopontin positive group (Rudland et al., 2002). The prognosis for S100A4+ group was less favourable than for S100A4− group; the median survival time was >204 months for S100A4− and 186 months for S100A4+ patients over an 18-year follow-up. The detection of both S100A4 and osteopontin seemed to presage shorter survival time (Rudland et al., 2006), but there is no clear identification of difference in survival of S100A4+/osteopontin+ and S100A4+/osteopontin− groups. Recently Lin et al. (2011a) found markedly enhanced expression of osteopontin in metastasising hepatocellular carcinoma than in non-metastasising tumours.

NF-κB are a family of transcription factors that activate many genes closely concerned with inflammation, cell survival, cell proliferation and apoptosis, angiogenesis, cell adhesion, invasion and metastasis (Table 9.2). The association of osteopontin with several activated NF-κB pathways gives osteopontin a predominant position in the search for therapeutic intervention in the biology of cancer. Berge et al. (2011) showed recently that S100A4 induced the expression and secretion of osteopontin in some osteosarcoma cell lines in an NF-κB-dependent fashion. Inhibition of osteopontin using siRNA at least partially counteracted the effects of S100A4 on uPA and MMP13, suggesting that osteopontin could be a downstream effector of S100A4. Equally, one ought to consider that S100A4 can activate NF-κB via the classical pathway mediated by MEKK/IKKβ (Grotterod et al., 2010; Hua et al., 2010) with

Table 9.2 NF-κB Target Genes Relevant in Cancer Biology

Cellular Feature	Activators/Target Genes
Inflammation	TNF, interleukins, chemokines
Cell survival, apoptosis	Bcl family, 14-3-3σ
Cell proliferation	Myc, cyclins, interleukins, PDGF, CSF
Angiogenesis	VEGF, NOS,TNF, interleukins
Invasion, motility	MMPs, uPA, vimentin
Adhesion phenomena	ICAM, VCAM, CD44 ELAM-1
Metastasis	S100A4, osteopontin, S100A6, S100P

Source: Collated from Pahl (1999), Nishikori (2005) Sherbet (2001, 2011a,b).

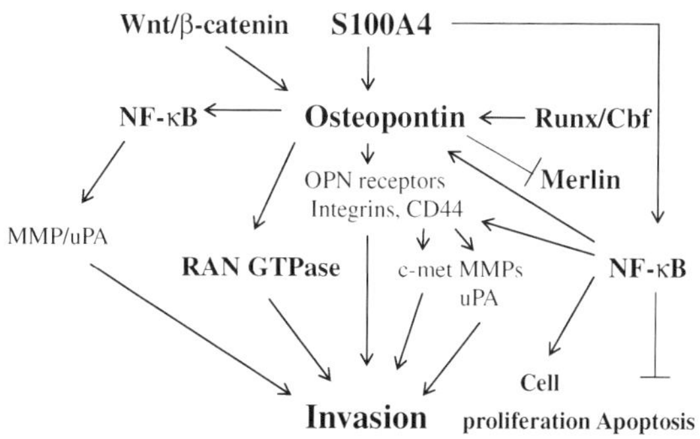

Figure 9.1 Osteopontin intermediary function in promoting cell proliferation and invasion with potential onward progression of tumours. This figure focuses on how osteopontin co-ordinates the signalling pathways involving S100A4, Wnt/β-catenin, NF-κB and Merlin in generating the invasive and metastasising phenotype. This representation is based on references cited in the text.

the same phenotypic outcome. This can occur quite independently of osteopontin. One would recall here that S100A6 and S100P also are capable of exerting pro-metastatic effects by activating the NF-κB pathway (Figure 9.1).

The activation of NF-κB can upregulate the expression of MMP and uPA and in this way promote cell motility. NF-κB can inhibit apoptosis and so promote cell population expansion (Matusan-Ilijas et al., 2011). Equally, osteopontin has been shown to promote the motility of SMMC-7721 hepatocellular carcinoma cells *in vitro*. This occurs via the agency of CD44*v*6. Antibodies against CD44*v*6 inhibited the invasion. In parallel the osteopontin overexpressing cells also showed enhanced MMP-2 and uPA levels (Chen et al., 2011c).

Since inhibition of osteopontin is known to be able to inhibit tumour development and angiogenesis, inhibition of both might result in synergistic suppression of tumour progression. A conceivable strategy would be to suppress osteopontin expression, which is regulated by the transcription factor Runx2 and the co-factor Cbf. Not only does Cbf expression correlate with cell motility but also that it is a necessary ingredient of Runx2-mediated induction of osteopontin expression and that of MMPs and osteocalcin (Mendoza-Villanueva et al., 2010).

Osteopontin function can itself be suppressed by targeting its receptors or its downstream signalling components. One of the surface molecules of interest is CD44. Osteopontin and its 45kDa fragment contain the RGD (Arg-Gly-Asp) sequence which binds to a number of cell surface integrins. Notably, osteopontin binds to α5β1, αvβ3 and αvβ5, and this determines its ability to promote intercellular adhesion and cells spreading (Yokosaki et al., 2005). Ovarian clear cell carcinomas strongly express osteopontin. Osteopontin upregulates integrin receptor expression in ovarian clear cell carcinoma cell lines, RMG-1 and TOV-21 G and increases *in vitro* cell motility. These effects are suppressed by Simvastatin (Matsuura et al., 2010). Osteopontin interaction with CD44 has been suggested as a putative route to the activation of c-met and so lead to cell invasion (Yoo et al., 2011).

One would recall here that the RAN (Ras-related nuclear) GTPase occurs at greatly increased levels in human clear cell renal carcinoma as compared with normal tissue and further enhanced expression was associated with tumour grade and has correlated with invasion and metastasis (Abe et al., 2008a). Transfection and forced expression of osteopontin in the non-invasive Rama37 cell line transforms them into an invasive and metastatic phenotype. These cells also show enhanced expression of the RAN GTPase. Furthermore, transfection of an expression vector for RAN GTPase is also able to generate the invasive phenotype of Rama37. Taken together these observations have been interpreted as suggesting that RAN GTPase is induced by osteopontin leading to the manifestation of invasive and metastasising ability (Kurisetty et al., 2008).

With the current sharp focus on miRNA in the biological behaviour of cells, one should note that transfection of miRNA-96 into HCC cells has markedly enhanced the expression of osteopontin in HCC cells. Reducing miRNA-96 expression using anti-miRNA-96 oligonucleotides led to reduced migration and osteopontin expression (Chen et al., 2012a).

The clear influence of Wnt signalling on S100A4 expression together with the perceived involvement of both signalling systems in EMT and the function of osteopontin as an intermediary in S100A4 coheres these courses of signalling to seek and formulate a form of focus for therapy. Wnt pathway has been implicated in the regulation of osteopontin expression. In Wnt signalling β-catenin, the transcriptional factors Lef-1, TCF-4 and other HMGs function in collusion and these are known to be differentially expressed in some tumours. Osteopontin is activated by Wnt signalling and indeed the transcription factor TCF-4 not only can activate osteopontin expression but also promote cell motility (Ravindranath et al., 2011). The osteopontin promoter is responsive to β-catenin/Lef and the response is enhanced by Ets family transcription factors. Indeed osteopontin transcription is regulated by a complex that includes Ets, PEA3 of the Ets-domain transcription factor subfamily and β-catenin/TCF-4 (El-Tanani et al., 2004).

Osteopontin has been implicated in loss of the suppressor protein, Merlin encoded by NF2, the neurofibromatosis 2 gene on chromosome 22q12. Merlin is an ERM protein, which in an under-phosphorylated form functions as a suppressor gene and regulates many downstream signalling systems. It is able to link transmembrane proteins with the actin cytoskeleton and modulate cell behaviour. Merlin regulates these processes via the agency of CD44, integrins and the cadherins. Reverting to the story of osteopontin, expression of Merlin protein, not gene transcripts, inversely correlates with osteopontin in breast cancer. We know that Merlin inhibits PI3K/Akt signalling. It is also known to inactivate PAK (p21-activated kinase) 1, a serine/threonine kinase that participates in the regulation of the actin cytoskeleton. On the other hand, osteopontin has been linked with Akt-mediated phosphorylation and degradation of Merlin (Morrow et al., 2011). Furthermore, the regulatory link between osteopontin and Merlin is supported by the fact that CD44, which acts as an osteopontin receptor and routes its signals downstream, is antagonistic in function to Merlin. Either way osteopontin negates the function of Merlin and its negative regulation of growth via the downstream Hippo pathway to constrain cell proliferation and growth and promote apoptosis. So prevention of degradation of Merlin by PI3K or by inhibition of osteopontin could be a valuable approach. To complete the picture of the complexity of interactive signalling, it has emerged that active Merlin localises Wnt/β-catenin at the cell membrane which renders the proliferative signalling by Wnt ineffective (Zhou et al., 2011a) (Figure 9.1).

Since osteopontin function is subject to regulation by Wnt, S100A4 and NF-κB, here is provision of many options available for therapeutic intervention. The Hippo signalling system and its components, for example the transcriptional co-activator YAP (Yes-associated protein) and LATS2 (large tumour suppressor homolog 2) that can inactivate YAP are also potential targets in this network of interactions.

Many studies have implicated other signalling pathways, for example Ras/ERK and the Src kinases. At the risk of reiteration, osteopontin transcription is regulated by a complex that includes Ets, PEA3 and β-catenin/TCF-4. Besides, the activity of the PEA3 is accentuated by Ras-ERK signalling. Substantive proof of the operation of these pathways in the generation of the invasive phenotype is currently awaited.

RAGE/NF-κB Signalling in S100A4 Function

The receptor for advanced glycation end products (RAGE) has been implicated in human pathogenetic conditions such as diabetes, atherosclerosis and Alzheimer's disease. RAGE interacts with diverse ligands and activates NF-κB as well as binds to integrins, thus subserves several cellular functions of promoting cell proliferation, motility and metastasis. Quite appropriately therefore inhibition of RAGE and/or NF-κB activation has been viewed as viable options of therapy of tumour growth and progression.

The S100 family proteins too act as RAGE ligands. RAGE ligands interact with Amphoterin, a 30 kDa protein with HMG1-type sequence. Amphoterin is a heparin binding protein and shows particular localisation at the leading edges of motile cells and thus shares biological features with the HMG group. The disruption of this interaction is known to inhibit tumour growth. Kawada et al. (2010) have isolated a natural compound (NBRI17671) from a species of the fungus Acremonium, which is highly effective in inhibiting growth. It was also found to downregulate MAPK suggesting the inhibition of RAGE-Amphoterin might have led to inactivation of

NF-κB. For instance, S100P promotes cell proliferation and survival by activating ERK and NF-κB signalling. The effects of S100P were suppressed by Amphoterin-derived peptide that inhibits RAGE activation and also by anti-RAGE antibodies, and by expression of a dominant-negative RAGE (Aarumugam et al., 2004). Akin to this is the observation relating to S100A4. Activation by S1004 of the NF-κB pathway seems to involve RAGE. The induction of cell migration by S100A4 is inhibited by the inhibition of RAGE using siRNA. This also inhibited the phosphorylation of ERK and MMP-2 (Spiekerkoetter et al., 2009). Presumably this could suggest activation of NF-κB by the classical course of action. Exposure of cardiac myocytes to S100β increases VEGF mRNA and protein together with the activation of the classical NF-κB pathway. These effects were inhibited by caffeic acid phenethyl ester (CAPE). Myocytes expressing dominant-negative RAGE did not show induction of VEGF or NF-κB activation indicating that S100β functioned via RAGE (Tsoporis et al., 2012).

Caffeic acid (CA) and CAPE have been investigated for their anti-tumour effects and for inhibition of NF-κB signalling. Both were shown some time ago to inhibit *in vitro* invasion and *in vivo* metastatic spread of HepG2 HCC cells. Concomitantly these compounds inhibited MMP-9 activity and suppressed NF-κB signalling (Chung et al., 2004). Although Chung et al. (2004) suggested at the time there the biological effects resulted dual modes of MMP inhibition and inhibition of NF-κB, we know now that the two are indeed linked. NF-κB can signal via CD44 and also the osteopontin pathway to activate MMPs. Also as discussed earlier, the promotion of invasion and induction of MMP expression by S100A4 is mediated by NF-κB. Since then much progress has been made with these compounds. CAPE, a constituent of honey bee resinous product called propolis, inhibits NF-κB and activates the transcription factor NFAT and inhibits the classical pathway by delaying IκBα degradation and the translocation in to the nucleus of p65 (RelA) (Ang et al., 2009). At the phenotypic level, CAPE inhibits cell proliferation and promotes apoptosis, with parallel upregulation of the pro-apoptotic Bax expression and downregulation of the anti-apoptotic Bcl-2. *In vivo* tumour development upon implantation of cholangiocarcinoma (CCH) cells into Balb/c nude mice had a longer latency and showed growth inhibition (Onori et al., 2009). Similar growth inhibition occurs in CAPE-treated MCF-7 and MDA-231 breast cancer cells, with supplementary inhibition of VEGF and apparent inhibition of angiogenesis (Wu et al., 2011a). It would be worth noting here that CAPE was shown some time ago to be a highly efficacious inhibitor of NF-κB activation. CAPE-induced apoptosis was accompanied by the loss of IAPs (inhibitors of apoptosis protein) cIAP-1, cIAP-2 and XIAP (McEleny et al., 2004). Berger et al. (2007) have described another inhibitor of NF-κB, namely Bay 11-7085 which appears to be more efficient than CAPE in B-lymphoma cell lines, possibly functioning by a different mechanism from CAPE.

Now to return to the RAGE story, the RAGE inhibitor PF-04494700 is in clinical trials for the treatment of Alzheimer's disease. FPS-ZM1 is another putative RAGE-specific inhibitor that binds to the V domain of RAGE and blocks amyloid beta-mediated murine model of the disease (Deane et al., 2012). Yamamoto et al. (2001) reported some time ago that a putative AGE inhibitor,

(+/−)-2-isopropylidenehydrazono-4-oxo-thiazolidin-5-ylacetanilide(OPB-9195) prevented diabetic nephropathy in transgenic mice that overexpress human RAGE.

S100A4 Downregulates PRDM1 and VASH1 Suppressor Genes

The PRDM2 gene encodes two proteins, namely RIZ1 and RIZ2. PRDM2 is a tumour suppressor gene. RIZ1 referred to here as PRDM2 is a zinc finger protein belonging to the nuclear protein-methyltransferase superfamily. PRDM2 has a PR domain at its N-terminus, which is absent in RIZ2. PRDM2 is capable of binding Rb protein and ER. It is postulated to play a part in regulation of transcription of Rb and activates ER function. The gene is located on 1p36.21, a region which is often deleted in cancers (NCBI database). The gene is silenced by hypermethylation in many tumours. The loss of PRDM2 shows obvious links with disease progression. In ovarian cancer decreased expression of PRDM2 has been found to correlate with tumour grade, clinical stage and cell proliferation (Akahira et al., 2007). The progression of CML from the chronic phase into blast crisis occurs with parallel molecular changes that result in enhanced cell proliferation and survival and arrest of differentiation. This blast crisis phenotype requires molecular alterations of Bcr/Abl, which are deregulated in progression. These alterations affect the cell cycle control genes, namely p53 and Rb. Lakshmikuttyamma et al. (2009) have reported that CML progression was associated with decreased expression of PRDM2 together with reduction in apoptosis and increase in cell proliferation.

VASH1 (vasohibin 1) inhibits cell migration, proliferation and tumour growth and angiogenesis (Dessen and Le Minor, 2002; Watanabe et al. 2004; also see EMBL-EBI database). Recombinant adenovirus encoding VASH1 inhibits cell proliferation *in vitro*. Growth of tumours formed *in vivo* by subcutaneously injected H22 cells was inhibited by VASH1. Microvessel density associated with the tumours was also reduced, although not quantified (Li et al., 2010a). Unfortunately there is no information about metastatic spread. However Heishi et al. (2010) found that VASH1 not only inhibited corneal vascularisation induced by VEGF-A, VEGF-C, FGF2 and PDGF-BB, but also inhibited lymhangiogenesis and spread tumours to regional lymph nodes in a murine tumour model.

The inhibition of angiogenesis involves EZH2 (Zeste homologue 2) protein. The EZH2 belongs to the Polycomb family. These proteins are involved in maintaining several genes in transcriptional repression. The Polycomb group genes form two important multi-protein complexes, namely PRC1 (Polycomb repression complex 1) and PRC2. PRC1 is said to stably maintain gene repression, whilst PRC2 contains HDACs and methyltransferases and initiates silencing (Kohler and Villar, 2008; Han L et al. 2009). EZH2 is a histone-lysine *N*-methyltransferase PRC2 complex and controls epigenetic modifications and so maintain genetic repression. EZH2 mediates repression of VASH1. Suppression of VASH1 by methylation upregulates VEGF expression and promotes angiogenesis (Lu et al., 2010). Compatible with this mode of function is the ability of S100A4 to downregulate VASH1. Suppression of

S100A4 using interference RNA upregulates the expression of PRDM2 and VASH1 (Tabata et al., 2009). Saito et al. (2011) injected VASH1 intravenously using an adenovirus vector into diabetic mice and noticed suppression of VEGF induction together with increase in expression of P-cadherin. Interestingly they also noticed the suppression of S100A4 expression and of Slug/Snail. This might suggest that VASH1 might be effective in suppressing the activation of EMT. These observations support the attribution to S100A4 with the ability to promote angiogenesis.

S100A2 Suppressor Gene and S100A4 Function

S100A2 has long been regarded as a tumour suppressor, despite some dissenting findings. It shows diminished or loss of expression in many forms of cancer and the loss of expression has been linked with cancer progression and prognosis. A few investigations have revealed an inverse relationship between expression of S100A2 and S100A4. Loss of expression of the gene is mainly due to hypermethylation of its promoter.

Many S100 family proteins both tumour suppressor and promoters interact with members of the cell cycle regulator p53 family and in that way exert their influence on cell proliferation. Tetramerisation is required for p53 function, including binding to DNA, interaction with other proteins function and its degradation. P53 occurs in a monomeric form and tetramerisation is preceded by an intermediate state of dimerisation which is believed to be a rate-limiting step in achieving tetramerisation. S100 proteins bind different tetramerisation domains of p53 proteins and with differing affinities. Also different S100 proteins seem to bind different oligomeric forms of p53. Since they bind p53 in both monomeric and tetrameric state, it has been suggested that binding to the monomer is inhibitory whilst binding to the tetramer has an activating effect. These contrasting effects depend upon the degree of expression of the proteins (Van Dieck et al., 2009, 2010). Equally, S100A4 is said to bind preferentially to the p53 tetramerisation domain in its lower state of oligomerisation and effectively disrupt the tetramerisation and function of p53 (Fernandez-Fernandez et al., 2005). This would be expected to inactivate the cell cycle regulatory function of p53.

The involvement of RAGE is a shared feature in the functioning of S100 proteins. As noted earlier, S100P promotes cell proliferation and survival by activating ERK and NF-κB signalling. These effects can be prevented by inhibiting RAGE activation. S1004 activates NF-κB via RAGE. The induction of cell migration by S100A4 is inhibited by the inhibition of RAGE. RAGE functioning has been implicated in the suppression of cell motility by S100A2. An inverse relationship was noticed between RAGE expression and that of S100A2 (Nagy et al., 2001). Subsequent investigations by the same group encountered RAGE expression in many human cancers. In breast and lung cancers, that expression of S100A4 and S100A6 correlated with RAGE (Hsieh et al., 2003). However, doublestaining for S100A4 and RAGE does not appear to have been carried out; hence it is difficult to state with confidence that S100A4 expression and RAGE were related. But quite obviously there is a link between S100A4, not S100A2, and activation of NF-κB and this is aided by RAGE (Haase-Kohn et al., 2011).

10 MTAs in Cancer Invasion and Metastasis

The Biology of Metastasis Promotion by MTAs

The MTA family genes are a well-studied family of metastasis associated genes. The MTA1 occurs as a long and short isoforms which are generated by alternative mRNA splicing and the family contains two other members MTA2 and MTA3. The MTAs are expressed in many forms of cancer. Evidence is emerging of the involvement of MTAs in EMT. Overexpression of MTAs has been correlated with invasion and aggressive behaviour of many forms of cancer. They promote cancer cell migration by regulation of cell adhesion proteins in consort with membrane-cytoskeletal linker proteins such as Ezrin. MTAs also have been known to induce the expression of angiogenic agents. MTA2 is believed to interact with p53 and interfere with its cell cycle regulatory function and apoptosis pathway (Toh and Nicolson, 2009; Pencil et al., 1993; Toh et al., 1994, 1995; Mahoney et al., 2002).

MTA proteins repress genetic transcription by virtue of their function in histone deacetylation and are involved with nucleosome remodelling and histone deacetylase (NuRD) complex. MTA1 is a component of and engage with specific NuRD complexes and interacts with HDAC-1. The MTA1-NuRD has been found to regulate many genes of significance in the present context. Among them are ERα, c-myc and the tumour suppressors BRAC1 and p53. The short isoform of MTA1 contains an ER binding motif (LRILL) (Kumar et al., 2002). It is said to sequester cytoplasmic ERα and in this way inactivates its transcription function. MTA1 has been reported to be able to repress ERα activation by chromatin deacetylation of the ER-response element of responsive genes (Toh and Nicolson, 2009). However, ERα and ERβ have opposing functions. The former promotes cell proliferation, but ERβ inhibits cell proliferation and invasion. It has also been apparent from recent work that ERβ2 can inhibit ERα-mediated transactivation via oestrogen response element (Zhao et al., 2007b). The long isoform of MTA is a co-repressor of ER (see UniProtKB Q13330). Also increased expression of ER/MTA1-NuRD can influence ER/HER2 signalling which is crucially important in TNBC. Therefore, the effects of MTA on ER need further exploration.

Whether enhanced non-genomic responses of ER might occur is a point for consideration. The rapidity of response by oestrogens suggests that the signalling might originate with ER at the membrane representing non-canonical pathways distinct from conventional signalling by nuclear receptors (Ravelli et al., 1998; Nadal et al., 2000; Ropero et al., 2002). Membrane bound ER, which may be relocated nuclear ER, have been identified and these have been postulated to be GPCR30, the so-called orphan receptor. Signalling by GPCR30 represents the non-genomic pathway.

Therapeutic Strategies in Cancer Biology and Pathology. DOI: http://dx.doi.org/10.1016/B978-0-12-416570-0.00010-X

Oestrogens do function by activating GPCRs (Sherbet, 2011a). Nonetheless, the importance of MTAs in tumour promotion and dissemination is amply supported by the finding that they are also capable of inducing the expression of neoangiogenic agents. Increased VEGF expression has been reported in conjunction with MTA proteins in many tumours but lack of accessibility of the reports has prevented full discussion of the results here.

Modulation/Inhibition of MTA Expression

The significant biological effects exerted by MTAs have prompted investigations designed to inhibit or modulate their expression. Among notable regulators of MTA is miRNA-146a. According to Li et al. (2010f), the expression of this miRNA is reduced in pancreatic cancer cells and experimental enhancement expression levels by treating the cells with 3,3′-diindolylmethane (DIM) or isoflavone has been found to inhibit cell invasion *in vitro*. In parallel the treatment downregulated EGFR, MTA2, IRAK-1 (interleukin-1 receptor-associated kinase 1) and NF-κB resulting in inhibition of pancreatic cancer cell invasion. This suggests possible interaction of MTA signalling with EGFR and NF-κB signalling pathways. Inhibition of MTA gene in neuroblastoma cells by using siRNA stimulates neurite extension (Singh et al., 2008). In B16F10 cells, inhibition of the MTA1 expression has been shown to reduce cell migration *in vitro* and lung colonisation upon introduction of the cells via the tail vein (Qian et al., 2007a).

MTA Signalling Intercalates with Wnt/Notch/Hh Signalling

The massive phenotypic changes occurring in the wake of MTA activation in terms of enhanced cell migration and aggressive behaviour of tumours have implicated a positive role for MTA1 in EMT activation. Whilst MTA1 induces a more aggressive phenotype, MTA3 exerts an opposite effect. It is able to inhibit the transcription factor Snail and lead to an upregulated expression of E-cadherin and inhibition of EMT. In contrasting function, MTA1 activates Snail and downregulates E-cadherin to bring about the expression of an aggressive cancer phenotype. It is becoming increasingly apparent that MTA3 influences the Wnt signalling following upon the finding that MTA3 directly represses Wnt4 transcription (Zhang et al., 2006a). MTA3 could be countermanding Snail activation induced by Wnt. At the biological levels, there is a solitary report of its downregulation in cancer. MTA3 occurred at reduced expression in endometrial adenocarcinomas and levels of expression were inversely related to tumour grade and lower in poorly differentiated tumours (Bruning et al., 2010). There are a couple of reports that suggest possible tumorigenic and EMT promoter effects of MTA3. However, the paucity of evidence does not allow a consensus.

The developmental pathways of Wnt, Notch and Hh signalling are often activated in cancer and have been directly or indirectly linked with EMT. A highly co-ordinated network of interaction in the signalling by these ligands functions and mutual regulation of phenotypic outcome has been encountered in developing systems as well as in the activation of EMT, cell proliferation and tumour progression. Therefore, targeting MTA3 signalling to modulate Wnt/E-cadherin pathway could provide a potentially powerful means to inhibit EMT.

As noted earlier, both the short and long forms of MTA1 possess the potential to modulate ER signalling. E-cadherin has been identified as the downstream target of MTA signalling mediated by ER/E-cadherin via the transcription factor Snail which regulates E-cadherin expression. As also noted above and illustrated in Figure 4.3, the Wnt/E-cadherin pathway also involves regulation by Snail. Certain miRNAs have been found to regulate EMT activation and preliminary studies do show that miRNAs might also regulate MTA expression. So MTAs might provide an alternative albeit complex approach to cancer treatment, especially to TNBC management.

Part 4

Genetic Determinants of Tumour and Metastasis Suppression

Several genetic determinants that induce the neoplastic phenotype have been identified, and some among these are regarded as the conventional tumour and metastasis suppressor genes and their mode of function has been studied in much detail. Many signalling systems are frequently deregulated in the development and secondary spread of cancer, and several components of these signalling cascades have been linked with the neoplastic process and have been described as tumour suppressors. Whether one can legitimately draw a distinction between the conventional tumour/ metastasis suppressors and modulators of signalling is debatable; these so closely impinge upon each other's function that distinction becomes arbitrary and artificial. Nonetheless, here these are treated as separate but not exclusive groups, with adequate emphasis on the nature and degree of mutual modulation of function.

11 Metastasis Suppressor nm23 and Manipulation of its Expression

The nm23 gene is now unquestionably established as a metastasis suppressor. Allelic loss of nm23 is an important feature of many forms of tumour. Loss of the normal regulatory function of nm23 by methylation has been linked with tumour progression. The overriding influence of the gene is unmistakable and patently obvious from its ability to counteract the influence metastasis promoter genes, interfere with growth factor receptor signalling, regulate cell proliferation and modulate cell membrane components that function as determinants of cell motility and invasion. Nm23 was originally identified as an NDPK (nucleoside diphosphate kinase). Histidine kinase activity has been putatively associated with its function in the suppression of motility. Nm23 was recognised as a metastasis suppressor on account of its differential expression in tumour with high and low metastasising abilities and positively demonstrated as a suppressor by gene transfer studies. The human nm23 family is composed of nm23-H1, nm23-H2, DR-nm23, nm23-H4 and nm23-H5 genes, which play important roles in cell proliferation, differentiation and in tumour development and metastasis. Of these, nm23-H1 has been persuasively and clearly linked with many human cancers. Nm23-H1 and nm23-H2 encode NDP kinase subunits. Nm23-H1 has been implicated in monitoring the integrity of DNA repair and replication.

Metastasis suppression by nm23 is well established in many tumour types but is unproven in some forms of cancer. But this has to be gauged in the background of activity of promoters of tumour growth and metastatic spread. The downregulation of expression of nm23 and of E-cadherin correlates with invasion and the presence of nodal metastasis. Besides, nm23 is a growth suppressor and possibly annuls the abrogation of cell cycle checkpoint control mechanisms by metastasis promoter gene such as S100A4. Oestrogen and progesterone do influence nm23 expression and do bring about changes in cell motility and invasion. Thus oestrogen appears to downregulate nm23-HI expression together with promotion of cell migration and invasion by activating the PI3K/Akt pathway, whilst progesterone opposes this effect. Furthermore, *in vitro* oestrogen enhances and progesterone decreases cell motility (Hua et al., 2006; Sherbet, 2011a).

Oestrogen exerts its physiological effects via ERα and ERβ mediation. Both the receptors bind 17β-oestradiol with high affinity and in similar if not identical manner to the oestrogen response elements (EREs). But there are major differences: 17β-oestradiol has selective affinity for ERα, but 17α-epiestriol is an ERβ selective agonist. Besides they differ in respect of their transcriptional activities (Barkham et al., 1998; Dechering et al., 2000). Possibly as a consequence of this, ERα and ERβ modulate the expression of different genes and exert different effects on cell proliferation and invasion. Whereas ERα promotes cell motility and invasion, ERβ

Therapeutic Strategies in Cancer Biology and Pathology. DOI: http://dx.doi.org/10.1016/B978-0-12-416570-0.00011-1

inhibits these processes, albeit by different signalling systems and ligand dependent or independent manner (Lazennec et al., 2001; Sanchez et al., 2010). But 17β-HSD1 (17beta-hydroxysteroid dehydrogenase type 1), which is involved in oestradiol synthesis and activation of dihydrotestosterone, enhanced the levels of both ERα and ERβ and promoted cell growth and migration. ERα levels had increased to a far greater extent than did ERβ (Aka et al., 2012). Obviously here promotion of cell growth and migration by ERα might have prevailed against the negative effects of ERβ. Hence, the relationship between oestrogen/ER and progesterone with invasive behaviour is not quite as straightforward as it would appear.

Sometime ago Lin et al. (2002) found that oestrogen induced the expression of nm23-H1mRNA and protein and this corresponded with levels of ERα. Also these effects were accompanied by suppression of invasive behaviour *in vitro*. The mediation of ERα was evident from the fact that BCM1 cells lacking the receptor showed no response to oestrogen. Antagonists of oestrogen inhibited nm23-H1 transcription. According to Sahab et al. (2010), ER− cells derived from breast cancer also show a marked reduction in nm23-H1. Therefore, in the environment of nm23-H1, it is essential to note that it interacts with and increases the engagement of ERα with ERE and suppress transcription of oestrogen-responsive genes such as cathepsin D and Bcl-2. Inhibition of nm23-H1 by using siRNA enhanced the expression of these genes generating its suppressor effect (Curtis et al., 2007). Compatible are also the findings of Rayner et al. (2008) who most convincingly demonstrated that nm23-H2 co-localised with ERβ but not with ERα. In MCF-7 cells over expressing ERβ, nm23-H2 nearly doubled ERE related transcription. Furthermore, oestrogen and nm23-H2 synergistically reduced cell migration *in vitro*, obviously an ERβ-mediated effect. Also relevant in relation to cancer is that similar co-ordinated expression is found in vascular cells associated with atherosclerosis and inflammation (Rayner et al., 2007).

Also there is much circumstantial and inferential evidence that the nm23-H1 promoter responds to corticosteroids and hence the suggestion that steroids could be deployed to activate nm23 function. In fact dexamethasone does induce increase of nm23-H1 expression and this increase is glucocorticoid receptor dependent (Ouatas et al., 2003).The perceived differences might be due to the fact that the two ERs differ in biological function and also one is dealing here with two different nm23 proteins. Nm23-H2 is identical in sequence to PuF (the Purine binding factor), an activator of c-myc and also implicated in the activation of several genes involved in myeloid differentiation, whilst it is uncertain if nm23-H1 has transcriptional functions. A function ascribed to nm23-H1 is monitoring the integrity of DNA repair and replication. So by and large the two proteins could be influenced differently by oestrogen and might differentially impact disease prognosis.

Manipulation of nm23 Expression as a Therapeutic Approach

Gene therapy can be administered in many ways. Where the appropriate path is silencing of genes as in the case of tumour and metastasis promoter genes, suppression of gene expression by methylation of the promoter of the genes, inhibition of expression with siRNAs or interference with their signalling downstream have been advocated and tested in experimental systems. Targeting of suppressor genes requires their reactivation and gene or suppressor protein transfer. Gene transfer would be accomplished incorporating the therapeutic genes into viral (bacteriophage) and plasmid vectors. These can be delivered into the cells using nanotechnology. Virus or plasmid vectors can be incorporated into nanocapsules. Attempts have also been made to permeate the cell membrane with therapeutic proteins to allow the cell to regain suppressor function. Again the therapeutic protein is subject to intracellular degradation and also administration via enteric and injectable modes is impractical. With both therapeutic gene and protein transfers a major problem would be achieving target specificity. So gene therapy obviously faces many strategic as well as tactical logistic problems to be effective in patient management.

Gene transfer studies relating to nm23 are not many. Transfer of a plasmid carrying nm23-H1 by electroporation into T24 cells has been reported to reduce *in vitro* cell motility and increased the sensitivity of the transfectants to cisplatin (http://www.medical-res-papers.com/urologic-surgery/5741). A full description of this work is not available for proper assessment. Nm23-H1 protein that can permeate the cell membrane has been designed. Li et al. (2006) used an adenovirus vector to transfer nm23-H1 into a high metastatic variant ovarian cancer cell line and found that this resulted in a 60% reduction in the number of animals developing liver metastases when cells carrying nm23-H1 were introduced intraperitoneally. They claimed a reduction in metastatic deposition as estimated from the extent of involvement of liver tissue. The reliability of this mode of determining extent of spread could not be evaluated in the absence of adequate illustration. However, the host animals did survive considerably longer with nm23-H1 than the controls. More recently poly-L-lysine-modified iron oxide nanoparticles (IONP-PLL) have been used to transfer the gene into host cells. Injection of transfected cells by the intravenous route reduced their lung colonisation as well as enhanced host survival time when gene transfer was combined with cyclophosphamide treatment (Li et al., 2009f). It is needless to say nm23-H1 does in fact influence intercellular and substratum adhesion, so the observed effect on lung colonisation does not necessarily reflect a true and proper diminution of the metastatic process. Tumour cell lines into which the protein was introduced showed reduction in the incidence of lung metastases, reduced pre-existing metastatic load and enhanced survival of hosts carrying tumours (Lim et al., 2011). These studies are laudable *per se*, but it is a long shot to pre-clinical or clinical testing.

Upregulation of nm23 by Medroxyprogesterone Acetate

With the identification of such a broad spectrum of biological effects of nm23, upregulating the expression of nm23 has to be considered as a viable therapeutic option. As noted earlier, both oestrogen and progesterone regulate and influence the expression nm23-H1 and nm23-H2. Progesterone upregulates the expression of nm-23H2 (Syed et al., 2005). MPA (medroxyprogesterone acetate) was suggested for clinical testing for its ability to upregulate nm23 and inhibit metastatic dissemination of murine tumour models (Palmieri et al., 2005, 2006). MPA might also inhibit cell proliferation (Tanaka et al., 2008). An added advantage is that MPA seems able to modulate cytoskeletal dynamics; this possibly explains the inhibition of invasion by nm23 which is also known to target the cytoskeletal machinery.

Deployment of MPA as an anti-metastatic and anti-angiogenic agent is being clinically evaluated in postmenopausal patients with ER−/PR− breast cancer and in locally recurrent and metastatic breast cancer in combination with cyclophosphamide plus methotrexate (NCT00577122). Treatment of endometriod adenocarcinoma by MPA is also being studied (NCT00064025); its effect as a chemopreventive agent (NCT00003179) is also being assessed. Another phase III study is looking into how effective MPA might be in preventing endometrial disorder in postmenopausal women with ductal carcinoma *in situ*, lobular carcinoma *in situ*, Paget's disease, and stage I or stage II breast cancer and who are on tamoxifen therapy (NCT00002920).

Targeting S100A4 to Restore nm23 Function

In the framework of upregulating nm23 expression, targeting S100A4 and in this way enabling nm23 function would appear to provide a novel therapeutic approach. This has exceptional relevance in tumours types which significantly display an inverse relationship between the expression of nm23 and S100A4 and EGFR and ER/PR. Although this has not been examined thoroughly at present, currently available evidence indicates that nm23+ tumour cell lines tend to be EGFR− and ER/PR+, whilst in contrast S100A4+ ones are more often than not EGFR+ and ER/PR−. That EGFR, nm23 and S100A4 might be co-ordinately regulated is supported by the demonstration sometime ago by Parker and Sherbet (1992) that verapamil downregulated S100A4 in B16 murine melanoma cells and not in the opposite direction as incorrectly stated elsewhere (Sherbet, 2011a). Verapamil significantly reduced EGFR expression at both the mRNA and protein levels in A549 cells and also significantly enhanced the expression of nm23 protein, but nm23mRNA remained unchanged suggesting that here the effect was exerted at the translation level or later (Zhang et al., 2009a). Recently nm23-H2 has been found to negatively regulate EGF- and Ras-induced activation of the ERK pathway and cellular proliferation (Lee et al., 2009). This situation is fully compatible with the findings of Lin et al. (2002) who reported that oestrogen induced the expression of nm23-H1mRNA and protein and that this corresponded with levels of ERα. Furthermore, ER antagonists inhibited nm23-H1 transcription.

An area of vibrant interest covers TNBCs. TNBCs are most refractory to treatment by their very nature of being ER−/PR−/HER2− and therefore show no response to tamoxifen, aromatase inhibitors or to Herceptin. Around a tenth of breast cancers turn out to be triple negative, TNBCs are highly aggressive and associated with enhanced recurrence and metastasis and poor prognosis. Thus the problem of managing these patients is a seriously tactical proposition. The relationship that seems to subsist between S100A4/nm23 and ER/EGFR (possibly also HER2, not established todate) might lend itself to further exploration of potential approach to inhibit S100A4 as a means of upregulating nm23 in TNBCs.

Promoter methylation has often been seen as a route to silencing gene expression. SiRNA are able to target genes that possess sequence homology with siRNAs (discussed in pp. 3 and 4) and these seem able to promote methylation of CpG sites (Jones et al., 1999; Wassenegger, 2000; Suzuki et al., 2008). The heterochromatic siRNAs are involved in the maintaining genomic integrity by DNA methylation and modification of histones (Katiyar-Agarwal et al., 2006; Matzke et al. 2009).

As discussed previously siRNA can be appropriately targeted to inhibit S100A4. With directed methylation, one can suppress S100A4 expression and in this way control the aberrant behaviour of cells. This approach could downregulate S100A4 and effectively restore nm23 function (Sherbet, 2009). Indeed this might constitute a bimodal suppression of tumour invasion and metastasis. There is currently little activity aimed at testing HDAC inhibitors and DNA methylation for inactivation of metastasis promoter genes.

Is Tumour Suppressor p53 a Route to nm23 Manipulation?

The suppressor gene p53 is generally accepted as a negative regulator of the cell cycle. The possibility that p53 might provide a means of manipulating nm23 expression has been seriously considered. Several attempts were made in early years to establish a connection between the expression of p53 and nm23. No firm conclusions could be drawn from most the studies due to flawed experimental design and the lack of understanding for the need to demonstrate simultaneous or co-expression of these proteins and also from the lack of appreciation that p53 protein was detectable by immunohistochemical procedures due to the enhanced half-life of the protein either by conjugation and stabilisation with other cellular proteins or the protein occurred in the mutated form. So overall, most of these studies were poorly informative and inadequate to establish any notion of correlation between the variables.

Sometime ago, Chen et al. (2003a) linked p53 signalling and nm23-H1in MCF-7 and J7B cells with the demonstration that nm23-H1is upregulated by wild-type p53 and paralleled by inhibition of the invasive ability of the cells, but then nm23-H1 is downregulated by p53 in RKO and H1299 cells together with increase of invasion. This differential effect seen in different cell lines could be due to the involvement of other factors associated with p53 signalling. STRAP (serine–threonine kinase receptor-associated protein; stress-responsive activator of p300) is known to be a regulator of p53 signalling. STRAP is co-ordinately regulated by ATM (ataxia telangiectasia mutated) and Chk2 (Checkpoint kinase 2) kinases upon DNA damage (Adams et al., 2008). STRAP seems to be able to promote recruitment of p53-coactivators and could also antagonise mdm2, which is itself a negative regulator of p53. Now nm23-H2 can interact with mdm2 (Polanski et al., 2011). Jung et al. (2007) have shown that nm23-H1 and STRAP directly interact with p53 and enhance p53 activity. Also p53 activation by nm23-H1 and STRAP occurs as a result of the removal of mdm2 from its complex with p53. A general conclusion would be that the interaction between nm23 isoforms with p53 signalling might result both in the inhibition of invasion and activation of the apoptotic pathway. It is worthy of note here that nm23-H1 has been postulated to possess 3′-5′ exonuclease activity and ascribed a role in the induction of apoptosis independently of the caspase pathway.

12 The Metastasis Suppressor KiSS-1 Gene

The KiSS-1 gene was identified as a suppressor of invasion and metastasis some while ago (Lee and Welch, 1997a). The gene occurs on chromosome 1q32, has two non-translated and two partially translated regions and four exons encoding a 145 amino acid primary product or precursor (West et al., 1998; Ohtaki et al., 2001) and shorter peptides. The peptide initially called metastin is the secreted and active form (Harms et al., 2003). The precursor peptide with 145 amino acid residues is cleaved to form the Kisseptin-54 with C-terminal 1–54 amino acid residues. The C-terminal peptides Kisseptin-14 (41–54), Kisseptin-13 (42–54) and Kisseptin-10 (45–54) are also biologically active. Kisseptin-54, believed to be most effective, functions as the ligand for GPCR54 receptor (Ohtaki et al., 2001; Kotani et al., 2001; Kirby et al., 2010) given the somewhat ungainly name metastin, referred to here with the more acceptable nomenclature of Kisseptin. The Kisseptin family peptides possess a common Arg–Phe–NH_2 motif at their C-terminus, which characterise them as members of the RF-amide superfamily (Roa et al., 2008a,b). Besides the obvious relevance to cancer biology, Kisseptins play a prominent role in neuroendocrine control of the hypothalamic-pituitary-gonadal (HPG) axis (Pineda et al., 2010). They have also been associated with possible cardiovascular abnormalities.

The Tumour Suppressor Function of Kisseptin

The suppressor function of Kisseptin requires the mediation and binding to GPCR. Activation of GPCR54 results in signal transduction via G_q/G_{11} G-proteins. The GPCRs co-ordinate and mediate several pathways of signalling that regulate neuroendocrine function, cell proliferation, survival and apoptosis and cell motility (Liu et al., 2008; Castano et al., 2009; Sherbet, 2011a; Davenport and Mead, 2010). GPCRs liaise with EGF and PDGF in mitogenic function and possibly also with VEGF. Their mediation may be seen in the functions of some cytokines, including interleukins and tumour necrosis factor (TNF). The SMO component of the Hh signalling cascade is recognised as a member of the GPCR family. So GPCRs participate in several important signalling systems such as Src, Ras, MAPK pathway involving ERK1/2, p38 and JAK/STAT pathways. They mediate signalling via PLC/IP3/Ca^{2+} mobilisation leading PKC and PKA systems to genetic transcription and promotion of cell proliferation. Cell survival or apoptosis takes place through PI3K/Akt, with the modulation of genes such as apoptosis-inducing Bax and inhibitory Bcl2 genes, and regulation of NF-κB function.

Therapeutic Strategies in Cancer Biology and Pathology. DOI: http://dx.doi.org/10.1016/B978-0-12-416570-0.00012-3

Lee and Welch (1997b) attributed the peptide with the ability to regulate cell-substratum adhesion possibly involving cytoskeletal proteins. Loss of both receptor and gene expression was reported to relate to and predictive of the presence of lymph node metastases (Ikeguchi et al., 2004) who also pointed out their potential clinical value. Recently, Kisspeptin-10 has been found to inhibit osteotropic invasion of breast cancers carrying GPCR54 (Olbrich et al., 2010). The metastasis and invasion suppressor function has been attributed to the suppression of expression of MMPs (Bilban et al., 2004; Hesling et al., 2004; Yan et al., 2001). At the phenotypic level, Kisseptin can inhibit invasive behaviour by inhibiting the DNA-binding activity of NF-κB and leading to a loss of MMP activity (Takeda et al., 2012). The inhibition of migration occurs via paxillin focal adhesion protein, FAK, MAPK and Rho A mediation. The focal adhesions and stress fibres induced by Kisseptin are inhibited by Rho kinases inhibitors (Shoji et al., 2009, 2010).

Historically, loss of Kisseptin with associated progression to the metastatic state has been reported in many human neoplasms (Shoji et al., 2010). Compatibly, its expression in metastatic tumours is downregulated in comparison with the corresponding primary tumours. More recently patients with endometrial cancers with high expression of GPCR54 showed higher overall survival and GPR54 expression was related to FIGO (International Federation of Gynecology and Obstetrics) stage and grade and myometrial invasion (Kang et al., 2011b). Earlier Nagai et al. (2009) had identified pancreatic tumours expressing Kisseptin with less frequent recurrence. High expression of the ligand and GPCR54 was predictive of significantly longer survival. But many dissenting voices have also been heard. No differences have been found in serum Kisseptin between normal subjects and NSCLC patients and between patients with and without metastatic disease (Karapanagiotou et al., 2011). Indeed, plasma levels were similar in prostate cancer patients and in gastric and pancreatic cancer plasma levels were higher than in normal subjects (Curtis et al., 2010; Ergen et al., 2012; Katagiri et al., 2009).

Kisseptin in Clinical Medicine

Some clinical trials involving the administration of kisspeptin 112–121 are currently being carried out to examine its effectiveness in subjects with reproductive disorders. Kisseptin agonists TAK-448 and TAK-683 have been developed for evaluation and are currently being tested for tolerability, toxicity and pharmacological features. Both these analogues have been tested in animal models for their biological properties, for example responsiveness of hypothalamic gonadotoropin-releasing hormone (GnRH) neurons (Matsui et al., 2010). Scott et al. (2013) have successfully tested TAK-683 for tolerability by normal subjects.

With the present background of the biology of the gene, possibly it would be premature to expect similar clinical trials with cancer patients. Whilst the primary target is clearly defined by various studies, we are at the observation stage rather than one of active intervention. In any event, this has to be preceded by both *in vitro* and *in vivo* testing. It would also be helpful to address the question of potential secondary targets.

13 KAI1 (CD82) Suppresses Metastasis, Cell Proliferation and Invasion

The KAI1 was identified as a metastasis suppressor gene by Dong et al. (1995), who demonstrated that upon transfer into rat AT6.1 prostate cancer cells suppressed metastatic spread and also that it was downregulated in metastatic prostate cancer. It is now known to function as a suppressor in many tumour types. Over the years, much has been learnt about the biology of KAI1 and the wide ranging influences that it exerts are increasingly being appreciated so much so that it could be deemed now as potential target for cancer therapy. The KAI1 gene product (CD82) is a tetraspan transmembrane protein. KAI1 also markedly interferes with the expression of ECM components such as β1 integrin and fibronectin involved in intercellular adhesion (Lee et al., 2011) and in this way influences cell motility and invasion. Abe et al. (2008b) have implicated KAI1 in E-cadherin/β-catenin signalling, and with E-cadherin closely associated cell–cell adhesion this does seem distinctly possible. The HGF/c-met signalling system is a major activator cell migration. One should also recall here that HGF can activate Wnt/β-catenin signalling and promote proliferation (Apte et al., 2006). KAI1 has been shown to inhibit the promotion of invasion by HGF (Figures 13.1 and 13.2). On the other hand, KAI1 could be functioning via the agency of integrins, whose functions it does influence. Sridhar and Miranti (2006) showed that it markedly suppressed integrin-dependent activation of c-met and its activation by HGF. Thus it could be involved in inhibition of both migration and cell proliferation promoting effects of HGF. HGF can activate several pathways of signalling. Binding to c-met can activate several pathways (Figure 13.2). Of these, KAI1/CD82 is said to selectively diminish Ras-Cdc42/Rac and IP3K/Cdc42/Rac pathways, but not IP3K/Akt and PI3K/MAPK (Takahashi et al., 2007). This seems to suggest that KAI1 selectively inhibits cell migration, not both migration and cell proliferation. However, PI3K/Akt signalling can function through mTOR/Rac influencing both cell proliferation and migration. Indeed Choi et al. (2009) found KAI1 downregulated Rac 1 expression through the PI3K/Akt/mTOR pathway. This aspect does need further study beneficial not only from the point of view of the effective outcome, but also from the viewpoint of possible strategic use of agents that could function additively with KAI1.

When read in this background, the recent report concerning the inhibition of VEGF and HIF-1 expression by KAI1 (Park et al., 2012) gains much significance. Park et al. (2012) have described a complex pathway by which KAI1 seems to do this, involving VHL which is known to upregulate HIF-1α that stimulates VEGF

Therapeutic Strategies in Cancer Biology and Pathology. DOI: http://dx.doi.org/10.1016/B978-0-12-416570-0.00013-5

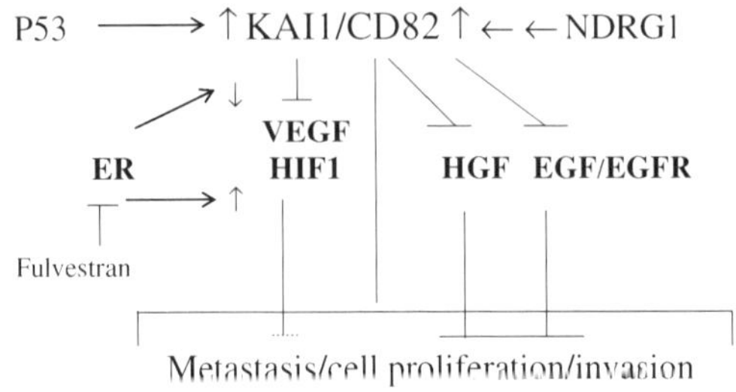

Figure 13.1 Postulated routes of suppression of metastasis, cell proliferation and motility by KAI1. Also shown are some interacting genes such as the suppressor p53 and NDRG1. NDRG1 can inhibit the metastasis promoting transcription factor ATF3. Further details of HGF signalling are shown in Figure 13.2 and the effects of KAI1 on that pathway are described in legend to that figure and also in the text.

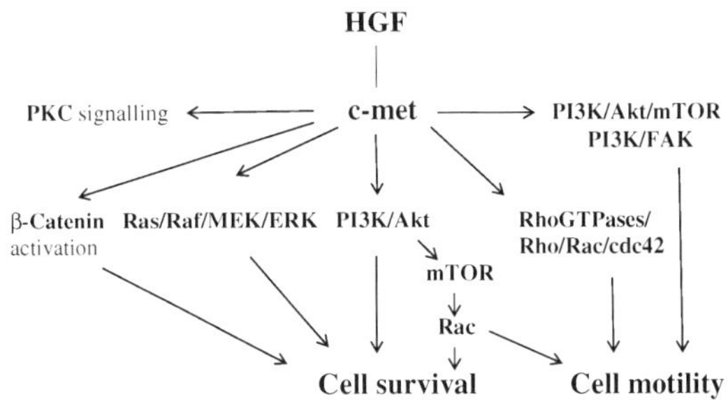

Figure 13.2 The potential pathways HGF mediated effects on cell survival and motility regulated by HGF. KAI1 might affect signalling involved with both cell proliferation and survival, but there are also suggestions that it can selectively inhibit Rho GTPases/Rho/Rac/cdc42 signalling cell motility.

expression. Unfortunately, no attempt has been made to quantify angiogenesis *in vivo*. Nonetheless, inhibition of angiogenesis is an assured mode of modulating metastasis.

The p53 protein is said to directly activate the KAI1 gene by interacting with its promoter via a consensus-binding sequence (Mashimo et al., 1998). Loss of both p53 and KAI1 has been associated with poor survival in prostate cancer. These findings assume great importance with p53 being mutated in a large majority of human neoplasms and in the context of their relevance to the deregulation of cell proliferation. It would be relevant to cite here the work of Christgen et al. (2008) who noticed a link between KAI1 expression and the expression of ER. ER positivity closely correlated with loss of KAI1. KAI1 transcription showed a far greater downregulation in ER-positive than in ER-negative tumours. These effects were suppressed by ER antagonists accompanied by KAI1 upregulation and inhibition of cell proliferation and motility.

Another metastasis suppressor gene apparently involved in regulating the activity of KAI1 is NDRG1 (N-myc downstream-regulated gene). The latter can induce KAI1 and suppression of NDRG1 reduces KAI1 expression. In fact loss of KAI1 diminishes the metastasis suppressor function of NDRG1. The loss of both NDRG1 and KAI1 occurs in prostate cancer progression NDGR1 also inhibits the metastasis promoting transcription factor ATF3, which can modulate KAI1 transcription (Liu et al., 2011d). ATF3 is a stress-inducible transcriptional repressor. The stress response is mediated by JNK with NF-κB, a component of a transcription complex containing ATF3, and MAPK or p53-dependent signalling pathway. The p53 gene can induce transcription of ATF3. The involvement of p53 with NDRG1 signalling is discussed in detail elsewhere in this book (pp. 111–117). Thus taken together, a complex network of signalling seems to function in bringing about a co-ordinated control of cell proliferation, intercellular/substratum adhesion, invasion and metastatic dissemination.

Reactivation of KAI1

The reactivation of KAI1 by p53 was cited earlier. Mashimo et al. (1998) identified a tandem repeat of the p53 consensus-binding sequence in the KAI1 promoter and the expression of the two genes showed a strong correlation in prostate cancer and loss of both resulted in poor prognosis. Mashimo et al. (2000) further showed that the topoisomerase II inhibitor etoposide activated KAI1 and markedly inhibited cell motility in human prostate cancer cell lines and a lung carcinoma cell line. These changes occurred by the mediation of p53 or c-Jun. As one would recall, in addition to p53, binding sites for c-Jun/transcription activating factors AP1 and AP2 have been identified in the KAI1 promoter (Marreiros et al., 2003).

14 14-3-3 Proteins in Normal and Tumour Cell Biology

The 14-3-3 proteins are ubiquitously expressed in many normal tissues and have been associated with cell signalling systems, cell proliferation, cell cycle regulation and apoptosis, cytoskeletal dynamics, to name those cellular functions crucial in tumour development and dissemination. Their wide ranging function can be attributed to the numerous interacting proteins that interact with 14-3-3s. Several isoforms of 14-3-3 have been identified, seven β, γ, ϵ, ζ, η, σ and τ encoded by seven genes in mammals. These appear to display much diversity of phenotypic outcome ascribable with no uncertainty to the gamut of their targets. Some are tumour suppressors and others promoters. The inclusion of the discussion of this protein family with tumour suppressors is therefore merely a matter of expediency. The 14-3-3s operate by binding to phosphorylated-serine/threonine motifs on their target proteins and modulate their function. In doing so they effectively modulate several signalling pathways to generate positive or negative regulation of cell motility and invasion, cell proliferation and apoptosis, EMT and metastasis (see Neal and Yu, 2010, for a highly erudite review). The ability of 14-3-3 to complex with and function as scaffold protein affects subcellular localisation of the targets. This faculty of 14-3-3 has been exploited as a therapeutic tool.

Expression of 14-3-3σ in Tumour Progression

The 14-3-3 proteins are differentially expressed in tumours. The isoform 14-3-3σ, an epithelial marker, is the most attributed with tumour suppressor ability on account of its downregulated expression by promoter methylation in many tumours. Early work on breast cancer revealed that the expression of 14-3-3σ decreased from 92% in breast ductal hyperplasia to 65% in ductal carcinoma *in situ* and further down to 23% in invasive ductal carcinoma (Simooka et al., 2004). This downregulation occurs by promoter methylation and this has been encountered in around 9/10 patients as compared with about a fifth of hyperplastic tissue and none being found in normal breast tissue (Luo et al., 2010). But this is not in agreement with the early effort of Moreira et al. (2005) who found equivalent levels of 14-3-3σ in non-malignant and malignant breast cancer and only 3/68 tumours showed loss of its expression. Albeit in samples from a small number of patients, Li et al. (2009b) found that14-3-3σ was downregulated in metastatic lesions of squamous cell carcinoma of the lung, suggesting cell subpopulations expressing low levels of 14-3-3σ might possess a greater ability or propensity to form metastases. They transfected

Therapeutic Strategies in Cancer Biology and Pathology. DOI: http://dx.doi.org/10.1016/B978-0-12-416570-0.00014-7

lung cancer cells with siRNA directed against 14-3-3σ and demonstrated that inhibition of 14-3-3σ increased *in vitro* invasion. Promoter methylation has been found in around a third of nasopharyngeal carcinomas (Chan et al., 2010). The promoter was more frequently partially methylated in node positive than in node negative tumours (Yi et al., 2009), but in the latter study there was little of interest in respect of correlation with distant metastases. It was downregulated in oesophageal squamous cell carcinomas and indeed reduced expression was related to invasion and lymphatic and distant metastatic spread and reduced 5-year survival (Ren et al., 2010a).

A most telling link between 14-3-3σ expression and metastasis may have been established by Schultz et al. (2009) with the demonstration of enhanced methylation of the gene in lymph node and cutaneous metastases of melanoma than in primary tumours, with attendant decrease in mRNA expression potentially indicating a clonal link of the metastases with subpopulations of the primary lesion. Ling et al. (2012) used another approach. They demonstrated that loss of one or both alleles of the 14-3-3σ gene leads to increased cell motility *in vitro*, tumour growth and metastasis and further that with the loss of 14-3-3σ MAPK but not PI3K/Akt signalling is activated together with increase in proliferation but without any changes in apoptosis. Finally, Ling et al. (2012) show that loss of 14-3-3σ increases both invasion and metastatic ability. However, in respect of the last parameter, only the lungs of tumour-bearing animals were checked for metastases.

It is needless to say there are sharp disagreements too. Shiba-Ishii et al. (2011) found higher expression of 14-3-3σ in invasive than in *in situ* adenocarcinoma of the lung. Okumura et al. (2010) studying a large group of patients with oesophageal squamous cell carcinomas concluded that nuclear, rather than cytoplasmic, 14-3-3σ correlated with clinical stage and lymphatic invasion and poor prognosis. The latter study employed immunohistochemical methods to assess the expression. It is needless to reiterate the problems intrinsic to this form of assessment of degree of expression. Although methylation frequency was higher in sporadic breast cancer than in hyperplastic lesions, methylation status was unrelated to 5-year survival (Luo et al., 2010).

Further disagreements and divergence in views have surfaced recently in relation to the expression status of 14-3-3σ. Contrary to expectation, augmented expression has been reported in many tumour types. As stated below, 14-3-3σ expression in cholangiocarcinomas directly correlated with lymph node involvement (Wu et al., 2012a). 14-3-3σ expression was greater and more frequently encountered in gastric cancers and furthermore, high expression correlated with disease progression and poorer prognosis with poorer overall survival and progression-free survival (Zhou et al., 2011b). Radhakrishnan et al. (2011a) found higher levels of 14-3-3σ in NSCLC with attendant hypomethylation of the gene and they also emphasise that there were no alterations in gene copy numbers or mutations that could account for higher or aberrant expression of 14-3-3σ. In oral squamous cell carcinoma exfoliative cells, the 14-3-3σ gene was hypermethylated in patients as well as in control subjects (Cordeiro-Silva et al., 2011). There was no obvious correlation between 14-3-3σ and p53 in oesophageal squamous cell carcinoma. In fact from a study of a large group of patients, Okumura et al. (2010) concluded that nuclear 14-3-3σ correlated with clinical stage and lymphatic invasion and poor prognosis, in total contradiction to the report by Ren et al. (2010a). Gastric carcinoma cell lines have also shown upregulation of 14-3-3σ (Kuramitsu et al., 2010). No relationship subsisted between 14-3-3σ and various clinical parameters of tumour stage, grade and response to therapy or overall survival or disease-free survival in patients with ovarian epithelial carcinoma (Mhawech-Fauceglia et al., 2009).

Naidoo et al. (2012) detected 14-3-3σ in a majority of pancreatic ductal adenocarcinoma. Metastatic epithelia showed upregulated expression in the cytoplasm. Interestingly, even the metastatic tissue of primary tumours not expressing the protein showed upregulated expression. This work involved immunohistochemical assessment of expression and the reliability of the scoring system of summing up intensity and frequency of staining requires much scrutiny. One ought to query the rationale for choosing 14-3-3σ, a postulated suppressor protein but excluding 14-3-3γ and 14-3-3θ which have been reported to promote tumorigenesis and possibly metastatic progression. Also it is somewhat intriguing that suppressor proteins such as NDRG1 and Raf kinase inhibitor protein (RKIP) which have been linked with the progression of pancreatic tumours to the metastatic stage do not occur in the list of differentially expressed proteins.

An important point that has lapsed in many investigations is that very little attention is paid to regulation of expression at the translational levels. Most studies have concentrated upon measuring mRNA or protein expression. Holm et al. (2009) found no relationship between 14-3-3σ mRNA levels and protein levels. This emphasises the difficulties inherent in attempting correlative relationship between protein expression and clinical features and prognostic outcome.

Nakamura et al. (2012) found 14-3-3σ in about a third of breast cancer specimens examined. Expression of 14-3-3σ negatively correlated with resistance to paclitaxel/5-FU/epirubicin/cyclophosphamide therapy. 14-3-3σ may regulate p53 and stabilise its expression by suppressing its ubiquitination. It has also been shown to promote p53 oligomerisation and increase its transcription activity (Yang, 2003b). A specific loss of function mutations of p53 has been attributed with the ability to confer multidrug resistance on cells. Mutant p53 can transactivate the mdr-1 gene and this is accompanied by increased P-glycoprotein levels (Chan and Lung, 2004). This could be an isolated finding. However, Nakamura et al. (2012) found no correlation of drug resistance with wild-type or mutated p53. But as noted in a later section, 14-3-3 proteins are known to be induced by p53 in response to DNA damage. This suggests that they might be involved with DNA repair and therefore can be implicated in inducing drug resistance. This has been discussed in the following section.

How Do Other 14-3-3 Isoforms Perform in the Clinical Settting?

In pleasant contrast, one detects much agreement regarding the effects of other isoforms of 14-3-3. The isoforms 14-3-3ζ, β, ε, θ, τ and γ have all been linked with tumour promotion. 14-3-3ζ targets many oncogenic proteins and promotes growth by inhibition of apoptosis and promoting cell survival pathways. 14-3-3ε, ζ and θ were expressed in meningioma in a tumour grade related fashion (Liu et al., 2010), and their expression levels increased with the increase of pathological grade of meningioma. Ko et al. (2011) examined primary hepatocellular carcinomas and matching extra-hepatic metastatic lesions. Both primary and secondary tumours showed enhanced 14-3-3γ expression. Also the metastases stained more strongly that the corresponding primary tumours. High expression of 14-3-3β was found in a majority of primary tumours and this correlated with potential for extra-hepatic metastases. Furthermore, high expression was also associated with poor survival (Liu et al., 2011c). Similar findings have been reported for vulvar carcinomas with higher levels of expression of these isoforms linked with aggressive disease (Wang et al., 2011d). According to Radhakrishnan et al. (2011b), 14-3-3γ expression is higher in NSCLC and correlated with p53 mutations thus linking them in terms of cell proliferation. They appear to imply that wild-type p53 suppresses 14-3-3γ expression. This may be so, but mutated p53 would have also lost its cell cycle control property with attendant phenotypic effects with the loss of this provision and independently of the 14-3-3γ expression status. Earlier they had come up with the notion that it activated MAPK and PI3K signalling (Radhakrishnan and Martinez, 2010). Wu et al. (2012a) found 14-3-3σ, γ, β and η are all expressed in cholangiocarcinomas; also 14-3-3σ and 14-3-3β expression positively correlated with lymph node metastases and poor survival rates. In the relevant table in the publication only, lymph node involvement is indicated but in the description they refer to distant metastases. Unfortunately patient data was not available for analysis of these findings.

14-3-3 Proteins in Regulation of Cell Proliferation

The regulation of cell proliferation and cell motility and invasion has been held for some time now to be in the realm of 14-3-3 protein function. Several pathways have been implicated in 14-3-3 protein signalling. These proteins interact with many target proteins mutually affecting their cellular functions. The tumour suppressor p53 family of proteins as well as certain growth factors and their receptors has been identified as potential targets of interaction with 14-3-3 proteins with a real and active bearing on cell cycle regulation and cell proliferation.

Several lines of evidence have also been adduced to support the view that these proteins actively regulate the cell cycle traverse. 14-3-3σ is upregulated by the suppressor genes p53 and BRCA1, which are both closely allied to the upregulation of cdk inhibitors $p21^{waf1}$ and $p27^{kip1}$ leading to the inhibition of cell proliferation and tumour suppression. The cell cycle regulatory role via p53 has been seriously investigated in the recent past, which has included the participation of growth factors. The functions of 14-3-3 proteins have also included the mTOR pathway, which as discussed else in this book (pp. 11–13) and transduction of growth factor signals from activated RTK to PI3K/Akt and Ras/MAPK axis (Figure 14.1).

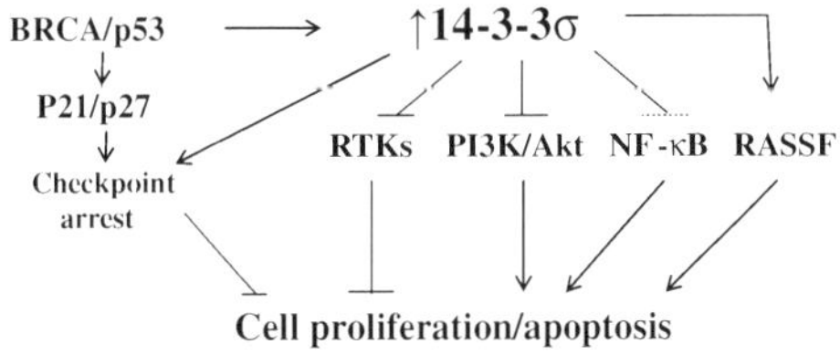

Figure 14.1 The postulated 14-3-3σ suppressor signalling pathways. Other isoforms of 14-3-3 which promote cell proliferation and inhibit apoptosis also by regulating the signals in the opposite direction to that of 14-3-3σ. 14-3-3σ signalling might also engage mTOR, Wnt and Hh pathways. The isoforms that can stimulate cell motility have been suggested to function via the MMP system.

P53 in 14-3-3 Function

The generally postulated suppressor function of 14-3-3σ would only be compatible if it were shown to inhibit cell proliferation and this would not be consistent with self-renewal capabilities of stem cells, whether of embryonic or adult origin. Indeed, contrary to conventional thinking embryonic stem cell survival and maintenance may be an important area in developmental and cancer stem cell biology, where 14-3-3s might play a role. The whole array of 14-3-3 isoforms has been detected in embryonic stem cells. Inhibition of 14-3-3σ with siRNA reduces stem cell proliferation. Furthermore 14-3-3σ has also been found to interact with Wnt signalling (Chang et al., 2012a), which as noted elsewhere collaborates with Hh signalling.

In a series of breast cancers not only was 14-3-3σ expression reduced but this correlated with that of p73 (Geng et al., 2011), which ties in with both p53 and p73 directly targeting 14-3-3σ with a pro-apoptotic outcome. Indeed p53 protein binds to certain consensus binding sequences that have been identified in the promoter of 14-3-3γ which is not only highly expressed in lung tumours but high expression correlates with p53 mutations with the implication that wild-type p53 might regulate 14-3-3γ expression (Radhakrishnan et al., 2011b). Wild-type p53 bound to the consensus sequence. However, when this motif was deleted, the ability of p53 to suppress 14-3-3γ expression also disappeared. This provides a conceptual nexus between 14-3-3 proteins and p53 family proteins with cell proliferation, but the caveat remains that loss of function mutation of p53 would influence proliferation status without reference to 14-3-3γ expression. Nonetheless, a causal link does exist between expressions of 14-3-3σ and p53. 14-3-3σ is a p53 inducible protein and may regulate p53, and stabilise its expression by suppressing its ubiquitination. It has also been shown to promote p53 oligomerisation and increase its transcription activity (Yang, 2003b). Su et al. (2011) has been shown most persuasively that DNA damage results in the diminution of COP1 (constitutive photomorphogenic 1), an E3 ubiquitin ligase. COP1 is a repressor of photomorphogenesis in plants which targets p53 for proteasomal degradation via ubiquitination (Dorman et al., 2004). The reduction of COP1 would be expected to result in the prevention of ubiquitination of p53 and lead to its stabilisation. In contrast in 14-3-3σ (−/−) cells, COP1 was not downregulated so allowing the ubiquitination and destruction of p53 (Su et al., 2011). This does not in any way exclude the stabilisation of p53 also by the direct binding of 14-3-3 proteins to specific C-terminal motifs of p53. Rajagopalan et al. (2010) have identified specific phosphorylation sites in p53 which serve as potential 14-3-3 binding sites. Phosphorylation at these sites facilitated the binding of 14-3-3γ and 14-3-3ε, but not essential for the binding of 14-3-3σ and 14-3-3τ.

The Function of 14-3-3 via PI3K/Akt Survival Pathway

The 14-3-3 proteins activate the PI3K/Akt survival signalling in many ways. They have been putatively linked with the activation of PI3K/Akt survival pathway by direct interaction with the pathway as well in collaboration with non-canonical SMO-mediated activation of Hh signalling (see Figure 4.1). 14-3-3ζ binds to the regulatory unit of PI3K and can activate it and promote its translocation to the cell membrane (Neal et al., 2012a) and in this way generate a response to activated RTK. Interestingly, these authors have reported the association between 14-3-3ζ and Akt activation in human breast cancer and linked this with tumour recurrence. Phosphorylation at Ser-58 is essential for interaction between the two (Powell et al., 2002). MAPK-activated protein kinase 2 (MAPKAPK2) has also been shown to phosphorylate 14-3-3ζ (Powell et al., 2003). Interaction with substrates such as Raf-1 and Bax require phosphorylation at other residues and these phosphorylations are carried out by casein kinase 1 and JNK, respectively (Dubois et al., 1997; Tsuruta et al., 2004). The phosphorylated-serine/threonine motifs facilitate the binding of 14-3-3 to its substrate proteins leading to alteration of their function.

The suppressor gene LKB1 (see pp. 179–194) is said to be a mediator in the activation by 14-3-3 of Akt. Akt seems to phosphorylate LKB1 at a specific serine residue and appears to promote 13-3-3 binding to the suppressor and block its growth suppressor function (Liu et al., 2012b). On the other hand, suppression of apoptosis and promotion of cell survival may be mediated by miRNAs. The Akt survival pathway is suppressed by miRNA-375 which inhibits PDK1 and prevents Akt activation, leading to the promotion of apoptosis. In gastric carcinomas, miRNA-375 is downregulated and this could result in increased cell survival and tumorigenesis. Another reason for this is the involvement of the anti-apoptosis 14-3-3ζ. Transfection of the miRNA-375 into gastric carcinoma cells inhibited the anti-apoptosis 14-3-3ζ and promoted apoptosis (Tsukamoto et al., 2010). Another route to apoptosis via TRAIL (TNF-related apoptosis-inducing ligand) is also influenced by Akt and 14-3-3. PACS-2 (phosphofurin acidic cluster sorting protein 2), which mediates apoptosis induced by TRAIL, is phosphorylated by Akt in collaborative function with 14-3-3 (Aslan et al., 2009).

Growth Factors and Their Receptors in 14-3-3 Function

The signalling of many growth factors and their receptors is modulated by 14-3-3 proteins, among those modulated are the EGFR family, insulin receptor substrate

(IRS) and TGF-β/Smad. EGF/EGFR do stimulate cell migration besides promoting growth. EGFR interacts with 14-3-3σ, possibly this leads to sequestration of EGFR. Cells transfected with 14-3-3σ showed a reduction in EGFR expression and cells carrying transfected siRNA against 14-3-3σ showed reduction in the expression of the latter and enhanced EGFR expression. *In vitro*, enhanced 14-3-3σ expression and reduction in EGFR led to reduced cell motility. In the opposite scenario with reduced 14-3-3σ and attendant increase of EGFR expression cell motility was enhanced (Huang et al., 2010). These transfectants do not appear to have been tested in metastasis assays. The presence of 14-3-3σ also inhibits the normal function of the EGFR family receptor HER2 (erbB2). Ling et al. (2012) have shown recently that loss of one or both alleles of 14-3-3σ lead to accelerated growth of mammary and salivary tumours in mice, in the background of activated HER2. They demonstrated that with the loss of 14-3-3σ MAPK signalling, but not PI3K/Akt, was activated and compatibly there was an increase in proliferation and no effect on apoptosis was seen. Finally, Ling et al. (2012) show that loss of 14-3-3σ increases both invasion and metastatic ability. However, in respect of the last parameter only the lungs of tumour-bearing animals were checked for metastases. 14-3-3σ expression downregulates HER2. Loss of 14-3-3σ upregulates HER2 as well as the transcription factor EGR2/CITED1. Indeed EGR2 binds to HER2 and leads to the relocation of the complex to the cytoplasm thus sequestering the regulatory transcription factor (Dillon et al., 2007).

The inhibitory effect of 14-3-3σ contributes to TGF-β1-mediated growth inhibition. But 14-3-3ζ negatively regulates growth inhibition by TGF-β (Hong et al., 2010), compatible with their known opposing effects on cell proliferation. However, one ought to be circumspect about experimental models operating in the backdrop of TGF-β, for this has a differential function in early transformation and tumorigenesis where it exerts an inhibitory effect, whilst promoting invasion and metastases at later stages of tumour development. Equally, EMT is activated by TGF-β, EGFR and HER2 receptors.

Regulation of Cell Cycle Checkpoints by 14-3-3 Proteins

The maintenance of genomic integrity requires that cells are prevented from entering into mitotic or meiotic division when DNA replication is incomplete. DNA replication checkpoints control cell cycle progression. DNA damage response is regulated by the ATM protein, a kinase similar in homology to PI3K. ATM is activated by MRE11 (meiotic recombination 11)/Rad50/NBS1 (Nijmegen Breakage syndrome 1) complex linked with DSB (double strand break) damage recognition and repair and checkpoint regulation (Carney et al., 1998) or 53BP1. The latter is involved with G2-M and S-phase checkpoint regulation (Wang et al., 2002).

Three DNA damage surveillance checkpoints have been identified and defined. These are the G1-S transition checkpoint, which restrains cells with damaged DNA from entering the S-phase. The second checkpoint, namely the S-phase checkpoint, monitors the progress of cells through the S-phase and regulates the rate of

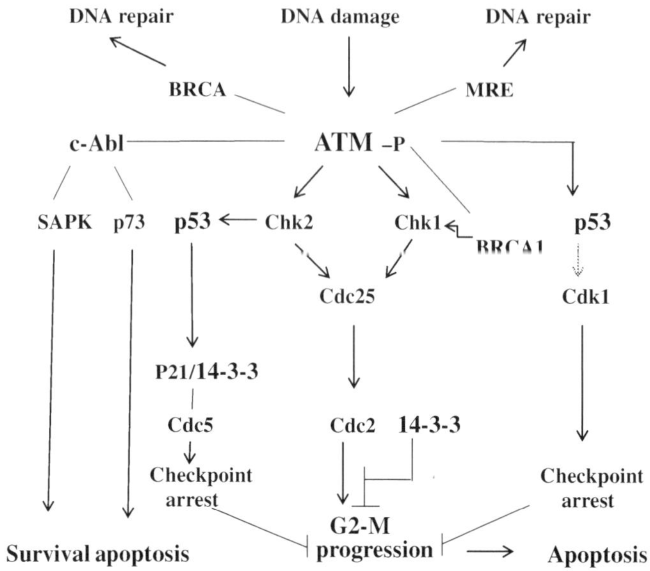

Figure 14.2 Pathways of ATM-mediated checkpoint regulation, apoptosis and cell survival. 14-3-3 might participate in G2-M arrest by the p53/21 and also Cdc25/Cdc2 axes of signalling. Cdc25 dephosphorylates and activates Cdc2 kinase. ATM signalling via NF-κB is not shown here.

DNA synthesis. The third checkpoint monitors the G2-M boundary. DNA damage response involves many signalling pathways that regulate cell cycle checkpoints, DNA repair and apoptosis programmes; these help to maintain genomic integrity. The checkpoint kinases Chk1 and Chk2 function downstream in DNA damage response. They are important participants in the network that maintains genomic integrity. Chk2 has an essential role in p53-dependent apoptosis. Indeed, checkpoint kinases have been regarded as tumour suppressor genes. These modes of genetic defence would be counterproductive to cancer management by chemo- and radiotherapy (Sherbet, 2003, 2011a). The ATM pathway involves the downstream checkpoint kinases Chk1 and Chk2, the suppressor proteins p53 and BRCA1 as well as c-Abl and NF-κB. The checkpoint kinases in turn target Cdc25/Cdc2. Thus emerges a composite picture in which ATM regulates many signalling systems that culminate in DNA repair, apoptosis or cell survival (Figure 14.2).

With this prelude a discussion of the role of 14-3-3 proteins in the repair of DNA damage and the regulation of cell cycle checkpoints seems highly appropriate. Indeed, the diversity of pathways that engage these proteins would make this an imperative from the point of view of therapeutic approaches with 14-3-3 because the checkpoint kinases have also been regarded as potential tumour suppressors. The fission yeast proteins Rad24 and Rad25 have been identified as homologous with mammalian 14-3-3 proteins (Ford et al., 1994). As stated earlier, Chk1 targets and phosphorylates Cdc25 and in this process creates a Rad24/25 (14-3-3) binding site and seems to determine its intracellular disposition (Lopez-Girona et al., 1999). Toyoshima-Morimoto et al. (2002) implicated the phosphorylation of a serine residue (Ser198) in a nuclear export signal sequence of human Cdc25C in promoting its nuclear localisation. One can speculate that possibly Rad24/25 (14-3-3) might interfere with Cdc25 localisation. As pointed out by Ford et al. (1994) in Rad24 (−/−), mutants cells show premature G2-M progression.

Do 14-3-3 Proteins Participate in DNA Repair?

The expression of some 14-3-3 proteins is inducible by p53 in response to DNA damage, suggesting that they might possess DNA repair function. From this arises the possibility that 14-3-3 proteins might be involved in generating drug resistance. The cytotoxicity of DNA damaging drugs is alleviated by DNA repair of which there are several pathways. Inhibition of repair of DNA damage potentiates the effects of cytotoxic drugs. DNA damage induces PARP (poly [ADP-ribose] polymerase) activation which then participates in DNA repair. PARP participates in base excision repair and inhibition of PARP prevents the repair of DNA lesions and enhances cytotoxicity. PARP mediates an important pathway to apoptosis. This could conceivably function either dependently or independently of p53. Caspases can break down PARP, prevent DNA repair and induce apoptosis. On the other hand, there is a view that hyperactivation of PARP could activate AIF (apoptosis inducing factor) and apoptosis (Yu et al., 2002), although lately PARP-independent activation of AIF has been advocated (Kondo et al., 2010). PARP has also been implicated in stalled replication fork restart. It is believed to bind to and activated by stalled forks and possibly attracts MRE11 to promote the restart (Bryant et al., 2009). A connection with the tumour suppressor BRCA2 also seems to subsist. Increased MRE11 levels are associated with BRCA2 deficiency (Ying et al., 2012). This also forges a link between BRCA2 deficiency and PARP inhibition in cancer management. BRCA1 has also implications for 14-3-3 expression, in that the former induces 14-3-3σ expression in the backdrop of wild-type p53. Both loss of BRCA1 and reduced 14-3-3σ expression are not conducive to G2-M arrest upon exposure to ionising radiation (Aprelikova et al., 2001). Quite clearly, PARP and 14-3-3σ do suggest a potential synergistic route to the regulation of cell cycle progression.

DNA repair requires the assembly of nucleosomes, a process that involves CAF-1 (chromatin assembly factor-1). The latter is able to interact with several cellular proteins, notably for instance the ASF-1 (anti-silencing factor-1) involved in chromatin dynamics. CAF-1 can also silence specific target genes such as Notch, tumour suppressor and cell cycle regulator Rb proteins, and PCNA (proliferating cell nuclear antigen) (Goodfellow et al., 2007; Ridgway and Almouzni, 2000; Moggs et al., 2000), among others. In the present context, 14-3-3 seems to take part in many components of the DNA repair apparatus. For instance 14-3-3ζ forms a complex with CAF-1 (Hoek et al., 2011).

Exonuclease 1 (Exo1) which participates in the processing of stalled replication fork and checkpoint induction is a member of Rad2 family of nucleases involved in DNA repair, replication and recombination. Engels et al. (2011) showed that Exo1 is also an interacting partner of 14-3-3 proteins in the yeast and in mammalian cells. Deficiency of the protein leads to lack of hyperphosphorylation of Exo1 and persistent stalling at the replication fork leaves lesions of DNA damage as single strand gaps as seen by EM. Exo1 is phosphorylated at three Ser/Thr motifs and is a potential target of 14-3-3 proteins. Interaction between the latter and Exo1 occurs *in vitro* (Andersen et al., 2012). Thus 14-3-3 proteins may modulate Exo1 phosphorylation status and possibly affect DNA repair and cell cycle progression. It would not be out of place to mention here that in the fission yeast *Schizosaccharomyces pombe* Dss1p DNA recombination repair protein has been found to recruit Rad24, which is a 14-3-3 homologous protein, to DSBs (Selvanathan et al., 2010). Rad24 probably also takes part in cytokinesis (Mishra et al., 2005).

Response of cells to DNA damage involves induction of G1 arrest by p53-mediated transcription of target genes such as $p21^{waf1}$ and the inhibition of cdk activity. 14-3-3σ is upregulated by the suppressor genes p53 and BRCA1, which are both closely allied to the upregulation of cdk inhibitors $p21^{waf1}$ and $p27^{kip1}$ leading to the inhibition of cell proliferation and tumour suppression. The expression of p73, another member of the p53 family, also correlates positively with 14-3-3σ. Overexpression of the latter upregulates p73 and inhibits tumorigenicity (Geng et al., 2011). Compatible with this, the expression of 14-3-3σ has been associated with drug resistance in breast cancers (Liu, 2006). 14-3-3σ might collude with p53 in the development of drug resistance. For, as noted earlier not only is 14-3-3σ a p53 inducible protein but it may also regulate p53, stabilise its expression by suppressing its ubiquitination. 14-3-3σ has also been shown to promote p53 oligomerisation and increase its transcription activity (Yang, 2003b) to occur quite independently of wild-type or mutated p53. However, mutation of p53 does contribute to drug resistance instigated by 14-3-3σ. A specific p53 mutation has been claimed to confer multi-drug resistance on cells and this effect seems to be a result of transactivation of the mdr-1 gene and increase of P-glycoprotein levels (Chan and Lung, 2004). Although these findings suggest a direct link of 14-3-3σ with DNA repair, one cannot exclude other means of activating $p21^{waf1}$ transcription. The tumour promoter MTA (see pp. 77–79) is capable of repressing the transcription of $p21^{waf1}$ independently of p53 (Li et al., 2010b). On the other hand, 14-3-3σ is said to stabilise p53 and enhance the sensitivity of 5-FU-resistant MCF7 cells to 5-FU, mitoxantrone and cisplatin. The stabilisation of p53 would effectively inhibit Akt and downstream of survivin, Bcl-2, NF-κB and NF-κB1 (p50) (also see below) reducing cell survival and aiding the cytotoxic effects of 5-FU (Zheng et al., 2012). Besides, p73 can activate the transcription of genes that are targeted by p53 and acts quite independently of p53 (Zawacka-Pankau et al., 2010).

14-3-3σ and NF-κB Survival Pathway

NF-κB are a family of transcription factors that activate many genes that are closely concerned with inflammation, cell survival, cell proliferation and apoptosis, angiogenesis, cell adhesion, invasion and metastasis (see Table 9.2). The NF-κB survival pathway has been invoked in the suppressor function of 14-3-3σ. The latter can inhibit

NF-κB and suppress proliferative potential of cells. In breast cancer cells (MCF7, MDA-MB-231, BT-474, SK-BR-3 and T47D), 14-3-3σ expression was downregulated in comparison with other isoforms. 14-3-3σ is associated with reduced invasion and metastasis *in vivo* experimental models. Of interest is that in breast cancer tissues absence of 14-3-3σ correlated with the presence and activation of NF-κB and markedly reflected in poor prognosis (Ingles-Esteve et al., 2012). These authors suggest a direct interaction between 14-3-3σ and NF-κB on the basis that three 14-3-3σ binding sites have been reported in NF-κB protein (Aguilera et al., 2006). Whilst they do show the effects on the expression of several NF-κB target genes, they have made no attempt to target their attention on specific genes that regulate the phenotypic effects of NF-κB (see Table 9.2). Nonetheless, this is a very creditable demonstration of the inhibitory function of 14-3-3σ. NF-κB is activated by many cytokines and is important in the production of MMPs. In certain cell lines, for example the lung fibroblast IMR-90 cell line, MMP-1 is produced in response to 14-3-3β and 14-3-3α but not to 14-3-3σ (Asdaghi et al., 2012). It seems possible therefore that 14-3-3σ has no effect on cell migration unlike other 14-3-3 because of its inability to generate MMPs.

Does 14-3-3σ Influence Wnt Signalling?

The Wnt system is a major developmental signalling pathway whose activation is an important requisite for many biological phenomena such as stem cell maintenance and maintenance of potential for cell replication. Wnt signalling seems to be able to interact with NF-κB signalling in the activation of EMT. Hence it is of interest here that attempts have been made to draw together 14-3-3σ with Wnt signalling. However, in somewhat sharp contrast to what one would expect, Chang et al. (2012a) have attributed the faculty of promotion of cell proliferation of embryonic stem cells to 14-3-3σ. They have allied together the functions of 14-3-3σ with Wnt signalling. One would recall the characteristic feature of Wnt signalling is the formation of a multiprotein complex of Axin, GSK-3β, CK1, and APC with β-catenin when Wnt receptors are not bound by the appropriate ligand. Axin promotes the phosphorylation of β-catenin with the mediation of GSK-3β leading to the degradation of β-catenin. The turnover of β-catenin is interceded by β-TrCP (β-transducin repeat containing protein), which is responsible for its ubiquitination and degradation. The degradation of β-catenin enables repressor factors to bind to the TCF/Lef transcription factors to block the transcription of β-catenin responsive genes. In the presence of the Wnt ligand, the multiprotein complex breaks up and this leads to the stabilisation and accumulation of β-catenin in the nucleus. The binding of Wnt ligand to the receptor inhibits GSK-3β and stabilises β-catenin. β-catenin now forms a complex with the transcription factor TCF/Lef, and the complex is translocated into the nucleus leading to the transcription of responsive genes (see Figure 6.1).

Chang et al. (2012a) have in fact argued that in embryonic stem cells 14-3-3σ promotes cell proliferation. They have claimed that suppression of 14-3-3σ using siRNAs suppressed embryonic stem cell proliferation. This is annulled by β-catenin knockdown. 14-3-3σ bound and increased GSK-3β phosphorylation and dislocated beta-catenin association with the multiprotein destruction complex. 14-3-3σ enhanced Wnt3a-induced β-catenin level. This would possibly suggest a deregulated activation of Wnt signalling. The expression of 14-3-3ζ has been reported to correlate with tumour grade and progression to metastasis in NSCLC, in which higher 14-3-3ζ levels also corresponded with higher β-catenin levels (Zang et al., 2010).

14-3-3 and Hh Signalling

Hh signalling can be distinguished into the canonical and non-canonical systems which both seem to eventually target genes whose activity or suppression is required in generating phenotypic outcomes. The canonical pathway involves Gli activation and the non-canonical SMO-dependent and -independent routes. In the non-canonical pathway, SMO can activate Gli via PI3K/Akt and MAPK/ERK route or Gli can be activated by PI3K/Akt independently of SMO (see pp. 35–39). The Gli transcription factors have been found to bind 14-3-3ε and this reduces their transcription activity. This interaction requires phosphorylation of Gli by PKA, which is known to negatively regulate Hh (Asaoka et al., 2010).

14-3-3 Proteins Interact with RASSF Signalling

The RASSF (Ras-association domain) gene family has received much attention on account of frequent methylation and inactivation of its members in many human neoplasms. The RASSF pathways of apoptosis signalling have been discussed at length in a later location (pp. 163–167). The wide diversity of 14-3-3 effects includes interaction with RASSF signalling to construct some of its phenotypic outcome.

A mode of involvement of 14-3-3 proteins has been proposed in the regulation of RASSF1A mediation of apoptosis. 14-3-3σ and 14-3-3ε to a lesser extent are closely bound to RASSF1A and this association is dissolved upon stimulation by TNF-α or TRAIL. RASSF1A can now activate apoptosis via TNF-R1 or TRAIL-R1. Furthermore, phosphorylation of certain serine residues seems to be essential for RASSF1A to associate with 14-3-3, since mutations at these residues prevent the association and lead to premature activation of apoptosis. To witness, the presence of three mutations greatly enhanced apoptosis (Abu Ghazaleh et al., 2010). However, RASSF can potentially interact with Hippo and Hh signalling, thus vitiating any attempts to forge a direct link between RASSF and 14-3-3.

Do 14-3-3 Proteins Employ mTOR Signalling?

The mTOR signalling pathway has now pre-eminently associated itself with several cellular processes such as cell proliferation, growth, apoptosis, angiogenesis, cell motility and invasion. So its aberrant activation provides cancer cells with a huge proliferative and invasive advantage and this contributes significantly to the process of cancer metastasis. Several cellular functions, such as growth factor function, cell proliferation, angiogenesis, response to hypoxia among others, see the participation of mTOR (see Figure 3.1) signalling. Inevitably a link-up of 14-3-3 proteins with mTOR signalling has also been sought. It has been proposed that14-3-3σ might regulate the mTOR-RAPTOR signalling cascade. An inverse relationship has been noticed in the expression of 14-3-3σ and mTOR activation in prostate cancer (Evren et al., 2011). However mTOR signalling is also influenced by Wnt activation and as noted earlier, there are suggested links of14-3-3σ with the Wnt pathway.

Effects of 14-3-3 Proteins on Cell Motility and Invasion

14-3-3 proteins influence cell motility and cell shape and cytokinesis among other biological features by having recourse to Rac (Rac GTPase-activating protein)-1, Cdc42 and other Rho-GTPases-mediated modulation of cytoskeletal dynamics. 14-3-3 proteins regulate cytoskeletal reorganisation and the interaction between ECM components and the cytoskeletal machinery to bring about the phenotype of invasion and motility.

14-3-3ζ has been reported to promote cell motility and diapedetic behaviour of prostate carcinoma PC3 cell by activating the Rac signalling pathway. Mutants of 14-3-3ζ that are unable to form dimers inhibited Rac function and PAK1 and PAK2 activity (Goc et al., 2012). In certain cell lines, for example the lung fibroblast IMR-90 cell line, MMP-1 is produced in response to 14-3-3α and 14-3-3β but not to 14-3-σ (Asdaghi et al., 2012). It seems possible therefore that 14-3-3σ has no effect on cell migration unlike other 14-3-3s because of its inability to generate MMPs.

Some interesting cross connections have emerged recently from the work of Yoon et al. (2012) who studied the cytoskeletal dynamics of *Xenopus* retinal ganglion cell axon elongation. They noted that expression of 14-3-3ζ transcripts and proteins in retinal growth cones was greater during the phase of rapid axonal extension. Inhibition of 14-3-3 (all isoforms)/14-3-3ζ expression resulted in a marked diminution of axonal length. They also found that 14-3-3ζ co-localised with cofilin in the growth cones and cofilin is an essential requirement for axon growth of neurons. Now cofilin is a plasma membrane associated protein, promotes actin-depolymerisation, but primarily it is an F-actin binding protein. Its activity and subcellular localisation is regulated by phosphorylation (Bamburg, 1999). Phosphorylated inactive cofilin does inhibit cell migration (Popow-Wozniak et al., 2012). LIM kinase regulates cofilin activity and is itself activated by Rho GTPases, namely Rho, Rac and Cdc42, which are closely associated with cytoskeletal dynamics.

Therapeutic Approach with 14-3-3

With the broad spectrum of biological effects and the known ability to activate and participate in several signalling systems, the 14-3-3 proteins have been often advocated as therapeutic targets. This coupled with the differential effects attributed to the many isoforms of the family has led to many thoughtful studies into the possibility of appropriately influencing the status of their expression to control the biological parameters that they regulate.

Drug resistance is a major challenge in cancer management. Several pathways that lead to the development of drug resistance have been identified. Multi-drug resistance is a consequence of reduced drug efflux via the function of ABC transporters such as the P-glycoprotein. Induction of drug detoxification mechanisms, DNA repair and cell survival pathways, for example NF-κB and Akt are also involved in drug resistance (Tsuruo et al., 2003; Gottesman, 2002). The striking association of some 14-3-3 proteins to modulate apoptosis and cell proliferation and DNA repair has also instigated investigation into their role in the development of drug resistance and whether the proteins provide a route to neutralising drug resistance. Several studies have shown certain isoforms can actively modulate drug resistance. A salient example is the association of cisplatin resistance with 14-3-3σ expression. Pancreatic ductal adenocarcinoma show higher levels of 14-3-3σ expression than normal pancreatic duct cells. When the carcinoma cells were forced to express 14-3-3σ at high levels, they became resistant to cisplatin induced apoptosis but displayed increased invasion in response to EGF and IGF-1 (Neupane and Korc, 2008). Similarly HCT116 human colon cancer cells showed greater resistance to cisplatin than those in which 14-3-3σ had been blocked. However, intriguingly 14-3-3σ appeared to inhibit p53 activation and activation of p21 and p21-dependent senescence (Han et al., 2009c). Earlier Han et al. (2006) showed that siRNA-mediated inhibition of 14-3-3σ in prostate cancer cell lines enhanced their sensitivity to mitoxantrone and Adriamycin. Sang et al. (2006) found that induction of 14-3-3σ increased the sensitivity of resistant human breast cancer cell lines to Adriamycin. Both p53 and p73 directly target and activate 14-3-3σ leading to a pro-apoptotic outcome. Indeed, it seems more likely that 14-3-3σ and p53/p73 compound each other's effect and augment drug resistance. This would harmonise with the tumour suppressor function of 14-3-3σ, but would not be compatible with increase in cell motility intrinsic or induced by biological response regulators. In fact one has to entertain the possibility that the growth factors could have exerted their effects independently of 14-3-3σ.

Signalling systems such as HER2/EGFR/MAPK are particularly of interest. For, this signalling axis can be upregulated by miRNAs. MiRNA-21 has been shown to lead to the development of resistance to Trastuzumab therapy of breast cancer (Gong et al., 2011). Possibly, this is a consequence of lack of saturation binding of HER2 by Trastuzumab. High miRNA-21 could be counteracting the blockage of EGFR by Trastuzumab by activating EGFR/STAT signalling with apparent increase of resistance to Trastuzumab. MiRNA-125a-5p seems to suppress cell proliferation by directly targeting HER2. This inhibitory effect was compounded by Trastuzumab (Nishida et al., 2011). Now enhancing the expression of miRNA-451 effectively reduces 14-3-3ζ, together with increase in apoptosis, suppression of cell proliferation and reduced activation of HER2, EGFR and MAPK signalling (Bergamaschi and Katzenellenbogen, 2012). Tamoxifen is administered in ER+ breast cancer, but these authors have found that it downregulated miRNA-425 and upregulated

14-3-3ζ. Inhibition of 14-3-3ζ restored sensitivity to CHOP (cyclophosphamide, hydroxyl doxorubicin, vincristine/Oncovin®, prednisone) chemotherapeutic regimen and inhibition of Akt also has a similar effect (Maxwell et al., 2011). The advantages of targeting 14-3-3ζ to control and inhibit cell proliferation and drug resistance is obvious from this. A proteomic analysis of chemotherapy-sensitive and chemotherapy-resistant ER+ invasive ductal carcinoma tissue samples has shown that several 14-3-3 proteins, for example γ, θ, δ, ε, may also be associated with drug resistance (Hodgkinson et al., 2012). Although the mechanisms involved are not well understood, it has been recognised that other pathways might be switched on in the course of development of resistance to tamoxifen. Hh and PI3K/Akt signalling have been suggested as possibilities (see p. 36). The activation of Hh signalling linked up with PI3K/Akt has been demonstrated in tamoxifen-resistant breast cancer cell lines (Ramaswamy et al., 2012).

Among other therapeutic approaches being attempted is the development of 14-3-3 inhibitors that might directly attenuate the cell proliferation and tumour promoting activity of these proteins or synergistically sensitise tumour cells to conventional therapeutic agents. The strategy is based on interference with the interaction of 14-3-3 proteins with their interacting counterparts. Zhao et al. (2011) have shown that 4-[(2Z)-2-[4-formyl-6-methyl-5-oxo-3-(phosphonatooxymethyl) pyridin-2-ylidene] hydrazinyl] benzoate inhibited the binding of 14-3-3ζ to some of its substrates. It may be noted here that 14-3-3ζ (also 14-3-3ε) can bind to many substrates including IRS1 (Ogihara et al., 1997) and PKC and inhibit the function of both.

Corradi et al. (2010) and Mancini et al. (2011) have described another non-peptide small molecule called BVO2 (Asinex, Moscow), which is said to inhibit the docking of 14-3-3 to their targets. It can release c-Abl from binding with 14-3-3σ, which allows nuclear translocation of c-Abl leading to the activation of pro-apoptotic genes. BVO2 has been found to inhibit the proliferation of both imatinib-sensitive and -resistant leukaemia cells. Equally this can lead to oncogenic activation of c-Abl. Another way this can be achieved is by using competitive inhibitor R18 to dislodge 14-3-3 association with the pro-apoptotic transcription factor FOXO3a and induce apoptosis (Dong et al., 2007b, 2008). The R18 peptide specifically inhibits 14-3-3 function but it is not isoform specific. The similarity in binding affinities suggests a common binding site for R18 on 14-3-3 proteins (Wang et al., 1999). So the use of this inhibitor would depend upon whether the suppressor or promoter isoform is expressed. The potential value of R18 is underscored by the finding that R18 inhibits the interaction of 14-3-3 with Raf-1. Raf-1, which is activated by phosphorylation in response to mitogens, is a MAP3K kinase. It binds directly to membrane associated GTPases and regulates many cellular functions, importantly promotes cell survival by inhibiting apoptosis and promotes cell migration and invasion.

The 14-3-3 proteins function by recognising and binding to phosphorylated-serine/threonine motifs on their target proteins and modulate the function of the target proteins. Dimerisation is essential for their interaction with substrate proteins. Also dimerisation diminishes 14-3-3 susceptibility to phosphorylation (Ying et al., 2003). Now the provision of liability to be phosphorylated at the dimer interface blocks signalling function. Sphingosine and its analogue FTY720 both promote phosphorylation of 14-3-3 and promote apoptosis (Woodcock et al., 2010). Among other small molecule inhibitors are the natural extracts of the beetle *Blaps japanensis*, which Yan et al. (2012) have claimed to inhibit 14-3-3 interaction with target proteins. Arrendale et al. (2012) have designed a phosphoserine threonine mimetic pro-drug with a non-hydrolysable difluoromethylenephosphoserine which displays pronounced cytotoxic effects at very low concentration. They have shown that the inhibitory effects directly involve 14-3-3 proteins. Most of these are preliminary studies but should be deemed as suitable for further investigation. The strategy of developing agents that promote or impede and disrupt 14-3-3 interaction as the case may be with their target proteins is indeed a laudable one. It has been speculated that there might be over a couple of hundred targets and so the chances of identifying targets of value in terms of deploying 14-4-3 to modulate their phenotypic outcome may not be unrealistic. One cannot gainsay that there is a long but profitable way of beneficial exploration.

15 Suppressor Function of NDRG1

The NDRG1 (also known as of Cap43 and Drg-1) gene, as the name implies, is regulated, indeed suppressed, by N-myc gene. This is one of possibly a four member family of genes, NDRG1, -2, -3 and -4. Some of them show organ/tissue-specific distribution (Zhou et al., 2001) and possibly subserve different biological functions. NDRG4 which is expressed almost exclusively in the heart and brain may be essential for the survival of primary astrocytes and might indeed support the viability of components of glioblastoma multiforme (Schilling et al., 2009). The NDRGs may be involved in the pathogenesis of many human conditions. NDRG1 was identified and originally described as a stress-regulated and hypoxia-inducible and possibly a differentiation-related gene and has been attributed putatively with tumour and metastasis function. It is downregulated in many forms of cancer. Loss of expression of the gene has inversely correlated with tumour invasion, nodal dissemination and generally predictive of poor prognosis. Silencing of the gene has been linked with histone acetylation rather than with hyper-methylation of the promoter (Li and Chen, 2011). Equally, NDRG1 may be upregulated even in tumours on account of the prevailing hypoxic environment.

NDRG1 is able to inhibit tumour growth and cell migration, both essential biological elements of metastatic spread. The NDRG1 suppressor signals appear to be activated by many ligands and transduced through signalling channels such as the p53, NO and Wnt signalling cascades. Aberrant Wnt/β-catenin signalling is an important aspect of promotion of metastasis and discussed in an earlier context. The Wnt pathway is effectively inhibited by NDRG1. NDRG1 also inhibits ATF3 which supports metastasis, and Wnt/β-catenin/TCF signalling has been implicated in this process (Liu et al., 2012d). ATF3 is a stress-inducible transcriptional repressor. The stress response is mediated by the JNK/MAPK or p53-dependent signalling pathway. The p53 gene can induce transcription of ATF3. The myc gene in contrast inhibits NDRG1 but can activate ATF3. A new dimension added to the signalling picture is that EMT induced by TGF-β is inhibited by the upregulation of NDRG1 in HT-29 and DU145 cells by iron chelators. They also appear to preserve E-cadherin and β-catenin at the cell membrane, also thus implicating Wnt signalling (Chen et al., 2012b). In prostate cancer, reduced expression of membrane-associated NDRG1 closely corresponded with reduced expression of E-cadherin and the loss of both these correlated with poor prognosis (Song et al., 2010b). Thus one can envisage a complex regulatory cycle involving NDRG1, p53, myc, TGF-β and Wnt/β-catenin signalling together with ATF3 participation in the inhibition of EMT as well the metastatic process (Figure 15.1).

Therapeutic Strategies in Cancer Biology and Pathology. DOI: http://dx.doi.org/10.1016/B978-0-12-416570-0.00015-9

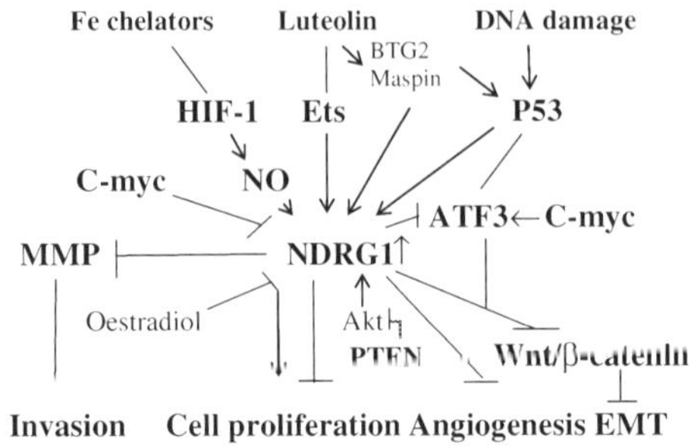

Figure 15.1 The several pathways by which NDRG1 exerts its inhibitory effects on invasion, cell proliferation, EMT and also possibly angiogenesis and so inhibit metastatic spread (discussed in detail in the text).

NDRG1 Suppresses MMP Activity and Invasion

There is a large body of evidence supporting the view that NDRG1 actively suppresses migration and cell proliferation, but the precise mechanisms involved in the generation of these biological effects have not been investigated. In gene transfer, NDRG1 exerts marked inhibitory effects on cell migration *in vitro*, which have been ascribed to reduced expression of MMP-9 (Maruyama et al., 2006). Human gastric adenocarcinoma cells in which NDRG1 was suppressed using siRNA showed enhanced *in vitro* migration and gelatinolytic activity due to MMP-2/MT1-MMP (membrane type 1 MMP), but when NDRG1 expression was restored, the cells lost the acquired invasive ability (Liu et al., 2011f). One would recall here that MT1-MMP is a transmembrane MMP capable of remodelling the ECM and influence cell motility.

Upregulation of NDRG1 Suppresses Cell Migration and Proliferation

Effects of Iron Chelators

NDRG1 is upregulated by free radical NO and this suppresses cell migration (Hickok et al., 2011). The single unpaired electron of NO enables it to bind strongly to the iron in heme groups and other iron proteins. Hickok et al. (2011) go on to show that free radical NO interacts with the chelatable iron pool generating complexes that lead to NDRG1 regulation and to the inhibition of cell migration. Raising NDRG1 levels might be a mechanism by which iron chelators exert their anti-proliferative effect. Iron chelators do indeed upregulate NDRG1 by the mediation of HIF-1 (Le and Richardson, 2004) and NO synthesis. The anti-proliferative effects of iron chelators may occur by other mechanisms, for example p21 and cyclin D1 among others (Richardson, 2005). Richardson et al. (2006) have described a series of di-2-pyridyl ketone thiosemicarbazone anti-proliferative compounds which selectively inhibit tumour growth *in vitro* and *in vivo*. Of these Kovacevic et al. (2011a) found di-2-pyridylketone 4-cyclohexyl-4-methyl-3-thiosemicarbazone hydrochloride to be a highly effective tumour inhibitor as studied in xenografts of pancreatic cancer cells into BALBc nu/nu mice together with upregulation of NDRG1. Furthermore, this compound appeared to be tumour-specific without inflicting significant changes in normal tissues. Although the putative metastasis inhibitor was upregulated, there are no indications whether the xenograft control tumours had metastasised and whether the treatment regime besides completely inhibiting tumour growth also rendered the host animals totally devoid of metastases. Another iron chelator that appears to upregulate both NDRG1 and BTG2 (B-cell translocation gene 2) is L-Mimosine which arrests cells at G1-S transition point, quite obviously the outcome of induction by L-Mimosine and HIF-1α of both genes (Chung et al., 2012).

Luteolin Upregulates NDRG1

The flavonoid luteolin (3′,4′,5,7-tetrahydroxyflavone) is reputedly an inhibitor of tumour growth and metastasis. It seems to be capable of inhibiting EMT (Lin et al., 2011b). Hypoxia-induced EMT is inhibited by luteolin. It reversed EMT-associated suppression of E-cadherin and increase of N-cadherin and vimentin. Luteolin also inhibited the expression of integrin β1 and FAK, features closely related to EMT (Ruan et al., 2012). The flavone possibly also functions by inhibiting PI3K/Akt and NF-κB signalling to suppress cell proliferation and growth and invasion. Whether NDRG1 is involved in the modulation of signalling by this cell survival pathway is still unclear. The suppressor gene PTEN inhibits the PI3/Akt pathway and promotes apoptosis. Unoki and Nakamura (2001) found that introduction of extraneous PTEN, presumably by gene transfer although not so stated, transactivated several growth inhibitory genes that included NDRG1. Ostensibly NDRG1, which was said to have been strongly induced, was investigated further. However no data are provided, except that the increase in NDRG1 had had no effect on cell proliferation. The uncertainty of PTEN involvement heightens when read together with the reported findings that NDRG1 upregulation correlated with the downregulation of PTEN (Chen et al., 2008a). Nonetheless a clearer picture emerges from Klawitter et al. (2010) showing that Lovastatin upregulated PTEN thus effectively inhibiting Akt resulting in enhanced expression of NDRG1. Also earlier Bandyopadhyay et al. (2004) had demonstrated the upregulation of NDRG1 by PTEN.

Treatment of LNCap prostate cancer cells with luteolin inhibits cell proliferation by cell cycle arrest at G1-S. This is accompanied by the enhanced expression of many genes including NDRG1 (Tsui et al., 2012). The others upregulated are BTG2, an NGF-inducible gene coding for a protein with anti-proliferative properties, and Maspin, a serum proteinase inhibitor and a putative tumour suppressor. Maspin was reported to inhibit motility of breast cancer and prostate cancer cells *in vitro* and in fact it is epigenetically silenced in a number of human tumours (see pp. 129–132). Both BTG2 and Maspin were shown some time ago to be regulated by p53 (Rouault et al., 1996; Hendrix, 2000; Zou et al., 2000). Thus many of these agents that upregulate NDRG1 seem to function by p53-mediation, as discussed below.

Vandetanib (ZD6474) Upregulates NDRG1

Vandetanib (ZD6474) is an inhibitor of VEGFR and EGFR kinases and reasonably successful in clinical trials with some forms of cancer. Efficacy of this drug in combination therapy is currently being investigated in clinical trials. Of much interest here is the demonstration that this drug upregulates several hypoxia-inducible genes, among them NDRG1. *In vivo* tumour models, the drug has been found to inhibit metastasis (Arao et al., 2006). These authors did not look at the expression of VEGFR function, so there is no suggestion that the drug inhibited metastasis by this route.

Regulation of Cell Proliferation by NDRG1 Mediated by p53

NDRG1 appears to be associated with regulation of cell proliferation. NDRG1 might be one of the targets of the cell cycle regulator p53 gene. HCC cells transfected with NDRG1 show cell cycle arrest at G0/G1 (Akiba et al., 2011), indicating the involvement of functional p53. Stein et al. (2004) showed some time ago that NDRG1 is upregulated by DNA damage. DNA damage also induces p53 expression. Indeed NDRG1 upregulation is p53-dependent. They identified a p53 binding site in the NDRG1 promoter. Obviously here one is looking at wild-type p53 being induced in response to DNA damage. Intriguingly, however, mutated p53 is also apparently able to target NDRG1. Polyamine-mediated regulation of proliferation of intestinal mucosal epithelia involves p53-induced expression of NDRG1. Polyamine depletion of intestinal epithelial cells is said not only to induce p53 but also upregulate NDRG1 expression by activating its transcription (Zhang et al., 2007). Lerner et al. (2012) induced apoptosis of HCT-116 expressing wild-type p53 gene and Colo-320 cells carrying a mutant form of the gene by exposing the cells to 3,3′-diindolylmethane. Both cell lines displayed induction of apoptosis, but NDRG1 was upregulated only by Colo-320 cell line that expressed mutant p53. Mutations abrogate cell cycle control of p53, so it would seem that in these experiments the upregulation of NDRG1 could be due to *de novo* transcription of wild-type p53 in response to stress factors. There is also the possibility that stress-induced p21 might be involved here. Indeed NDRG1 is said to upregulate p21 quite independently of p53 (Kovacevic et al., 2011b).

Oestradiol and NDRG1 Expression

Oestradiol seems to downregulate NDRG1 expression as shown by Fotovati et al. (2006). These authors transfected breast cancer cell lines with ERα cDNA and generated cells expressing high or low levels of the receptor. Upon treatment with oestradiol, those cell lines which expressed ERα at high levels showed reduced NDRG1 expression. Fotovati et al. (2006) also found an inverse correlation between NDRG1 and ERα expression in breast cancer samples. This is compatible with the fact that ERα promotes cell motility and invasion. A further useful exercise would have been to see how NDRG1 might have behaved in the backdrop of ERβ.

Metastasis Suppression by NDRG1

The metastasis suppressor function credited to NDRG1 flows from many accounts of its ability to suppress invasion besides reports of its suppression in many forms of cancer, EMT and angiogenesis. Cell lines overexpressing the gene showed downregulated expression not only of MMP-9 but also of VEGF and IL-8, two major angiogenic factors. In biopsies of pancreatic ductal adenocarcinoma, there is a significant inverse relationship between NDRG1 expression and microvascular density and depth of tumour invasion. Reduced NDRG1 was also associated with poor prognosis (Maruyama et al., 2006). Fan et al. (2012) also found a similar inverse relationship between microvascular density and NDRG1, but surprisingly NDRG1 expression was higher in advanced stage of the disease than the earlier T1 stage. In a majority of the tumours examined, NDRG1 was found in the cytoplasm, nuclear in 20% and approximately a tenth of the specimens showed membrane located NDRG1. This would have provided useful clues if only one knew what the distribution was like in normal tissues. Enhanced expression has been reported by others (Hakan, 2004; Chang et al., 2005) as Fan et al. (2012) have pointed out, but this might be related to hypoxic conditions. With reduced vascularisation, the development of a hypoxic environment is a possibility, but Fan et al. (2012) did not assess the state of hypoxia. In an illuminating publication, Sibold et al. (2007) found NDRG1 expression in both the cytoplasm and the cell membrane in HCC xenografts and further that NDRG1 expression increased in HIF-1α-dependent fashion. External stimuli alter the cellular distribution of NDRG1 from the cytoplasm to the membrane. But whereas hypoxia increased cytoplasmic levels, no relocation occurred in response to hypoxic signal. However, relocation to the plasma membrane may indeed have implications for signalling, cell adhesion and motility and tumorigenesis.

As noted earlier, Vandetanib (ZD6474), an inhibitor of VEGFR and EGFR kinases, increased disease-free survival time of patients with NSCLC. The complexity in evaluating situations of this type arise from the fact that NO is an inducer of angiogenesis and also incidentally of NDRG1. The latter is also induced by hypoxia, as discussed earlier. So interpretations of increased expression of the suppressor would be subject to certain caveats.

Views contrary to the suppressor function have been aired. Wang et al. (2004b) claimed that NDRG1 is correlated with metastatic spread in colorectal cancer. They found greater expression in primary tumours with metastases than those without metastases. So they have suggested that NDRG1 might indeed promote metastatic spread. However they found no difference in expression between primary tumour and its metastases. Given that tumours are heterogeneous, if cells with higher NDRG1 expression were more prone to metastasise one would have expected the metastases to show higher NDRG1 expression of the corresponding primaries. Nonetheless quite clearly one has to accept such a possibility. However, it ought to be mentioned here that a diametrically opposite view has been expressed in respect of colorectal cancer itself. Strzelczyk et al. (2009) indeed found that patients with reduced expression of NDRG1 showed shorter 5-year survival than those with unchanged or increased NDRG1 expression. In fact reduced NDRG1 expression was markedly more frequent in and associated with TNM (tumour/node/metastasis) stage of progression.

In support of those who advocate metastasis promoter function, one can cite reports of apparent induction of angiogenesis. In NSCLC cells suppression of NDRG1 expression using siRNA resulted in growth inhibition of xenografts *in vivo*, but induced the expression of VEGF and IL-8. In resected tumours, microvessel density correlated with nuclear NDRG1 (Azuma et al., 2012). Nishio et al. (2008) studied cervical adenocarcinomas for possible relationship between NDRG1 and clinical features of aggressive behaviour of the carcinomas. They found a close correlation between NDRG1 expression and lymphovascular invasion and lymph node metastasis and it was associated with poor prognosis. NDRG1 also correlated with angiogenesis leading them to suggest induction of angiogenesis was the reason for the apparent link between NDRG1 expression and poor prognosis. A similar situation seems to exist in hepatocellular carcinoma. NDRG1 was overexpressed in the carcinomas as compared with normal liver tissue, and it was accompanied by vascular invasion. The expression status correlated with tumour stage and grade and with shorter overall survival (Chua et al., 2007). Recently Nagai et al. (2011) investigated a large series of breast cancer patients for NDRG1 and 10-year follow-up for prognosis. Around half of nearly 600 patients expressed NDRG1 and this directly correlated with shorter disease-free and overall survival. Toffoli et al. (2009) examined the effects of intermittent hypoxia which causes endothelial dysfunction and reported to increase endothelial cell migration. Intermittent hypoxia increased NDRG1 protein expression and suppression of NDGR1 by NDGR1-siRNA transfection decreased endothelial cell migration estimated by the qualitative scratch assay in EAhy926 endothelial cell cultures.

The problems associated with and inherent in interpreting NDRG1 expression and tumour progression derive from the intervention of hypoxic stress and HIFs which confer marked growth advantage, as well as the tumour-associated angiogenesis which is vital for tumour spread. Adaptation to the hypoxic environment via the HIFs confers upon tumour cells, the faculties of cell proliferation and survival, renewal and maintenance of CSCs, and invasion. The phenotypic outcome would depend upon the target genes and these might inhibit or promote progression. To complete the picture, genetic or environmental stimuli besides hypoxia can induce HIFs, which in turn would influence tumour growth and progression. The evidence is overwhelmingly in favour of tumour promotion, so much so that HIFs are regarded as legitimate targets for therapeutic intervention.

Some clinical studies are under way to study the effects of HIF suppression. Of note are studies of antisense HIF-1α oligonucleotide (NCT01120288), cardiac glycoside digoxin in breast cancer (NCT01763931), and rapamycin derivative RAD001 in colorectal carcinoma (NCT01047293), PX-478 (*S*-2-amino-3-[4−-*N*,*N*,-bis(2-chloroethyl) amino]phenyl propionic acid *N*-oxide dihydrochloride) was reported to inhibit the growth of xenografts HT-29 colon cancer cells by inhibiting HIF-1α and VEGF

(Welsh et al., 2004). It has also been found to be effective inhibitor of growth of xenografts models of human SCLC and adenocarcinoma of the lung (Jacoby et al., 2010) and is currently being evaluated in advanced solid tumours and lymphoma (NCT00522652). Liu (2012c) recently demonstrated that Sorafenib inhibited HIF-1α and VEGF expression and also interfered with mTOR signalling in HCC cells and HCC xenografts. A word of caution however is required since inhibition of HIFs may have adverse effects in terms of promotion and enrichment of cancer stem cells (Conley et al., 2012). A phase II trial of Sorafenib in ovarian epithelial cancer is being carried out (NCT00522301).

16 The ING (Inhibitor of Growth) Suppressor Gene

The ING (inhibitor of growth) gene, which exerts wide-ranging biological effects on cell proliferation, regulation of the cell cycle, apoptosis, ageing, DNA repair and tumorigenesis. Its products are putatively thought to be tumour suppressors by virtue of their ability to modify chromatin structure and regulate genetic transcription. ING proteins contain the PHD plant homeodomain, a highly conserved zinc finger motif implicated in chromatin mediated regulation of genetic transcription, in the C-terminal end. ING proteins have N-terminal sequences required in their interaction with HAT and HDAC which enables them to regulate promoter activity of target genes. ING is downregulated in many tumours. Reduced expression may occur together with genetic changes that might compound the inhibitory functions of ING proteins.

ING3 downregulation has been reported in HCC tissues and this correlated with tumour grade with inhibited expression being associated with Edmondson–Steiner grade II–III tumours (Lu et al., 2012). In head and neck squamous cell carcinomas, there is a differential relationship between intracellular location of ING and tumour grade. Nuclear expression of ING4 decreased from normal epithelium to dysplasia and on to carcinoma and with TNM stages. ING4 levels in the nucleus also positively correlated with apoptosis. But cytoplasmic levels positively correlated with lymph node metastasis and consistent with this also correlated with 14-3-3η, a tumour promoter (Li et al., 2011d). This appears to be a consequence of ING4 binding to the 14-3-3 and being localised in the cytoplasm and prevented from performing its function. Such a functional inactivation by being restricted to the cytoplasm has been shown to occur, for example, in the case of ING1b (Gong et al., 2006). However, Li et al. (2011d) only demonstrated the co-expression of 4-3-3η and ING4 in the cytoplasm, not their binding to each other. Nevertheless, reduced nuclear levels of ING4 were correlated with the tumour stage and lymph node metastasis. ING4 expression was lower in grade III than in grade I–II tumours. In lung cancer, reduced ING4 mRNA has correlated with tumour stage and lymph node metastasis (Jara-Lazaro et al., 2010).

Adenovirus mediated transfer of ING4 brings about a host of alterations in recipient MDA-MB-231 breast cancer cells, among them induction of G2-M arrest and apoptosis together with upregulation of P21, P27 and Bax, and downregulation of Bcl-2, IL-8. Intratumoral introduction of ING4 into tumours formed by MDA-MB-231 cells in nude mice led to growth inhibition (Li et al., 2010i). MCF-7 cells overexpressing ING4 by gene transfer showed suppression of cell proliferation and cessation of the cell cycle at G0/G1 and enhanced apoptosis. There was enhanced expression of p21 and Bax mRNA (Wei et al., 2012b). Mutations of ING4 have been recorded and these inactivated the suppressor protein and reduced its ability to inhibit proliferation, cell migration and to induce apoptosis (Moreno et al., 2010).

Therapeutic Strategies in Cancer Biology and Pathology. DOI: http://dx.doi.org/10.1016/B978-0-12-416570-0.00016-0

The NF-κB survival pathway has also been implicated in the functioning of ING4. Klironomos et al. (2010) found that ING4 is downregulated in association with grade in astrocytomas. They also noticed that the expression of p65 (RelA) subunit of NF-κB was higher in grade IV astrocytomas than in grade I/II tumours.

The suppressive effects of ING on cell proliferation and induction of apoptosis may be compounded by their ability to inhibit neovascularisation in their overall ability to suppress tumour progression. Ovarian cancers not only showed ING3 downregulation, but this was also associated with enhanced microvessel density (Liu et al., 2012h). Earlier Li et al. (2010i) showed that ING4 was able to suppress the density of microvessels associated with tumours formed by MDA-MB-231 cells implanted *in vivo*.

Allelic loss of all members of the family (ING1-5) occurs at high frequency in tumours. The frequency of LOH of ING2 correlated with progression of head and neck squamous cell carcinoma (Borkosky et al., 2009, 2010). A point mutation has been reported in ING4 and this can conceivably lead to functional inactivation of the protein (Moreno et al., 2010). ING5 mutations together with the downregulation of ING5 mRNA have been reported in oral squamous cell carcinoma (Cengiz et al., 2010). Splice variants potentially differing in function have been found in human gastric adenocarcinoma (Li et al., 2009e).

The ING isoforms seem to have well-defined suppressor functions, but Unoki et al. (2009) state that ING2 is overexpressed in colorectal cancer and supports invasion via the MMP route. They also found that the presence of ING2a and ING2b inhibited cell cycle arrest and apoptosis, and some ING proteins might indeed be oncogenic, contrary to the accepted view in relation to a majority of the ING proteins. ING2 levels in colon cancer were twice as high as in normal mucosa, and this enhanced expression was seen in nearly half of the cancer tissues examined. Besides, ING2 bound NK-κB and this could have led to increased invasive behaviour encountered in this study. Also encountered was increased MMP 13 expression seemingly mediated by NF-κB (Kumamoto et al., 2009). MMPs are an activation target of NK-κB (see Table 9.2). Allelic loss of 4q35.1, the ING2 locus, occurred in head and neck squamous cell carcinomas and this was associated with advanced stage tumours (Borkosky et al., 2009). By implication, the loss of ING2 could be occurring late in progression rather than initiate progression. The topographical aspect of distribution is an important factor here. ING2 bears significant sequence homology to ING1b and loss of ING2 function could result from being restricted to the cytoplasmic compartment. Indeed ING1b seems to be inactive because of localisation in the cytoplasm (Gong et al., 2011). One ought to note that other reports that have supported a suppressor function for ING2. It occurred at lower in HCCs as compared with matched normal liver tissue. Loss of ING2 was also related to poor prognosis (Zhang et al., 2008a). ING2 expression is lost in melanoma progression too as Lu et al. (2006) reported some time ago.

ING1 suppresses the expression of many genes. Among them of interest in the present context of the inhibitory effects of ING proteins on cell proliferation is the tumour suppressor and cell cycle regulator p53, which has been implicated in the induction of apoptosis by ING. Another gene directly affected by ING is DGCR8, which is also involved in the early steps of miRNA biogenesis. As noted earlier, the primary miRNA transcripts are processed by a complex called the microprocessor complex formed of an RNase III Drosha (the catalytic subunit) and the protein DGCR8 (the Pasha protein of Drosophila, the subunit that recognises the substrate) into pre-miRNAs, then processed into the substantive miRNAs. ING1 is able to bind to the DGCR8 promoter and controls its transcription through chromatin regulation. ING1 quite clearly affects DGCR8 expression. Induction of ING1 suppresses cell proliferation and suppression of DGCR8 also was anti-proliferative, which could suggest that effects of cell proliferation might have occurred via interference with miRNA biogenesis (Gomez-Cabello et al., 2010). Although these effects do occur in parallel, the data do not constitute evidence for the direct involvement of miRNAs. It is needless to say that a number of miRNAs are upregulated in ING-deficient cells. High levels of expression of miRNA-192 and also of certain other miRNAs have been noticed. Gomez-Cabello et al. (2010) found no evidence of p53 binding to the DGCR8 promoter. Nevertheless, miRNA-192 is p53-inducible (Georges et al., 2008). MiRNA-192 and others such as miRNA-194 can be transcriptionally activated by p53; these then modulate mdm2 itself a regulator of p53 (Pichiorri et al., 2010). So the possibility remains that ING effects can occur by p53 mediation. In this setting, also to be reckoned with is the cell cycle regulatory Rb protein, which is targeted and downregulated by miRNA-192 (Feng et al., 2011c). Also relevant here would be to note that ING members can and do interact with p53. ING4 was shown to induce G2-M arrest by increasing $p21^{waf}$ expression in a p53-dependent manner. Furthermore, the NLS (nuclear localisation signal) domain was suggested to be involved in the interaction of ING4 with p53, which was disrupted by mutations in NLS, and this led to disruption of p53 function (Zhang et al., 2004, 2005). Zhu et al. (2009) found that ING1 variants, for example ING1b and ING1c, altered p53 function and also induced apoptosis and cell cycle arrest at G0/G1 phase, ING1c upregulated the expression of p53 and also $p21^{waf1}$. The overall effect on cell proliferation seems to have been compounded by enhanced expression of Bax and downregulation of Bcl-2 expression. They do not rule out possibility that alterations in p53 expression could have arisen from changes in p53 stability. Equally however ING1b has been described as being able to act independently of p53 (Coles et al., 2008). They noted that whilst ING1b is capable of inhibiting cell proliferation, this effect can be enhanced by the absence of p53. The NF-κB survival pathway has also been implicated in the functioning of ING4. Klironomos et al. (2010) found that ING4 is downregulated in association with grade in astrocytomas. They also noticed that the expression of p65 (RelA) subunit of NF-κB was higher in grade IV astrocytomas than in grade I/II tumours.

To conclude there are several important considerations if ING proteins do arrive at the stage of clinical deployment. Least among them is that at least one isoform might support progression. The study of signalling systems that INGs activate is at a rudimentary stage. So exploration of uses in the clinical setting needs considerable work on the basics and any thoughts of beneficial deployment are in reality a long way away.

17 The BRCA1 and BRCA2 Suppressor Genes

The development and systemic spread of breast cancer are closely allied to the loss of BRCA1 and BRCA2 suppressor genes. BRCA1 and BRCA2 are described as breast cancer susceptibility genes in that familial breast/ovarian cancer are a consequence of a germ line mutation in one of the alleles and a subsequent loss of function mutation in the second allele. BRCA1 has been associated with cell differentiation and embryonic development. Somatic mutations have been encountered in breast and ovarian cancers. The predisposition to develop cancer is related to mutations and the abrogation of suppressor function of the gene. Inactivation of these suppressor genes enhances cell proliferation and tumour growth, and mutations in these genes correlate to a high degree with inactivation of p53. The wild-type genes function as negative regulators of cell proliferation and tumour growth, and these abilities are bound to impinge upon and influence tumour progression. BRCA1 shows reduced expression in association with transition of *in situ* breast cancer to the invasive phase. Loss of heterozygosity of both BRCA1 and BRCA2 is more frequent in metastatic tumour than in primary breast carcinoma. The BRCA1 protein is a zinc-finger protein and therefore might participate in the regulation of gene transcription. Germ line mutations of BRCA1 seem to result in total loss of transcription or lead to the production of a truncated product, which might be functionally impaired (Sherbet, 2003).

BRCA1 and BRCA2 proteins have been attributed with putative participation in DNA repair. They participate in homologous recombination DSBR (double strand break repair). The defect in this process arising from genetic changes makes carriers of BRCA gene mutations more sensitive to radiation damage. It has been known for some time that BRCA2+/+ but not BRCA−/− cells can survive radiation damage. It would appear from recent studies that miRNAs play a role in radiation sensitivity. Moskwa et al. (2011) found that inhibition of miRNA-182 not only raised BRCA1 protein levels but also provided protection from effects of ionising radiation. They also state that sensitivity to radiation results from impairment of homologous recombination repair which accompanies downregulation of BRCA1 protein and enhanced miRNA-182 expression. Chang et al. (2011c) have offered an alternative mode of action. They show that BRCA1 can interact with HDAC2 leading to the deacetylation of histones H2A and H3 on the promoter of the miRNA and in this way repress the oncogenic miRNA-155 and inhibit cell proliferation.

The BRCA1 suppressor function has ramification with another miRNA. MiRNA-335 seems able to regulate ERα, ID4, IGF1R and Sp1. The miRNA upregulates BRCA1 expression (Heyn et al., 2011). This is in agreement with the view that miRNA-335 is an inhibitor of metastasis. Compatibly, miRNA-335 reduces ERα

Therapeutic Strategies in Cancer Biology and Pathology. DOI: http://dx.doi.org/10.1016/B978-0-12-416570-0.00017-2

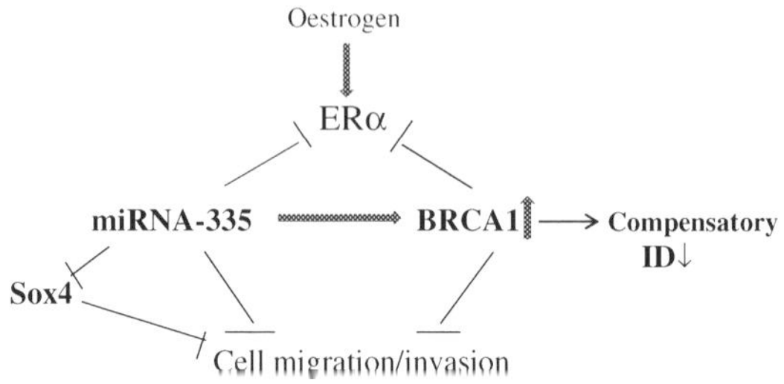

Figure 17.1 Regulatory loops of cell migration/invasion involving miRNA-335, BRCA1 and ERα. A second regulatory route not involving ERα is also shown. This involves Sox4. Suppression of Sox4 miRNA-335 would be expected to lead the suppression of cell migration/invasion.

level, but when the endogenous miRNA is suppressed using antisense oligonucleotides, ID4 expression increases, possibly a compensatory effect indicating a regulatory loop involving miRNA-335, BRCA1 and ERα. BRCA1 expression is said to inversely relate to both ERα and ID4 (Roldan et al., 2006), which insinuates a notion of co-regulatory link involving the miRNA, BRCA1 and ID4, but not ERα. Here miRNA-335 would be just as effective in suppressing aggressive behaviour of tumours. Indeed the miRNA does inhibit cell migration and metastasis as assayed by the less than ideal assay of lung colonisation upon tail vein injection even in the absence of ER. This in fact seems to occur by a different route, namely by inhibiting SOX4, an activator of EMT (Zhang et al., 2012c) (Figure 17.1).

18 BRMS1 (Breast Cancer Metastasis Suppressor 1) Gene

BRMS1 is reputedly a metastasis suppresses but does not affect tumorigenicity. It is downregulated by promoter methylation in breast cancer cell lines and primary breast cancers and their metastases (Metge et al., 2008). The conundrum is that transfection of BRMS1 leads to the development of invasive tumours, but these have low metastatic potential (Samant et al., 2000). Obviously, one is looking at the inhibition or non-operation of biological events occurring at later stages of metastasis. We do not know if the tumours formed from implanted transfected cells had induced angiogenesis, had or had not induced the expression of any angiogenic agents. Unfortunately, controlling the experiments using empty vector alone does not answer these questions. Therefore, the suggestion that BRMS1 suppresses metastasis by atypical means is open to debate even a decade later. As noted below BRMS1 seems capable of inhibiting CXCR4 and NF-κB signalling. Indeed, indirectly also BRMS1 can achieve this by inducing the expression of ING4 which is capable of inhibiting angiogenesis via suppression of NF-κB and IL-6 (Li and Li, 2010).

Sin3 are co-repressors that occur in complex with HDACs. These complexes deacetylate nucleosomes near Sin3-regulated genes to produce repressed chromatin structure. The BRMS1 protein is a component of the mSin3a HDAC complex and a repressor of gene transcription. In this way it might be involved in the downregulation of expression of anti-apoptotic genes. Suppression of its expression has led to increased cell adhesion accompanied by enhanced cell motility *in vitro* (Sheng et al., 2012). Inhibition of BRMS1 upregulated CXCR4 and activated NF-κB signalling. NF-κB notoriously enhances the expression of MMP and uPA and promotes cell motility. BRMS1 deacetylates the RelA subunit of NF-κB. By suppressing NF-κB, it is able to inhibit the tumour promoter osteopontin (Samant et al., 2007). Consistent with this, suppression of endogenous BRMS1 expression in Hep3B cells suppressed cell apoptosis. Here too BRMS1 seemed to suppress osteopontin expression and promoted apoptosis (Wu et al., 2012b), which might have occurred via suppression of NF-κB. As noted before osteopontin functions in conjunction with activated NF-κB pathways (see Figure 9.1 and Table 9.2). Conversely, NF-κB itself might bring about the downregulation of BRMS1 expression. RelA/p65 subunit of NF-κB seems to directly recruit DNMT1 (DNA (cytosine-5)-methyltransferase 1) to methylate and suppress BRMS1 (Liu et al., 2012g). But one must add a rider here that NF-κB transcriptionally activates a vast array of genes such as cytokines (e.g., TNF-α and interleukins), cell adhesion molecules and functions in conjunction with tumour promoters. This is the first attribution of methylation and downregulation of BRMS1 gene expression, so needs to be explored further.

Therapeutic Strategies in Cancer Biology and Pathology. DOI: http://dx.doi.org/10.1016/B978-0-12-416570-0.00018-4

Recently it has been suggested that the biological outcome of BRMS1 might relate to its subcellular localisation. Prominent cytoplasmic localisation has been related to increased disease survival of melanoma patients, inversely related to proliferation and Akt activation. Nuclear localisation was linked with enhanced invasion (Slipicevic et al., 2012). It is difficult to conceptualise these findings except by mobilising other invasion associated genes, but none has been nominated to date and endorse the findings as a point of departure in the study of melanoma malignancy. However, BRMS1 does influence growth factor receptor genes and tumour suppressors but not in the direction of invasion promotion.

Amplified expression of MiRNA-146a is known to reduce invasive ability and this occurs together with inhibition of EGFR, NF-κB signalling and suppression of MTA2 tumour promoter expression (Li et al., 2010c). BRMS1 upregulated both miRNA-146a and miRNA-146b in metastatic breast cancer cell lines. Transfection of these miRNAs into the cell lines suppressed *in vitro* cell migration and inhibited lung metastasis of cells introduced via the tail vein of athymic mice (Hurst et al., 2009a). The transfected cell lines also showed downregulation of EGFR, another feature these miRNAs share with BRMS1. Nonetheless, how this can conceivably selectively suppress metastatic spread without affecting tumorigenesis is unclear. BRMS1 is silenced by promoter methylation. It is conceivable that methylation of the gene might occur in the later phases of progression. This is not unheard of. EPB41L3, which is regarded as a tumour suppressor, has been reported to undergo tumour stage dependent methylation in NSCLC (Kikuchi et al., 2005). Now then Nagji et al. (2010) showed that in NSCLC, BRMS1 methylation and silencing were related to tumour stage. Further, it has now emerged that that miRNA-342 is capable of suppressing the expression of DNMT1. MiRNA-342 was downregulated in colon cancer tissues and cell lines, but when expression levels were restored a marked reduction of the expression of DNMT1 occurred. Therefore one does need to know the scope and identity of miRNAs that are modulated in consort with alterations in BRMS1.

It is now abundantly clear that BRMS1 functions in collaboration with other suppressor genes and, as discussed, with miRNAs with suppressor function. It is possible that the expression of some of the co-factors might be dependent upon degree of progression. It has been postulated that colonic and ovarian cancer progression is accompanied by a stage dependent expression of many genes, including suppressor genes producing cumulative effects leading to the manifestation of progression (Sherbet and Patil, 2006). Induction of angiogenesis is an essential requirement for metastasis to occur. BRMS1 is able to induce ING4 which has been shown to suppress the formation of blood vessels and the suppression of endogenous ING4 increases growth of human umbilical vein endothelial cells (HUVEC). Further support for the co-operation of ING4 and BRMS comes from the demonstration that repression of ING4 abrogates the suppression by BRMS1 of growth of HUVEC cells, but ING4 overexpression inhibited suppression of angiogenesis by BRMS1. In other words, the downstream ING4 target is a requisite for the suppressor function of BRMS1 (Li and Li, 2010). As discussed earlier, ING is known to influence DGCR8, which is involved in the early steps of miRNA biogenesis. ING1 affects DGCR8 expression. Induction of ING1 suppresses cell proliferation and suppression of

DGCR8 was also anti-proliferative, which could suggest that effects of cell proliferation might have occurred via interference with miRNA biogenesis (Gomez-Cabello et al., 2010). It is needless to say that a number of miRNAs are upregulated in ING-deficient cells; high upregulation of miRNA-192 have been noticed. Admittedly these events occur in parallel and so do not constitute evidence for the direct involvement between INGs and miRNAs. It would be useful to know how BRMS1 affects or is affected by miRNA-17-92 that is known to promote tumour angiogenesis *in vivo*, and Let7-f, miRNA-27b and miRNA-130a are believed to be pro-angiogenic *in vitro*.

The invasive faculty of tumour cells is another faculty that is associated with advanced stage of tumorigenesis and is linked with the functioning of MMP and uPA and other enzymes that aid in cell detachment and promote invasion. Intercellular adhesion molecules such as E-cadherin and CADM1 are inhibitory of invasion and are downregulated to assist invasion. In contrast, fascin, an actin-linking cytoskeletal protein, shows limited expression in normal cells, but it is overexpressed in cancer cells and so is regarded as a promoter of invasion and inhibits the expression of E-cadherin (Hayashi et al., 2011). Fascin has been found to downregulate BRMS1 (Al-Alwan et al., 2011). However, fascin expression itself is inhibited by miRNA-145, miRNA-133a and miRNA-133b, which are downregulated in some forms of cancer and which inhibit cell proliferation and invasion *in vitro* (Kano et al., 2010). In fact, Kano et al. (2010) have identified conserved miRNA sequence in the 3′UTR of Fascin. Here again we have effects that might be generated indirectly by intervening suppressors, albeit the latter could be regulated by different miRNA species.

19 Maspin (SerpinB5): a Postulated Tumour Suppressor

Maspin (SerpinB5) is a serine protease inhibitor, but uncharacteristic of the Serpin group does not possess serine protease activity. It is putatively regarded as a tumour suppressor since its expression is downregulated in many tumours. It was recognised some time ago that the loss of its expression was a result of promoter hypermethylation of one or both alleles of the gene (Ogasawara et al., 2004; Akiyama et al., 2003). Epigenetic silencing might involve promoter methylation and hypo-acetylation of associated histones H3 and H4 and conversely Maspin activation occurs with promoter demethylation and hyper-acetylation of the histones (Futscher et al., 2002; Fitzgerald et al., 2003).

Maspin is markedly downregulated in many tumours. An early report on breast cancers claimed that Maspin expression progressively decreased with breast cancer progression from *in situ* ductal carcinoma to invasive carcinoma and associated nodal involvement. Loss of Maspin also correlated with reduced disease-free survival (Maass et al., 2002). Reduced Maspin expression has been recorded in many tumours and it was related to tumour progression (Zhang and Zhang, 2002). Consistent with this is the finding that it inhibits angiogenesis. It effectively reduced microvessel density associated with tumours. *In vitro* it inhibited migration of endothelial cells towards bFGF and VEGF and blocked endothelial proliferation and tubule formation (Nickoloff et al., 2004; Zhang et al., 2000). The RSL (reactive site loop) is the major determinant of the function of Maspin of influencing cell adhesion, migration and apoptosis.

But suppression of angiogenesis is not believed to be totally dependent upon RSL. For, mutation of the RSL domain or its deletion does not fully prevent this inhibition (Zhang et al., 2000). In hindsight one would recall a second domain spanning from amino acid 139 to amino acid 225 is able to mediate adhesion (Cella et al., 2006). Whether this might be important for inhibition of angiogenesis is not known.

The RSL does mediate interaction with β1 integrins of the ECM. The expression of several β1 integrins is induced by angiogenic agents (see Avraamides et al., 2008; also see further discussion below). Maspin binds to mammary carcinoma cells and this is prevented by the RSL peptide, indicating that it participates in ECM-mediated adhesion of mammary carcinoma and is capable of inhibiting invasion (Ngamkitidechakul et al., 2003). It would appear that an intact RSL is required only for the invasion inhibitory but not anti-angiogenesis function.

In a small number of laryngeal carcinomas, nuclear expression of Maspin directly correlated with longer disease-free interval and good prognosis (Marioni et al., 2005). It is highly expressed in high grade intraepithelial neoplasia precursors and subsequently it is downregulated in low grade to high grade prostate cancers.

Therapeutic Strategies in Cancer Biology and Pathology. DOI: http://dx.doi.org/10.1016/B978-0-12-416570-0.00019-6

Paradoxically there was no Maspin detected in benign secretory cells (Pierson et al., 2002). The pattern of intracellular distribution might be of consequence. Invasive ovarian cancers tended to show high expression in the cytoplasm and high expression was seen in high tumour grade as compared with low malignancy tumours or benign tumours and poor prognosis (Sood et al., 2002). But curiously transfection of Maspin into ovarian cancer cell lines markedly reduced *in vitro* invasion. In these early investigations inducible expression vectors were not used. Transfection of vectors can produce collateral genetic activation. Recombinant Maspin was shown some years ago to be able inhibit *in vitro* invasion and this effect was blocked by antibodies raised against the RSL (Sheng et al., 1994). Solomon et al. (2006) found that Maspin translocation into the nucleus correlated with suppression of angiogenesis in ovarian serous carcinoma is associated together with VEGF expression. This is not unexpected or remarkable *per se*. There are several examples where cytoplasmic restriction of signalling ligands has affected biological function.

The presence of Maspin in the nucleus correlated with the expression of ER and PR. In contrast, cytoplasmic Maspin correlated with the absence of ER/PR and large S-phase fraction (Mohsin et al., 2003; Umekita et al., 2011). High Maspin expression in the nucleus has correlated with good prognosis in NSCLC (Berardi et al., 2012). Regrettably most studies have related to Maspin expression in tumour cell lines. Whilst loss of expression in them is important and significant to the establishment of a suppressor role for Maspin, many reports have subsequently cited them as if tumour derived cell lines are ipso facto representatives of tumour tissues themselves.

So a cautionary discussion of contrary verdicts is warranted here. Several early reports have indicated high Maspin expression in carcinomas. The tumour suppressor role has been the subject of intense scrutiny ever since Maass et al. (2001) detected its expression in 23/24 pancreatic cancers and also in high grade precancerous lesions. Ogasawara et al. (2004) found Maspin was expressed more frequently in undifferentiated and poorly differentiated than in well differentiated thyroid tumours. Also follicular adenomas and normal thyroid tissues did not express Maspin. Similarly it was expressed in invasive bladder cancer but not in tumour-free epithelium (Sugimoto et al., 2004). Maspin was detected in a majority of gastric carcinoma and also in foci of intestinal metaplasia (Akiyama et al., 2003); also in pancreatic ductal adenocarcinomas where expression was however seen mainly in low grade tumours (Ohike et al., 2003). Cytoplasmic staining was more frequent in stage III than stage I NSCLC. Maspin mRNA levels were also higher in stage III disease than in normal tissues. Poor survival corresponded with high Maspin expression (Hirai et al., 2005). Yatabe et al. (2004) noticed Maspin expression due to reduced methylation in virtually all the squamous cell lung cancers they tested, whilst half the number of adenocarcinomas showed no Maspin expression. Maspin is expressed in both the cytoplasm and nuclei of endometrial cancers (Li et al., 2007d) and papillary thyroid carcinoma (Shams et al., 2006). A recent study to this effect concerns TNBCs. Umekita et al. (2011) have detected Maspin expression in nearly 9/10 of TNBC cases and the expression was related to histological grade. Often investigations have focused on either detecting epigenetic silencing or detection of the protein expression. Hypermethylation of the gene promoter is always shown to correlate with loss of protein expression, but where the protein is not detected no effort is made to check if this is a consequence of translational control. It would be needless reiteration to say that translational control has a critical role to play in development, differentiation and cell proliferation. It may be recalled here that miRNAs repress translation of mRNAs of target genes. Maspin has shown inverse correlation with miRNA-21 expression.

Notwithstanding the evidence contradictory to the suppressor function of Maspin, it would be appropriate to emphasise the positive aspects. There is much experimental evidence linking Maspin to the suppression of cell proliferation and invasion. The p53 pathway was implicated early on with the demonstration that p53 bound directly to the Maspin promoter and furthermore mutant p53 inversely correlated with Maspin expression (Zou et al., 2000; Zhang and Zhang, 2002). Maspin and the apoptosis related gene GADD45A (growth arrest and DNA-damage inducible) are both regulated by p53 and the transcription factor ATF-2. Both activate Maspin transcription by binding to separate sites on Maspin promoter. ATF-22 binds to GADD45A via transcription factors Oct-1, NF-1 and BRCA1 (Maekawa et al., 2008). The operation of the MAPK signalling is obvious since ATF-2 is a target of p38/MAPK and JNK leading to apoptosis. In contrast, mutant p53 seems to relate inversely to Maspin (Cho et al., 2007). Possibly mutant p53 is capable of inhibiting Maspin and abrogate its ability to impose cell cycle arrest.

As noted earlier, the transcription factor Snail promotes EMT by suppressing E-cadherin. But Neal et al. (2012b) now show that Snail is recruited to and represses the Maspin promoter and this is linked with increased cell motility. Other suppressor might be involved with the functioning of Maspin. Suppression of Maspin alters the expression of many genes; notable among them is BRMS1 itself a suppressor gene (Liu et al., 2012fa). Whether this reflects a compensatory response by cells or a result of intervention by miRNA is a matter for further exploration. MiRNAs are known to be involved in the regulation of both Maspin and BRMS1. On the other hand, Maspin downregulation by promoter hypermethylation correlated with reduced Maspin protein levels and with increase of the tumour promoter protein MTA1 and VEGF (Sharma et al., 2011).

A mode of induction of apoptosis driven by Maspin involves the DNA repair protein Ku. Ku as a heterodimer (Ku70/80) participates in non-homologous end-joining and also in controlling transcription and apoptosis. Ku functions by interacting with many cellular proteins and nucleic acids. It is involved in Bax-mediated apoptosis. Ku protein levels are reduced in some neoplasms and high levels of Ku are associated with radiation resistance. In the cytoplasm, Ku70 occurs in a complex with Bax thus sequestering the latter from inducing apoptosis. This complex depends upon the state of acetylation of Ku70 and when it is acetylated Bax is released from the complex (Subramanian et al., 2011). Maspin has been found to inhibit HDAC1 and increase the acetylation of Ku70 and induce apoptosis (Lee et al., 2012).

The inhibitory effect appears to involve ECM components that determine intercellular adhesion and also migration, such as the trimeric glycoprotein laminin. Integrity of Maspin expression is essential for normal embryonic development. It seems to regulate both cell proliferation and migration in embryonic systems.

Maspin increases adhesive interaction between the endoderm and to ECM component laminin matrix. Maspin−/− endoderm proliferated more slowly than Maspin+/+ cells on laminin substratum (Gao et al., 2004). The engagement of integrin receptors by laminin regulates cell adhesion and migration. Consistent with this is the inverse relationship noticed between Maspin expression and expression of uPA (urokinase-type plasminogen activator) and uPAR (uPA receptor) in oral squamous cell carcinomas (Yoshizawa et al., 2011). Maspin might be localised and its adhesion function promoted by the uPA/uPAR system. Maspin, uPA/uPAR and β1 integrin might be functioning as a complex (Endsley et al., 2011). This could be a means by which Maspin regulates cell adhesion and inhibits cell motility. Maspin co-precipitates with α3β1 and α5β1 integrins. Binding of Maspin to α5β1 inhibited the activation of the integrin. Maspin inhibited migration of CHO (Chinese hamster ovary) cells overexpressing α5β1 but not of those that lacked it indicating that it mediated the inhibition (Bass et al., 2009). The enhanced adhesive effects of activation of the β1 integrin ECM component are transduced down the vinculin/actin cytoskeletal system (Qin and Zhang, 2010).

20 EPB41L3 and CADM1 Tumour Suppressor Function

The membrane proteins EPB41L3 (erythrocyte membrane protein band 4.1-like 3 also known as DAL-1 and 4.1B) and CADM1 have been focused upon in recent years. They occur as a complex and have been implicated in intercellular adhesion. EPB41L3 belongs to the family of proteins possessing N-terminal FERM (F for 4.1 ERM) domain. ERM proteins have a C-terminal F-actin binding region (Chishti et al., 1998; Pearson et al., 2000). The FERM domain proteins including EPB41L3 in this way link cytoskeletal proteins and the cell membrane.

Both EPB41L3 and CADM1 proteins are downregulated in cancers. Both have been linked with breast cancer invasion and progression (Takahashi et al., 2012). Repression of its expression has shown correlation with progression of NSCLC where its methylation has shown tumour stage-dependent increase (Kikuchi et al., 2005). Enforced expression of EPB41L3 in A549 suppressed cell migration, whilst suppressing the endogenous protein enhanced migration (Zhang et al., 2012d). These authors have portrayed its expression as related to metastasis for 5/7 established NSCLC cell lines showed no expression. Cells harvested from pleural fluid showed no expression of the protein. Nonetheless, it is disconcerting to note the somewhat promiscuous use of the term metastatic in cancer literature.

Dafou et al. (2010) approached this problem by transferring chromosome 18 into the ovarian cancer cell line TOV21G. This led to suppression of tumorigenesis. They identified several genes located on chromosome 18 in the hybrid cell line TOV21G. Whilst EPB41L3 was expressed in normal ovarian epithelial cell lines, it was methylated and suppressed in a majority of ovarian cancer cell lines. Methylation of the gene was also found in ovarian cancer tissue. Although these findings are interesting *per se*, one cannot fail to notice that among other genes that were overexpressed in the hybrid TOV21G cells is cadherin. Taking a quizzical view, it seems surprising that a near 17-fold increase in its expression would certainly have contributed to the suppression of tumorigenesis and alterations of *in vitro* cell motility. EPB41L3 is a cytoskeletal protein involved with the actin cytoskeleton and with intercellular adhesion phenomenon. So the transfection studies should also have included cadherin transfer. As the authors themselves have pointed out EPB41L3 is hypermethylated in many human cancers and possible co-ordinated functioning of EPB41L3 with cadherin cannot be excluded. Furthermore, EPB41L3 similarly co-operates with another membrane component CADM1 in the cell–cell adhesion process. CADM1 is said to interact with the actin cytoskeleton through the mediation of EPB41L3 (Yageta et al., 2002). It may be noted in this context that hypermethylation and silencing of EPB41L3 occurs in Nasal NK/T-cell lymphoma (NL) correlated with concomitant methylation of the CADM1 gene (Fu et al., 2009). CADM1 is an

Therapeutic Strategies in Cancer Biology and Pathology. DOI: http://dx.doi.org/10.1016/B978-0-12-416570-0.00020-2

integral transmembrane protein belonging to the immunoglobulin family that participates in cell–cell adhesion process independently of Ca^{2+} and has been ascribed with tumour suppressor function (www.ncbi.nlm.nih.gov/gene/23705 updated 2012). Mao et al. (2004) noticed that when re-expressed in NSCLC A549 cell line, CADM1 suppressed cell proliferation and induced caspase-mediated apoptosis and also suppressed tumorigenesis in animal models. CADM1 was downregulated by promoter methylation in many cell lines derived from human neoplasms. In NSCLC, promoter methylation seemed to be related to state of tumour progression, with larger proportion of advanced T2–T4 displaying methylation far more as compared to early T1 stage tumours (Fukami et al., 2003). In ovarian cancer, loss of CADM1 expression markedly correlated with metastasis to lymph nodes and with the presence of distant metastases and poorer survival (Yang et al., 2011b). Certain miRNAs have been suggested to be capable of directly targeting DNA methyltransferases to downregulate and inactivate suppressor genes. Such a mode of action has been ascribed to miRNA-342 (Wang et al., 2011b). EPB41L3 is modulated by miRNAs. MiRNA-223 is said to downregulate the expression of EPB41L3 by directly targeting its 3′-untranslated regions (Li, 2011c).

It may be noted that no link has been reported between miRNAs and other established suppressor genes including nm23, KAII and KiSS1, and the less convincingly postulated tumour suppressor MKK4 (mitogen-activated kinase kinase 4). This is not to say miRNAs are not involved with any of them or are not influenced by them, but possibly not many investigators are attracted to the proposition.

Deserving much attention is the thought that in some cases, miRNAs might be responsible for the patterns of metastatic spread encountered in human cancers. This is implicit in the finding that RKIP (the suppressor Raf kinase inhibitory protein), a putative metastasis suppressor discussed at another location in this book, seems to engage miRNA let-7 (Yun et al., 2011). They show that RKIP inhibits let-7 targets (HMGA2, BACH1) that are able to upregulate MMP1, OPN (osteopontin), CXCR4 involved in bone metastasis.

Part 5

Signalling and Transcription Regulators as Prospective Candidates in Cancer Therapy

Genetic determinants of phenotypic profiles transduce their signals through defined pathways characterised by distinctive components that carry the signal downstream to activate or inhibit target genes. The identification of these components of the signalling system and uncovering the links and nodes of the signalling pathways would be crucial if the biological effects of these genetic determinants were to be countermanded directly or indirectly for potential therapeutic deployment. The signalling routes activated may interact with one another and generate effects that could contribute in a positive or negative fashion to the functioning of the intended target. So specifically focusing on and targeting the nodes and links might be an unenviable undertaking. Nonetheless, it is important to examine this theme in the context of cancer whether modulation of the CSC phenotype, reversal of the EMT, and suppressing the invasive and metastatic behaviour are a practical proposition.

21 Is MKK a Metastasis Suppressor?

The MAPK Signalling Pathway

The Ras/Raf/MAPK signalling pathway is activated by many external stimuli, such as growth factors, mitogens, inflammatory cytokines, apoptosis inducing factors and others. Ras functions through its effector isoforms of Raf, namely Raf1 (C-Raf), A-Raf and B-Raf, to activate the signalling cascade. Ligand binding activates Ras/Raf1 which stimulates the signalling cascade involving MAP3K→MAP2K→MAPK transducing the signal down the cascade leading to cellular responses like regulation of transcription, cell proliferation, apoptosis, differentiation, development and angiogenesis. The MAP3K, of which there are around 20 in number, seems to integrate the signals transduced by the stimuli. Seven MAP2K have been identified. MAP3K→MAP2K regulate five MAPK families generating a diversity of sub-routes of flow of information towards achieving a wide range of phenotypic function.

External stimuli can differentially engage MAPK signalling. Thus growth factor mediated activation of RTKs, and hormonal and neurotransmitter signals imparted by GPCRs, for instance, activate Raf1 to engage MKK1 (MAP2K1/MEK1)→ERK1/ERK2 (MAPK3/MAPK1) route to lead to cell survival, proliferation and differentiation. Raf1 can stimulate signalling by another route by activating kinases such as ASK1 (MAP3K, apoptosis signal regulating kinase 1) and STK3 (the serine threonine kinase 3, MST2). This occurs independently of the kinase activity but by direct binding to ASK1. Stress factors in the form of ROS (reactive oxygen species), inflammatory cytokines, UV irradiation, deprivation of growth factors, etc. activate the ASK1 route via MKK4/7 (MAP2K4/MEK4; MAP2K7/MEK7)→JNK1-3 (stress-activated protein kinases SAPK1/MAPK8-10) or the MKK3/6 (MAP2K3/MEK3; MAP2K6/MEK6)→ p38MAPK (MAPK14) to inflammation and apoptotic outcome (Cuevas et al., 2007; Matallanas et al., 2011).

In essence the outcome of differential signalling might significantly impinge upon the phenotype. Thus, mutations of Ras and Raf, which are encountered frequently in cancers, deregulate extracellular signals stimulating cell proliferation and growth, whilst the p38MAPK/JNK route leads to opposite effects of apoptosis and suppression of the neoplastic process.

MKK in Tumour Biology

The MAPK signalling cascade leads to cellular responses like regulation of transcription, proliferation, apoptosis, differentiation and development in response to

Therapeutic Strategies in Cancer Biology and Pathology. DOI: http://dx.doi.org/10.1016/B978-0-12-416570-0.00021-4

a range of external stimuli. With ligands transducing their signal using the MAPK pathway to produce prominent phenotypic features crucial in tumour development and dissemination, MKK1 (MAP2K1) has quite deservedly enjoyed attention in respect of its relevance to the pathogenetic process. This has led to an extensive study of MAPK mediation of cell proliferation, migration and neovascularisation. MKK was described as a metastasis suppressor. Yoshida et al. (1999) reported that MKK4 (MAP2K4) behaved as metastasis suppressor gene in AT6.1 rat prostate cancer cells. AT6.1 cells carrying transfected MKK4 displayed greatly reduced metastatic spread as compared with control cells *in vivo*, without affecting the growth of the primary tumours. A more informative study is that of Stark et al. (2005) who found that in brain metastases of breast cancers, the suppressor genes KiSS1, KAI1, BRMS and MKK4 expression was reduced to one-tenth of the level of the primary tumours. This possibly suggests the component cells of the primary tumour with downregulated expression might be predisposed to form the metastatic clones. It is difficult to determine the relative importance of the individual genes that were downregulated in metastatic deposits. Interestingly they found nm23 expression was unaltered between primary tumours and brain metastases. In prostate cancer, MKK4 expression has shown an inverse relationship to Gleason score (Kim et al., 2001). On the other hand, there are reports that MKK4 induces rather than suppresses oncogenic changes. Experimentally enhanced expression of MKK4, in breast and pancreatic cancer cell lines not expressing the gene, induced cell proliferation and invasion, whilst inhibition of expression by siRNA in MKK4 expressing cell line MDA-MB-231 decreased anchorage-independent growth and suppressed tumour growth *in vivo* (Wang et al., 2004a).

The Inhibitory Effects of Anthrax Lethal Toxin on MKKs

The Anthrax lethal toxin (LeTx) is an inhibitor of MKKs. It specifically targets many members of the MKK family and has been used in many studies relating to tumour biology. LeTx is a binary or ternary complex of three proteins, namely the protective antigen, the lethal factor and the oedema factor. Both the lethal factor and the oedema factor are enzymes; the lethal factor is Zn^{2+}-dependent MMP and the oedema factor is adenylate cyclase. The protective antigen mediates the binding of the toxin to the cell and its translocation into the cells (Pannifer et al., 2001; Thoren and Krantz, 2011; Young and Collier, 2007). LeTx binds at the cell surface with specific receptors CMG2 (capillary morphogenesis gene 2) and TEM8 (tumour endothelial marker 8). The protective antigen binds to the integrin-like extracellular domain of the receptors and promotes the internalisation of the toxin into the cytoplasm. CMG2, TEM8 and integrin β1 contain the von Willebrand A (adhesive) domain which is essential for the binding of the toxin to the receptor. TEM8 is differentially expressed in normal and tumour associated vasculature and expression in the latter might be related to tumour stage. It is upregulated during vasculogenesis and formation of tumour vasculature and expressed copiously (St Croix et al., 2000; Rmali et al., 2005; Carson-Walter et al., 2001; Bradley and Young, 2003). Three isoforms have been identified; two of them are transmembrane proteins with the von Willebrand factor I adhesion motif and the third isoform contains only the extracellular domain and could be the secreted form (Bradley et al., 2001; Scobie and Young, 2005).

The receptors might also function synergistically in endocytosing LeTx (Martchenko et al., 2010). The latter cleaves and inactivates MEKs and inhibits p44/p42 MAPK (ERK1 and ERK2), p38 MAPK and the JNK arm of the MAPK pathway (Chopra et al., 2003; Duesbery et al., 1998; Young and Collier, 2007).

LeTx and Cell Invasion/Motility

LeTx inhibits two basic traits of the cancer cell, that is migratory behaviour and its ability to induce neovasculature. It appears to modify the cytoskeletal machinery, alter junction complexes and focal adhesions. These alterations might inhibit cell motility (Lehmann et al., 2009). It appears from a recent report that the oedema factor, namely adenylate cyclase might remove and internalise integrins from the cell surface and reduce cell adhesion (Martín et al., 2011). This is bound to have consequences for cell invasion. LeTx increases actin stress fibres and also vascular endothelial cadherin (VE-cadherin) distribution (Warfel et al., 2005). Normally VE-cadherin is selectively localised at intercellular contact points and forms adhesive dimeric bridges between monomers on the cell–cell contact surfaces (Brasch et al., 2011). Interestingly, LeTx seems able to bind to LRP6 and activate Wnt signalling as indicated by the nuclear accumulation of β-catenin. SiRNA mediated suppression of both CMG2 and TEM8 receptors inhibited Wnt signalling (Abrami et al., 2008). Wnt signalling has been closely implicated in the activation of EMT. It seems eminently worthwhile to investigate if LeTx does indeed activate EMT. There is no direct evidence of this at present. But it would be well to remember that LeTx has been shown to inhibit Akt signalling and also inhibit E-cadherin mediated intercellular adhesion and promote cell motility (Popova et al., 2009). Further elucidation is required here of the mechanism involved, for one would recall that growth factors are known to induce cell migration by activating Akt signalling. But there are no indications that LeTx alters the E-cadherin negative regulatory transcription factors.

LeTx Inhibits Angiogenesis

More important is the suppression of angiogenesis, which is an essential requirement of metastasis. Ras mediated transformation of murine 3T3 fibroblasts upregulates VEGF apparently by the intervention of MKK1 (Rak et al., 2000). As mentioned above, LeTx (lethal factor) is a Zn^{2+}-dependent MMP that cleaves and inactivates most isoforms of MKKs. LeTx mediated inactivation of MKK signalling reduces angiogenesis in tumours. It seems able to block the release of VEGF by endothelial cells. Intravenous administration of LeTx reduced growth of tumour xenografts together with loss of VEGF and VEGFR and reductions in microvessel density

(Depeille et al., 2007). Ding et al. (2008) found that LeTx treatment reduced the levels of phosphorylated ERK and p38 MAPK *in vitro* and prolonged treatment was anti-proliferative. LeTx also substantially inhibited the release of VEGF, IL-8 and bFGF. The *in vivo* effects were similar to those reported by Depeille et al. (2007). Ample confirmation of these effects has emerged from the study by Huang et al. (2008) that inhibition of MAPK kinase signalling inhibited the growth of renal cell carcinoma xenografts and inhibited cell proliferation also of endothelial cells and suppressed angiogenesis *in vivo*. Using a different model, namely the murine retinal model, Bromberg-White et al. (2009) have shown that LeTx treatment induces significant increase in the levels of secreted VEGF, but not interleukins or bFGF.

VEGF expression and release are enhanced by HIF in WM35 melanoma cell lines suggesting the possibility that HIF, a pro-angiogenic agent, could be promoting angiogenesis by this means. In cells carrying Ras or p53 mutations, HIF-induced VEGF release increases (Shellman et al., 2003). This latter finding further adds weight to the involvement of MKKs in the promotion of angiogenesis. FGF-2 stimulates VEGF release via p44/p42 MAPK (ERK1 and ERK2)/SAPK/JNK pathway, but activated p38 MAPK inhibits VEGF release (Tokuda et al., 2000). However, it ought to be borne in mind that inhibition of one can transactivate the other (Sharma et al., 2003).

Quite obvious is that different sub-routes of MAPK pathways might be engaged by different ligands and oncogenes. According to Shin et al. (2005), H-Ras activates the Rac-MAPK kinase (MKK)3/6-p38 pathway and promotes invasion and cell migration, which is probably attributable to the induction of MMP-2. But both H-Ras and N-Ras activate the Raf-MEK-ERK and PI3/Akt pathways to proliferation and differentiation. Now with the reasonably well-established role in promoting angiogenesis targeted inhibition of MKKs is worthy of clinical exploration. Of some clinical significance is the report that emerged some time ago that LeTx suppresses PR and ERα (Webster et al., 2004). It is rather intriguing why findings of this nature have not been confirmed and investigated further, especially in the light of the importance in TNBC. It would be interesting to assess the effects of LeTx on TNBC cell lines. One ought to recognise here that TEM8 has been implicated in endothelial cell migration and tubule formation and CMG2 in endothelial proliferation.

22 RKIP Suppresses Invasion and Metastasis

RKIP, as the terminology implies, inhibits Raf-1 kinase activity and regulates the MAPK signalling pathway (Trakul et al., 2005; Yeung et al., 1999). Besides MAPK signalling, RKIP function impinges upon GPCR and NK-κB signalling pathway. Much evidence has accumulated in recent years supporting the view that RKIP suppresses invasion and metastasis by regulating EMT, angiogenesis and apoptosis. RKIP is believed to accentuate apoptosis induced by chemotherapeutic agents and radiation. Its loss in tumours is said to afford protection against apoptosis (Woods et al., 2008). Loss or reduced RKIP activity has been linked with chromosomal instability and abnormalities. RKIP can mechanistically regulate the progression of the cell cycle by regulating the function of cell cycle checkpoints. RKIP thus coheres together and influences many aspects of the cancer phenotype.

RKIP Downregulation Creates Chromosomal Instability and Abnormalities

Chromosomal and cytoskeletal events in mitosis involve the function of mitotic kinases. Several protein kinases have been implicated in the formation and function of the mitotic spindle. Notable among them are Aurora kinases which are overexpressed in human cancers and whose enhanced expression correlates with DNA and chromosomal aneuploidy. Mitotic kinases have also been implicated in apoptosis. Aurora B kinase deletion produces defects in late anaphase and in cell division, for it seems to be required for sister chromatid separation and segregation. Aurora B deletion presumably affects microtubule dynamics. Mitotic kinases appear to be regulated by p53, which monitors the G1-S and the G2-M transition checkpoints (Sherbet, 2006). There is a large body of evidence that RKIP regulates Aurora B kinase and the spindle checkpoint of the M-phase via the Raf-1/MEK/ERK cascade. Loss of RKIP and increased Raf activity allows these cells to circumvent the spindle checkpoint resulting in chromosomal abnormalities (Al-Mulla et al., 2011; Eves et al., 2006; Rosner, 2007). Consistent with this, RKIP expression inversely correlates with chromosomal instability in colorectal cancer samples, and chromosomal loss has been correlated with loss of RKIP (Al-Mulla et al., 2008).

Therapeutic Strategies in Cancer Biology and Pathology. DOI: http://dx.doi.org/10.1016/B978-0-12-416570-0.00022-6

RKIP Inhibits Invasion and Growth of Cancer

In vitro studies show that RKIP markedly inhibits cell motility (Li et al., 2008, 2009d), accompanied in parallel by reduction of MMP-2 and MMP-9, cathepsin B and uPA (Xinzhou et al., 2011). Beshir et al. (2010) reported the induction of invasive behaviour together with enhanced expression of MMP-1 and MMP-2 resulted from silencing of RKIP and overexpression of RKIP led to decreased invasion and metastasis *in vivo*. No significance should be read into the different MMPs being affected by RKIP. The MMPs are corporately involved in ECM remodelling. Hao et al. (2012) have reported that RKIP transfected MDA-MB-231 cells showed reduced cell migration and this appeared to be related to and possibly attributable to the inhibition of MMP-1 and MMP-2. In accord with this, Xinzhou et al. (2011) and Martinho et al. (2012) found that suppression of RKIP increased cell survival and migration, but did not influence cell proliferation. Aside from the invasion promoting proteases, many other molecular entities that influence motility have been implicated in the absence of RKIP. The induction of β-catenin, vimentin, c-met receptor tyrosine kinase and PAK1 have been reported (Al-Mulla et al., 2011). As discussed below, RKIP might be regulating the initiation of EMT here. Beshir et al. (2010) believe that RKIP achieves MMP suppression by inhibiting NF-κB.

In the backdrop of the ability of checkpoint regulation and the integration of signalling by RTKs and GPCRs, it is rather difficult to comprehend and appreciate why there has been so little attention directed to the regulation of cell proliferation by RKIP or to appreciate the divergence of view in the scanty data currently at hand. Some while ago, Fu et al. (2003) transfected LNCaP and C4-2B cells with antisense and sense RKIP cDNA. The transfected derivatives showed no differences in proliferation *in vitro* or in colony-forming ability. According to Li et al. (2008), cell proliferation and anchorage-independent growth are also inhibited in ovarian cancer cells that overexpressed RKIP. Inhibition of proliferation was due to cell cycle arrest rather than to induction of apoptosis. But these authors found no suppression of cell proliferation in the case of breast cancer cell lines (Li et al., 2009d). Recent reports show that restoration of expression of epigenetically silenced RKIP by 5-azacytidine treatment and also by transfection of RKIP has led to the inhibition of proliferation of TE-13 oesophageal squamous cell carcinoma cells (Guo et al., 2012). Zhang et al. (2013) have shown that RKIP inhibited cell cycle progression and induction of apoptosis in the gastric cancer cell line SGC7901. RKIP also inhibited cell migration as shown by *in vitro* assay. Implantation into animal hosts showed RKIP inhibited the growth of tumour xenografts. More decisive evidence has recently emerged which shows that silencing of RKIP hastens G_1-S transition and DNA synthesis (Al-Mulla et al., 2011).

RKIP Downregulation is a Frequent Event in Cancer

RKIP is downregulated in many tumours. Loss of RKIP expression in oesophageal dysplasia and squamous cell carcinomas was due to promoter methylation and methylation corresponded with loss of transcription and protein expression (Guo et al.,

2012). RKIP showed downregulated expression in a small proportion of gliomas. Loss of expression was not related to histological type or malignancy grades, that is grade II and grade III/IV gliomas. The loss of expression has been claimed to be related to prognosis (Martinho et al., 2012).

The degree of downregulation seems to relate to the extent of differentiation or histological grade and stage in some tumour types. RKIP is downregulated in expression in the highly metastatic derivative cell line as compared with the parent LNCaP prostate cancer cell line with relatively low metastatic potential. Of further significance is that RKIP protein expression was lower in metastatic tumour than in the primary (Keller, 2004). This has been confirmed in a study of breast cancer, where RKIP expression was high in normal breast epithelia and expressed also in primary breast tumours. However, significantly RKIP expression was greatly reduced or absent in lymph node metastases (Hagan et al., 2005). Reduced RKIP expression was reported in 90% of metastatic prostate tumours as compared with 48% of primary tumours (Fu et al., 2006).

This suggests the RKIP negative component cells might show a greater propensity to metastasise. A most telling commentary on the importance of RKIP in prognosis is provided by the study of colorectal cancer with hepatic metastasis. RKIP loss was seen in 24.5% (12/49) of colorectal cancer without metastases, 47.1% (32/68) of colorectal cancers with metastases and 68% of hepatic metastases. There was a marked reduction in the median survival of patients with reduced RKIP expression as compared with RKIP positive patients (Kim et al., 2012a). It has also been claimed that restoration of RKIP expression reduced metastasis of prostate cancer C4-2B cells upon xenografting. In pancreatic carcinoma, loss of RKIP is related to metastasis to regional nodes and to distant sites. Also loss of RKIP correlated with poorer disease-free survival (Kim et al., 2010). So also an inverse relationship between RKIP expression and prognosis has been recorded for gliomas (Martinho et al., 2012) and oesophageal cancers (Birner et al., 2012). In some tumours, RKIP loss occurs only in a small proportion of specimens. Whether these subsets of tumours are more prone to aggressive behaviour is a moot point.

EMT is an important feature of neoplastic transformation. Several markers of EMT are currently available. As shown in Figures 22.2 and 22.3, suppression of RKIP is conducive to EMT. RKIP inhibits the transcription factor Snail, a negative regulator of E-cadherin and a consequence of this is EMT inhibition (also see below for a more detailed discussion). NPI-0052 induces RKIP expression and can upregulate E-cadherin expression by inhibiting Snail expression and EMT at the phenotypic level (Baritaki et al., 2009).

Pathways of RKIP Signalling

In the pre-eminent position occupied by MAPK and Ras/Raf signalling in cancer pathology, it is unsurprising that attention was focused on RKIP. RKIP is a member of the phosphatidylethanolamine-binding protein family. It has a functionally flexible pocket which can interact with and integrate the functions of the signalling ligand, Raf-1 and kinases that phosphorylate RKIP. Phosphorylation of RKIP disables binding

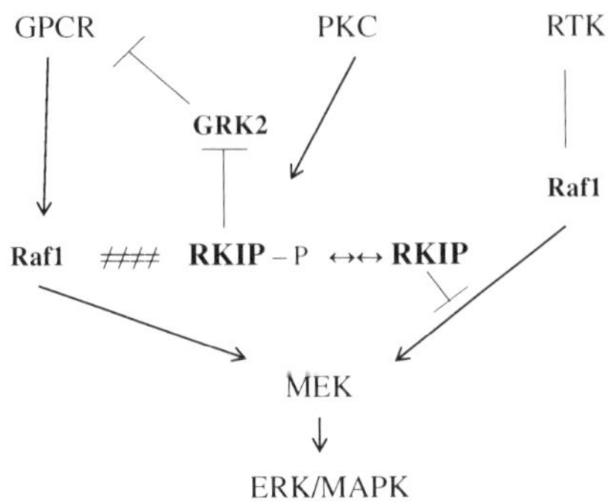

Figure 22.1 GPCR and RTK mediation of ERK/MAPK signalling and the involvement of RKIP in its regulation. Upon PKC-induced phosphorylation of serine-153, RKIP loses its ability to inhibit Raf-1, but it can now inhibit GRK2 and influence GPCR and the interacting EGF signalling pathways. This illustration is based on Lorenz et al. (2003), Kolch (2005) and references discussed in the text.

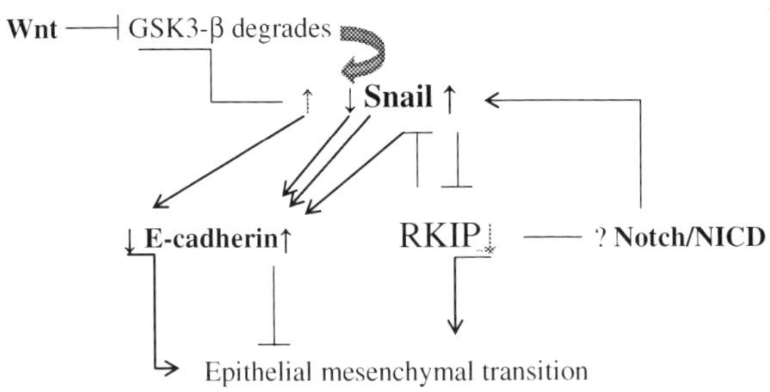

Figure 22.2 The potential interaction of Notch and Wnt signalling with RKIP function mediated by Snail. Loss of RKIP is said to stabilise β-catenin and induce expression of other markers such as c-met and vimentin. This is based on references cited in the text.

with Raf-1 and conversely ligand bound RKIP inhibits RKIP phosphorylation non-competitively and also prevents RKIP binding to Raf-1 (Granovsky et al., 2009).

RKIP operates in three distinctive ways, by inhibiting Raf-MEK-ERK, GPCR kinase and NF-κB signalling. It inhibits Raf kinase activation by binding to the phosphorylated form of Raf-1 and prevents Raf-1 kinase mediated phosphorylation and activation of MEK and also dissociates the RAF/MEK complex and said to competitively inhibit MEK phosphorylation (Rath et al., 2008; Yeung et al., 1999). In its phosphorylated form, RKIP also inhibits GRK2 (GPCR kinase 2) which is negatively regulates GPCRs. As shown in Figure 22.1 upon dissociation from Raf-1, RKIP couples with GRK2 and inhibits its activity. This occurs by PKC mediated phosphorylation of RKIP (Lorenz et al., 2003; see also Kolch, 2005). In this way, GPCR and EGF signalling pathways can interact and cross talk.

The third route is one by which RKIP targets NF-κB signalling. NF-κB is a survival pathway that is inhibited by RKIP. So the loss of the latter in cancers leads to promotion of cell survival. The core component of the NF-κB cascade is the IKK kinase

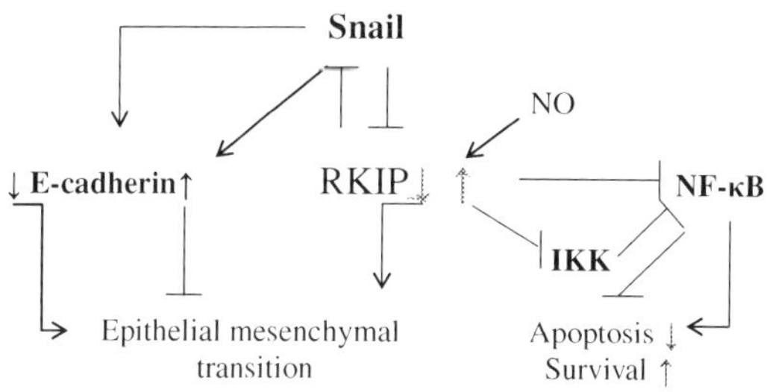

Figure 22.3 The RKIP/Snail/E-cadherin mediated regulation of EMT activation and the control of the normal NF-κB apoptosis/survival pathway by RKIP. The figure is based on references discussed in the text.

complex consisting of the kinases IKK-α and IKK-β and the regulatory subunit IKK-γ. RKIP inhibits both IKK-α and IKK-β and TAK1 (TGFβ-activated kinase 1), an IKK kinase that can activate NF-κB. In this way, RKIP can lead to the inactivation of the NF-κB cascade (Yeung et al., 2001). RKIP also interacts with TRAF6 (TNF receptor-associated factor 6) (Chatterjee et al., 2004; Tang et al., 2010), an essential element in NK-κB activation. As noted earlier, NF-κB is also involved with the activation of EMT in interaction with other pathways, for example of TGF-β and Wnt signalling, and effecting the downregulation of E-cadherin. Bonavida and Baritaki (2011) have found that NO (nitric oxide) induces apoptosis under certain conditions and this is mediated by its ability to upregulate RKIP expression leading to the inhibition of NF-κB, but this can probably occur independently of NO.

The transcription factor Snail, a negative regulator of E-cadherin, is also apparently involved in this regulatory loop. Snail seems able to downregulate RKIP (Wu and Bonavida, 2009) by suppressing transcription mediated by a proximal E-box (enhancer box palindromic motif) on the RKIP promoter (Beach et al., 2008) and in this way promote activation of EMT. In reciprocation, RKIP can conceivably suppress Snail (Bonavida and Baritaki, 2011) and this can result in upregulation of E-cadherin and suppression of EMT. To insert a sub-loop, NF-κB can itself regulate transcription of Snail.

One can visualise interaction of RKIP with Notch and Wnt signalling in influencing the activation of EMT. RKIP possesses the E-box motif in its promoter. E-boxes are binding sites for bHLH proteins such as Notch. The Snail family zinc finger transcription factors also bind to gene promoters with the E-box motif. The intracellular domain of Notch directly engages the Snail promoter and upregulates Snail expression (Sahlgren et al., 2008). This would lead to downregulation of E-cadherin and activation of EMT. Now GSK3-β can degrade Snail, but Wnt can suppress this activity of GSK3-β and stabilise Snail (Yook et al., 2005, 2006). Essentially this would be a negative regulation of the function of RKIP which itself is able to downregulate Snail expression. Therefore in the background of RKIP, inhibition of Notch and or Wnt would synergise with RKIP towards suppression of EMT. Ras and Notch signalling pathways do interact but no direct interaction between RKIP and NCID (Notch intracellular domain) has been reported (Figure 22.3).

A further thought is that certain miRNAs might be downstream targets of RKIP. MiRNAs have been strongly implicated in cancer pathogenesis, as discussed in detail in another location in this book. Many species of miRNA have been regarded as tumour suppressors. MiRNA Let-7 species are downregulated in tumours, which is compatible with their ability to inhibit the activation of EMT. Let-7d inhibits EMT and its expression is inversely related to Twist and Snail which are negative regulators of E-cadherin expression. Dangi-Garimella et al. (2009) showed that RKIP mediated suppression of MAPK signalling decreases Lin-28B transcription by Myc and leads to enhanced let-7 processing in breast cancer cells. Let-7 in turn inhibits HMGA2, an activator of the transcription factor Snail. These changes in Lin28/Let-7 suppress bone metastasis. In C5 molecular subtype of high grade serous ovarian cancer, marked changes have been reported in the expression of N-myc, Lin-28B, Let-7 and the high mobility group HMGA2. Characteristic amplification and overexpression of N-myc and overexpression of its targets Lin-28B together with loss of Let-7 expression and amplification and overexpression of HMGA2 were encountered (Helland et al., 2011). Among other targets of Let-7 that RKIP inhibits is Bach1. HMGA2 and Bach1 upregulate MMP1 conducive to invasion and metastasis as well as osteopontin and CXCR4 both implicated in bone metastasis (Yun et al., 2011) (Figures 3.4 and 3.5).

The transcription factor YY1 (Yin Yang 1) belonging to the Gli-Kruppel class of zinc finger proteins has been attributed with ability to repress and activate a diverse number of promoters by directing histone deacetylases and histone acetyltransferases to the promoters (www.ncbi.nlm.nih.gov/gene/7528). In hepatocellular carcinoma, YY1 and RKIP are inversely related (Notarbartolo et al., 2011). Induction of NF-κB increased YY1 expression and ablation of the RelA/p65 subunit of NF-κB reduced YY1 expression (Wang et al., 2007) again emphasising a close-knit relationship of RKIP with the NF-κB survival pathway. CD20, the B-lymphocyte antigen, is putatively linked with the regulation of B-cell activation and proliferation. Cross linking of CD20 with anti-CD20 antibodies induces apoptosis and probably also inhibits the cell cycle traverse. CD20 signalling does induce Src family kinases (Deans et al., 1995). ERK activation is a requisite for the induction of B-cell receptor mediated apoptosis and SAPK also a downstream effector of an arm of MAPK signalling might provide mitogenic signals (Healy and Goodnow, 1998). Quite compatibly, Jazirehi et al. (2005) showed that the anti-CD20 antibody Rituximab significantly increased RKIP expression and inhibited NF-κB signalling pathway. Vega et al. (2009) have broadly confirmed this using another humanised antibody which upregulated RKIP expression and inhibited constitutively activated NF-κB and p38 MAPK pathways.

Effects of Re-expression of RKIP on Metastatic Spread

Inhibition of NF-κB Signalling

In line with conventional wisdom, attempts have been made to reinstate RKIP expression. Many potential opportunities can be identified with RKIP operating by inhibiting Raf-MEK-ERK, GPCR kinase and NF-κB signalling. Upregulation of

RKIP would influence and modulate the phenotypic outcome of more than one mode of function and provide an additively substantive result.

Inhibition of the NF-κB survival signalling seems to upregulate RKIP. Anti-CD20 antibodies upregulate RKIP expression in this way (Jazirehi et al., 2005; Vega et al., 2009). The mechanism of the upregulation is unclear, but inhibition of NF-κB pathway occurs together with the downregulation of the anti-apoptotic Bcl-x(L) and Bfl-1/A1, a homologue of Bcl-2. As noted above, NO also inhibits NF-κB pathway and induces RKIP expression.

Proteasome Inhibitors

The proteasome consists of a cylindrical core particle with two identical regulatory particles one at each end of the cylindrical core. Some subunits of the regulatory particles can recognise ubiquitin. Ubiquitinated proteins bind to the regulatory particles. The ATPase components unfold the protein which then moves to the central core where it is degraded. Proteasome inhibition has been recognised for some time as valid approach in cancer treatment. Many inhibitors have attracted much attention since the successful introduction of Bortezomib in the treatment of multiple myeloma and mantle cell lymphoma and evaluated in many other forms of cancer, for example NSCLC, prostate cancer and B-cell neoplasms, follicular Non-Hodgkin's lymphoma, Waldenstrom's macroglobulinemia, chronic and acute lymphocytic leukaemia, gliomas and colorectal and pancreatic cancers. Bortezomib also counteracts chemoresistance and induces cell cycle arrest and apoptosis.

NPI-0052 (salinosporamide A; marizomib) is another inhibitor on which much attention has been focused. It binds irreversibly to inhibit proteasome action. Carfilzomib is also a proteasome inhibitor that is undergoing comparative evaluation for efficacy with Bortezomib. Omuralide is another inhibitor structurally related to NPI-0052. NPI-0052 and Bortezomib in combination induce apoptosis in several ways including caspase activation, PARP and inhibition of NF-κB signalling (Chauhan et al., 2008; Rocaro et al., 2008). Rocaro et al. (2008) also implicated inhibition of Akt pathway and noted abrogation of Akt signalling reduced the cytotoxic effects of NPI-0052. However, Sloss et al. (2008) found that NPI-0052 and Bortezomib indeed activated EGFR, ERK and inhibition of these enhanced the antitumour effects exerted by NPI-0052 and Bortezomib. NPI-0052 and Bortezomib treatment inhibited cell adhesion and migration. This could be accounted for by the inhibition of Snail and consequent upregulation of E-cadherin and of RKIP (Baritaki et al., 2009).

Four clinical trials are listed in clinicaltrials.gov/ct2/results?term=NPI-0052+. A phase I clinical trial to evaluate the safety and efficacy of NPI-0052 is under way in multiple myeloma (NCT00461045) and another non-randomised study into its efficacy, pharmacokinetics and pharmaodynamics in combination with the HDAC inhibitor Vorinostat in NSCLC, melanoma, lymphoma and pancreatic cancer (NCT00667082). Phase I trials in advanced malignancies (NCT00629473) and refractory lymphomas (NCT00396864) are in progress. No information about the outcome is available at present.

23 CRSP3 Metastasis Suppressor

The CRSP3 (also known as CRSP130, DRIP130, ARC130, among others) protein is encoded by the gene MED23. CRSPs (cofactor required for SP1 activation) bind to several nuclear receptors including vitamin D3 receptor (VDR), hence one of the aliases is DRIP (an acronym for vitamin D receptor interacting protein) and function as co-factor complexes enhancing ligand dependent transcription. CRSPs are co-factors required for gene activation by Sp1, a zinc finger transcription factor that binds to GC-rich motifs of many promoters. Sp1 transcription factor is an activator or repressor of transcription of numerous genes involved in cell proliferation, apoptosis, cell differentiation and immune response. It is hardly surprising that CRSP3 as a co-factor required for Sp1 activation should be linked with the cancer phenotype. Importantly, Sp family members participate in activation of VEGF gene transcription (Abdelrahim et al., 2004). The expression of Sp1 has been shown to correlate closely with VEGF expression in human NSCLCs, and further hypoxia increases Sp1 binding to the VEGF promoter (Deacon et al., 2012). Consistent with these findings, Sp1 levels have directly correlated with expression of VEGF and with microvascular density. Knockdown of Sp1 led to inhibited growth and metastasis of pancreatic cancer cells growing as xenografts (Yuan et al., 2007). So without much ado, one can conclude that Sp1 and related transcription factors are liable to be active participants in the tumour progression to the metastatic state. However, one should be mindful that VEGF is regulated by many suppressor genes including MKK and ID (see pp. 137–140; 204–208).

CRSPs interact with regulatory proteins and function as a scaffold for the formation of the pre-initiation complex with RNA polymerase II and the transcription factors. The recognition of CRSP3 as a metastasis suppressor follows ostensibly from attempts aimed at identifying potential upstream regulators and downstream effectors of KiSS metastasis suppressor gene. The attribution of the ability to suppress metastasis derives from the demonstration that cells transfected with CRSP3 showed upregulation of the metastasis suppressor genes KiSS1 and TXNIP (thioredoxin interacting protein). Also loss of CRSP3 correlated with loss of KiSS1 expression in melanoma samples (Goldberg et al., 2003). Loss of CRSP3 in melanoma cells led to a reduction of activation of the KiSS1 promoter. When Sp1 and CRSP3 were co-expressed, KiSS1 suppressor activity was reinstated in cancer cells leading to inhibition of invasive behaviour. Furthermore, Sp1 bound to a specific GC-rich region in the KiSS1 promoter to regulate expression of the gene, since deletion of this motif interfered with the regulatory process (Mitchell et al., 2007). Overall, the

Therapeutic Strategies in Cancer Biology and Pathology. DOI: http://dx.doi.org/10.1016/B978-0-12-416570-0.00023-8

involvement of CRSP3 in tumour biology is unremarkable. But from the point of view of its potential interaction with other molecular determinants, astonishingly little progress has been made in the context of its suppressor propensity in human neoplasm. Such an approach would perhaps make more valuable contribution than merely looking at its expression levels in tumours.

24 The Suppressor Function of TXNIP

TXNIP (also known as vitamin D3 upregulated protein 1) is an arrestin family protein (Alvarez, 2008). Arrestins are important signalling regulators. GPCRs, arrestins and G-receptor kinases (GRKs) are activated leading to the activation or inhibition of many signalling systems. β-Arrestin for instance is involved with ERK/p38/JNK, Akt and NF-κB pathways, and via GRKs MEK1 (MKK1), RKIP/Raf1 and PI3K/Akt pathways are activated (Lefkowitz et al., 2006; Reiter and Lefkowitz, 2006). TXNIP is localised in the nucleus. It is a negative regulator of thioredoxin (Nishiyama et al., 1999; Nishinaka et al., 2004). With possible involvement in a wide range of signalling systems, it is small wonder that TXNIP subserves several cellular functions, for example in cell proliferation (Han et al., 2003; Kuljaca et al., 2009), cell apoptosis (Chen et al., 2008b), immune response (Lee et al., 2005; Son et al., 2008), among others and significantly interferes with the function of some suppressor genes (Figure 24.1).

There is much more definable success with establishment of the suppressor function of TXNIP which was recognised to be important in the activation of KiSS1 metastasis suppressor by CRSP3. Goldberg et al. (2003) noticed that CRSP3 upregulated both KiSS1 and TXNIP in their quest to elucidate the upstream and downstream signalling cascade of KiSS1. In breast cancer, TXNIP and TXNRD1 (thioredoxin reductase 1) are both closely related to metastasis-free survival of node negative breast cancer patients. TXNIP expression indicated favourable prognosis (Cadenas et al., 2010). Not only this, these authors found that HER2 (ErbB2) + tumours expressed higher levels of TXNRD1. When HER2 expression was induced in MCF-7 cells, a rapid increase in TXNRD1 and a robust decrease in TXNIP occurred. Possibly, there is a practical benefit here.

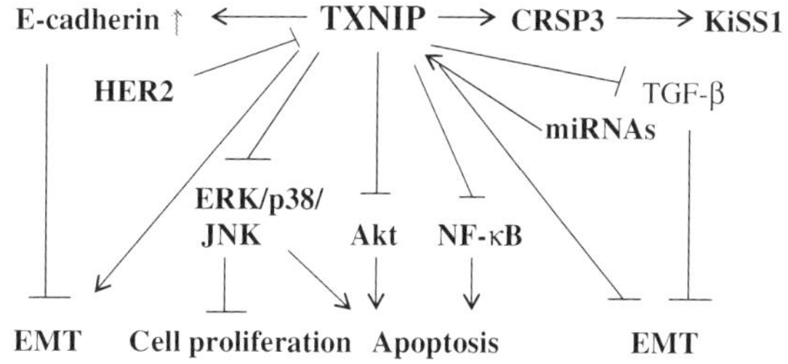

Figure 24.1 The signalling systems deployed in the suppressor effects of TXNIP. The Wnt/β-catenin pathway may be activated by miRNA-373 is not shown here.

Therapeutic Strategies in Cancer Biology and Pathology. DOI: http://dx.doi.org/10.1016/B978-0-12-416570-0.00024-X

TXNIP can block TGF-β/Smad mediated activation of EMT. TXNIP deficiency leads to increased expression of TGF-β and of the TGF-β inducible transcription factors Snail and Slug. This has led to the degradation of E-cadherin and activation of EMT, accompanied by a modest increase of Vimentin (Masaki et al., 2012).

TXNIP in Cell Proliferation and Apoptosis

The suppressive effects of TXNIP on cell proliferation and tumour growth are well documented. These occur mainly by cell cycle arrest and by the activation of apoptosis signalling. The suppression of metastatic behaviour is obviously related to the suppressive effects it exerts on cell motility and adhesion-dependent phenomena.

TXNIP engages many signalling systems to bring about these phenotypic outcomes (Figure 24.1). Minn et al. (2005) transfected a pancreatic islet beta cell line to over-express TXNIP and this resulted in the induction of apoptosis by the expression of Bcl2 family genes; the ratio of pro-apoptotic Bax/anti-apoptotic Bcl2 increased several fold together with increase of caspase-3. As shown in the figure, inhibition of the anti-apoptotic Akt by TXNIP induces apoptosis and this might involve regulation of PTEN. Embryonic fibroblasts derived from TXNIP knockout mice show an accumulation of inactive (oxidised) PTEN and enhanced Akt activation together with increased growth rates. TXNIP seems to be essential to maintain active PTEN, inhibit Akt phosphorylation and induce apoptosis (Hui et al., 2008). Wang et al. (2006) showed that dexamethasone induces TXNIP expression in a proper glucocorticoid-receptor mediated manner and induced apoptosis in a murine T-cell lymphoma cell line and also normal murine thymocytes. RAGE has been shown to induce the expression of TXNIP (Sbai et al., 2010) but incongruously RAGE binds several ligands including S100β, S100A4, S100A9, HMGB1, Amphoterin, amyloid fibrils, and others. The upregulation and activation of RAGE induces cell proliferation, invasion and metastasis.

TXNIP and Angiogenesis

Whether TXNIP influences angiogenesis, a process highly relevant in the metastatic process, is still an open question, but indications are that it does influence angiogenesis under given conditions. Kuljaca et al. (2009) showed that TXNIP inhibits endothelial cell migration and angiogenesis. They have adequately unravelled the molecular circuitry involving TXNIP in angiogenesis. They have demonstrated that cyclin-dependent kinase inhibitor $p21^{waf1}$ promoted angiogenesis by suppressing the transcription of TXNIP. Since TXNIP is a negative regulator of thioredoxin, this results in an increase of secretion of thioredoxin, which is pro-angiogenic. Conditioned media obtained from MCF-7 cells transfected with siRNA against p21 reduced endothelial cell migration and vascular sprouting. Knockdown of p21 led to an increase in TXNIP and a resultant decrease in thioredoxin. In agreement with this was their finding that siRNA specific for TXNIP promoted endothelial cell invasion.

Equally one has to consider other mechanism by which TXNIP might suppress angiogenesis. As discussed above, TXNIP is known to inhibit the Akt signalling pathway. Now in invasive ductal breast cancers, VEGF-A and -C isoforms show enhanced expression upon activation of Akt and this correlated with microvessel density (Tsutsui et al., 2010). Possibly then TXNIP could inhibit angiogenesis by suppressing Akt mediated upregulation of VEGFs. TXNIP also inhibits NF-κB pathway. VEGF expression is also enhanced by NF-κB induced Akt activation.

In Ras environment, TXNIP is upregulated leading to the inhibition of thioredoxin and increase in ROS and the induction of endothelial cell proliferation and tubule formation (Piao et al., 2009). But it is noteworthy that Ras upregulates VEGF expression as well as proteolytic systems such as MMPs and uPA which aid in the release of ECM-associated VEGF and in this way promotes angiogenesis. Hypoxia mediates the induction of VEGF; this also requires activated Ras signalling. Furthermore, Ras controls the function of the anti-angiogenic TSPs (Kranenburg et al., 2004). As noted earlier, RAGE upregulates TXNIP and interestingly RAGE can also upregulate the expression of VEGF (Ma et al., 2007). So an accurate definition of the molecular environment is most essential in ruling in or ruling out the function of TXNIP in the angiogenic process.

Finally equally important in terms of TXNIP participation in pathogenesis is that glucose can upregulate its transcription and expression. Minn et al. (2005) identified a distinct carbohydrate response element in the human TXNIP promoter of two E-box sequences and more precisely the effects of glucose are mediated via a region at −341 to −324 bp upstream of the translational starting point of the TXNIP gene (Pang et al., 2009).

miRNAs in TXNIP Function

The ubiquitous miRNAs have also been implicated in TXNIP suppressor function. Yan et al. (2011) claim that it is a direct target of miRNA-373, one of several putative candidates, and possibly route the signal to achieve E-cadherin expression as noted earlier. Equally, a target site has been provisionally identified in the promoter of E-cadherin that binds this miRNA-373. Binding of the miRNA has been found to induce E-cadherin expression (Place et al., 2008). This should result in the suppression of EMT and invasive behaviour. These findings are important whether they occur independently or in a TXNIP-dependent manner. However, recently totally contradictory results have been reported in respect of the effects of miRNA-373. The Wnt pathway, of which aberrant activation leads to E-cadherin inactivation and activation of EMT, has also been shown to be targeted by miRNA-371-373. The expression of these miRNAs has shown positive correlation of Wnt/β-catenin signalling in many human cancer cell lines. TCF/Lef1-binding elements occur in the promoter region required for Wnt-dependent activation of miRNA-371-373 and conversely miRNA-37-373 activate Wnt/β-catenin signalling and also target elements in the cascade (Zhou et al., 2012a).

Voorhoeve et al. (2006) believe that miRNA-372 and miRNA-373 function as oncogenes in testicular germ cell tumours and probably collaborate with Ras in the oncogenic process. MiRNA-520c and miRNA-373 upregulated the expression of MMP9 and promoted invasive behaviour of human fibrosarcoma HT1080 cells (Liu and Wilson, 2012). In this study, the effect of the miRNAs was not a result of directly binding to the MMP promoter, but via binding to mTOR and SIRT1 and suppressing their translation, resulting in activation of the Ras/Raf/MEK/ERK signalling pathway and of NF-κB. Of course this would result in enhanced cell migration and growth as the authors claimed. This is too complex a situation that warrants further study and confirmation. Nonetheless miRNA-373 and 520c are both able to downregulate many splice variants of CD44 and stimulate cell migration (Yang et al., 2009b). Iczkowski (2011) has confirmed that both miRNA-373 and miRNA-520c reduced the expression of CD44s (standard variant) in prostate cancer cells. Cell proliferation and invasion were restored upon re-expression of CD44s. But then it is necessary to recognise while relating CD44 to cell migration and proliferation that CD44 splice variants display markedly differential expression in different tumour systems.

Heightened expression of miRNA-373 in HCC samples and in HCC cells promoted cell proliferation. Apparently miRNA-373 targets the gene encoding the protein phosphatase 6 catalytic subunit, which negatively regulates the cell cycle (Wu et al., 2011b). The miRNAs-371/373 and miRNA-302 are overexpressed in all malignant germ cell tumours and serum levels of the same might potentially serve as disease markers (Murray et al., 2011). However, in colon cancer, miRNA-373 is downregulated by methylation (Tanaka et al., 2011). It is needless to point out that it is generally accepted that miRNAs can promote or suppress cancer-related processes as discussed earlier, but whilst this relates to different miRNA species equally it is conceivable that opposite outcomes might result upon the prevailing signalling conditions. Another suppressor gene that might be targeted by miRNA-373 is Lats2. Lats2 appears to be directly inhibited by miRNA-373 and their expression is inversely related (Lee et al., 2009). Hence much caution is required in designing experimental protocols and interpreting the data concerning miRNAs and CD44 variants in TXNIP function in tumours.

25 The Essence of the Hippo Signalling System

The Hippo signalling system has received much attention in recent years because of the part it prominently plays in development, morphogenesis and growth but also on account of its remarkable and conspicuous suppressor function in cancer pathogenesis. The mainstay of the system is a quintessentially conserved kinase cascade together with upstream activators and downstream effector transcription factors that regulate target gene expression. Here the discussion is centred round the suppressor function of Hippo and how the system interacts with several other developmental signalling pathways and those that transduce cell proliferation signals in the context of cell transformation, activation of EMT and possibly also metastasis. The Hippo system with its component cascade of kinases represents a unique tumour suppressor. The loss of any component element leads to unregulated growth, loss of differentiation, induction of EMT and cell transformation (Table 25.1). The significance of this signalling network has been encapsulated succinctly and aptly by Halder and Johnson (2011) by labelling the system as Hippo signalling: growth control and beyond.

Lats (Large Tumour Suppressor) Gene Signals via Hippo

The Lats gene mammalian homologues Lats1 and Lats2 constitute a major component of Hippo signalling and Lats are known to be tumour suppressors. Lats kinases are aligned with the inhibition of cell proliferation, promotion of apoptosis, and with chromosomal and cytoskeletal events that occur during mitosis. The kinases MST2/STK (serine/threonine kinase) 3 and MST1/STK4 are in complex with and regulated by Sav1 (human WW45). The pro-apoptosis MST2 and MST1 phosphorylate and activate Lats occurring as a complex with the regulatory protein Mob (Mps one binder) 1 phosphorylates and inhibits YAP1 and its translocation to the nucleus thus affecting target gene regulation.

The p53-dependent checkpoint prevents cell cycle progression in G1. Loss of p53 function is known to result in tetraploid and aneuploid tumours capable of metastatic progression. Lats also can activate this checkpoint. Lats and p53 both activate and control this checkpoint and maintain genomic integrity. Lats2 binds Mdm2, a negative regulator of p53 and activates p53 (Aylon et al., 2006). Lats2 can inactivate the transcriptional co-activator or co-repressor YAP, the downstream regulatory target of Hippo signalling system to control morphogenesis, cell proliferation and apoptosis. TAZ is also a downstream target of Lats.

Therapeutic Strategies in Cancer Biology and Pathology. DOI: http://dx.doi.org/10.1016/B978-0-12-416570-0.00025-1

Table 25.1 The Hippo Signalling Components, Their Possible Upstream Regulators and the Interacting Signalling Systems That Determine Phenotypic Outcomes

Upstream Regulators and Activators	Activator Identity	Function
Fat Fat4	Atypical cadherins	Cell adhesion, motility, proliferation, apoptosis
Dachsous Dchs1, Dchs2		
Four jointed		Fat signalling Ser/Thr kinase
Merlin/NF2	CD44	Transmembrane receptor implicated in cell adhesion, metastasis
Kibra		
		FERM-domain adaptor protein
		WW-domain adaptor protein
RASSF1–6		RA-domain adaptor
Signalling Component	**Kinase Identification**	**Function/Target**
Hippo	MST1, MST2	Ste20 family Ser/Thr kinase
Sav1 (WW45)		WW-domain adaptor
Lats1, Lats2	NDR Ser/Thr kinases	
Mats		Co-factor of Lats1 and Lats2
YAP/Yorkie/Taz		WW-domain transcription co-factors
Interacting Signalling Systems		**Phenotypic Outcome**
P53		Cell cycle regulation, apoptosis
Growth factors		Cell proliferation, EMT
TGF-β	YAP-mediated sequestration of Smads	Inhibition of TGF-β Signalling
AREG	YAP induces AREG expression	EGFR-mediated induction of cell proliferation and invasion
Connective tissue growth factor	YAP-dependent induction	AREG and HB-EGF via activation of EGFR
Hh	YAP	Inhibition of neuronal differentiation
Wnt/β-catenin	YAP and TAZ bind to β-catenin	Interference with nuclear translocation, Wnt signalling inhibited
Ras/ERK		

Source: This is based on information derived from Halder and Johnson (2011), Sherbet (2011a) and references cited in the text.

Signalling components of the Hippo system are characterised by the presence of the SARAH (Salvador 1/RASSF1A/Hippo) and WW domains that mediate interaction with other proteins. The SARAH domain mediates the interaction between RASSF1 and MST1 and MST2. SARAH possibly also mediates coupling of

MST1 and MST2 to Lats. YAP and TAZ possess WW domains composed of modules with 38–40 amino acid residues that may be required for interaction with other proteins with proline-rich motifs (see pawsonlab.mshri.on.ca; Hergovich, 2012). Lats2 phosphorylates YAP and leads to its cytoplasmic translocation and inactivation and so interferes with the expression of genes relevant to the processes of cell proliferation or apoptosis. On the other hand, inhibition of phosphorylation results in the promotion of growth (Zhao et al., 2007a). The deregulation of Hippo is evident from the increased expression of YAP in HCC cells and inhibition of YAP using siRNA reduces cell survival (Li et al., 2012c). Lats has also been shown to phosphorylate TAZ, lead to its retention in the cytoplasm and inactivation. Also ectopically expressed TAZ induces cell proliferation and activates EMT (Lei et al., 2008). These findings are confirmed overall by Zhou et al. (2011c) who showed that TAZ is overexpressed in NSCLC cells and forced overexpression in experimental situations induces cell proliferation and transformation (Figure 25.1).

Mats (the Mob as tumour suppressor) also acts as a co-activator of Lats (Lai et al., 2005). Overexpression of the human homologue Mob1 inhibits proliferation and induces apoptosis of MCF-7 and HeLa cells. Several human isoforms of Mob have been identified. These display tissue specific expression but the significance of this is unclear at present. Only Mob1A and Mob1b feature in interactions with Lats (Chow et al., 2010). Mutation or inactivation of Mobs does occur in human cancers and these naturally affect Lats and YAP.

The attribution of tumour suppressor property of Lats has flowed from the reported loss of heterozygosity, promoter methylation and mutations of the gene reported in human tumours. In certain tumour types, genetic polymorphism has been encountered but mutations might be infrequent (Hansen et al., 2002; Hisaoka et al., 2002; Ishizaki et al., 2002). Furthermore, Lats genes might be differentially

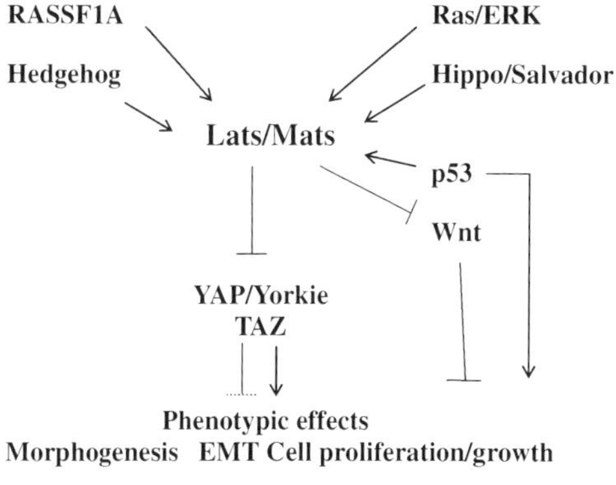

Figure 25.1 Several pathways intersect with the Lats/YAP axis in determining phenotypic effect of morphogenesis, cell transformation, cell proliferation and EMT. Many upstream modulators of Hippo are known to subserve tumour suppressor function. Note that ectopically expressed TAZ promotes EMT, but it is sequestered in the cytoplasm by phosphorylation by Lats and inactivated. Here only RASSF1A is shown. This figure is based on references cited in the text.

regulated. For instance, Zhang et al. (2008b) have shown that YAP interacts directly with and inactivates Lats1 but not Lats2. Suppression of Lats1 mimics the phenotypic effects of YAP overexpression. Possibly therefore Hippo signalling is transduced down different sub-routes. Besides, YAP might function in collusion with many transcriptional regulators such as Smads, Runx2 (Runt-related transcription factor 2), ErbB4, p53BP-2 (p53 binding protein-2), among others which clearly link it with cell proliferation and the function of growth factors and activation of EMT. This reveals the potential of cross talk and switching of signalling sub-routes by Lats/YAP between Hippo, Wnt, Hedgehog and growth factor signalling and provides a patently plausible scenario of the interaction of signalling pathways associated with developmental and pathogenetic processes (see Figure 25.1).

Hippo in Cross Talk with Growth Factor Signalling

Members of the TGF-β family transduce their effects through a relatively uncomplicated system of two types of receptor, the type I and type II (RI and RII) receptors. These receptors are transmembrane proteins consisting of a ligand-binding extracellular domain, a transmembrane domain and a cytoplasmic serine/threonine kinase domain. TGF-β signals are then transduced through the canonical Smad cascade. Of the Smad family transcription factors, Smad6 and Smad7 inhibit signalling downstream of TGF-β RI receptors. Smad7 is less discriminatory and can inhibit signalling via TGF-β RI-related receptors such as BMP type I and activin receptors (Sherbet, 2011a). Upon cytoplasmic localisation instigated by Hippo, YAP sequesters Smad complexes and inhibits TGF-β signalling (Varelas et al., 2010). Ferrigno et al. (2002) showed that YAP65 forms a complex with Smad7 and recruits Smad7 to the activated TGF-β receptor type I to inhibit Smad3/4-dependent gene transactivation by TGF-β. TGF-β signalling is also implicated by Fujii et al. (2012) in the induction of CTGF (connective tissue growth factor) expression and modulating the growth of mesothelioma cells. It may be recalled here that TGF-β/BMP/Smad signalling also functions via the Runx2 to regulate cell proliferation. Runx2 is also the effector of Wnt/β-catenin signalling in determining cells fate and pattern formation in embryonic development. This signalling pathway involves Groucho4/TLE4 (transducing-like enhancer of split 4) in a complex network of interaction with FGF (Burks et al., 2009). EGFR-mediated signalling similarly involves Runx2. EGF/EGFR inhibits osteoblast differentiation not by its proliferation inducing power but by inhibiting the expression of Runx2 and osterix in differentiating osteoblasts (Zhu et al., 2011). To complete the narration and the direct relevance to Hippo and Wnt interactive signalling, Runx2 associates with YAP65 and in complex increase anchorage-independent growth more effectively than each component individually (Vitolo et al., 2007).

Direct evidence of EGFR involvement with Hippo is provided by the finding that YAP-expressing MCF10A breast cancer cells can induce proliferation of neighbouring cells not expressing YAP. YAP seems to target amphiregulin (AREG) gene and increase its expression and AREG seems to promote cell proliferation via activation of EGFR (Zhang et al., 2009b). It may be noted here that AREG participates in cell proliferation and adhesion-related biological function, namely cell motility and invasion. There is ample evidence that AREG is aberrantly expressed in many human neoplasms and AREG expression correlates with both in disease progression and prognosis (Sherbet, 2011a). Of much interest is the induction of CTGF by AREG and HB-EGF that activate EGFR. This was dependent upon YAP. Also this study has identified elements in the CTGF promoter that binds to YAP complexes (Urtasun et al., 2011).

Hippo Interacts with the Wnt Signalling System

The activation of Wnt signalling begins with the binding of Wnt proteins to the Frizzled family of protein receptors and co-receptors. Many Wnt ligands are known. These bind to its receptor Frizzled or Frizzled and LRP5/6complex and the signal is transduced to the cytoplasmic hosphoprotein DVL (Dishevelled), which in turn relays the signal to non-canonical and the canonical β-catenin pathway. β-Catenin is localised in the membrane as a part of a multi-protein complex. The ligand/receptor complex releases β-catenin from the Axin/APC/GSK3 complex. In the absence of Wnt, GSK3 (glycogen synthase kinase-3β) phosphorylates β-catenin leading to its degradation and TCF forms a repressor complex with the transcriptional repressor Groucho and HDAC to inhibit target gene transcription. With the activation of Wnt signalling, GSK3 is inactivated and unphosphorylated β-catenin accumulates, enters the nucleus and displaces Groucho from its binding to TCF and the complex turns into an activator transcription complex and transcribes the target genes (Habas and Dawid, 2005; Sherbet, 2011a).

YAP has been viewed as a co-ordinator of cross talk between Hippo and Wnt. Xin et al. (2011) found that YAP inhibition suppressed proliferation of murine cardiomyocytes, whilst active YAP produced the opposite effect. They further showed that YAP can function via the IGF pathway, inactivate GSK3 and promote accumulation of β-catenin, clearly suggesting an interaction with Wnt signalling. YAP and TAZ have been found to be able to bind to and interfere with nuclear translocation of β-catenin thus blocking Wnt signalling. Hippo kinases phosphorylate YAP inducing its translocation to the cytoplasm thus assisting the sequestration of β-catenin in the cytoplasm and aiding the inhibition of Wnt/β-catenin signalling (Imajo et al., 2012). Not only this, but β-catenin/TCF4 complexes bind to an enhancer element within the first intron of the YAP gene and induce YAP expression (Konsavage et al., 2012). On the other hand, YAP-mediated promotion of cell proliferation and inhibition of neuronal differentiation intersects with Sonic Hh signalling rather than Wnt (Lin et al., 2012). How YAP might switch signalling in this way is yet to be elucidated. The subtlety of the Hippo system is explained by the possibility that the loss of any one signalling component is compensated by the activation of another component (Hergovich, 2012). So it is conceivable that a similar compensatory activation of one or more interacting signalling systems might occur.

Upstream Regulators of Hippo Are Tumour Suppressors

Suppressor Function of Atypical Cadherins Dachsous and Fats

Hippo signalling is modulated by many upstream elements which have themselves been attributed with tumour suppressor function. Prominent among them are the atypical cadherins, Fats and Dachsous (Dachs). The Fats seem to act downstream of Dachs. Dach1 gene encodes a chromatin associated protein that binds to transcription factors and regulates gene expression. Dach1 contains the highly conserved ski domain. Dach1 is said to regulate the Six family genes involved in organogenesis and also implicated in the control of cell proliferation and cell motility in embryonic development. It probably acts as a co-repressor of Six6 to regulate cell proliferation via repression of cyclin-dependent kinase inhibitors. Interestingly it is also believed to inhibit TGF-β signalling by interacting with Smad4 receptor and NCOR1 (see UniProtKB/Swiss-Prot: DACH1_HUMAN, Q9UI36). In essence, the Fat/Dachs/Hippo axis is closely associated with these biological features, essential in tumour development, invasion and secondary spread. In murine mammary epithelial cells, loss or suppression of Fat4 expression by promoter methylation promoted tumorigenesis, whilst induced re-expression inhibited tumorigenesis. In human breast cancer cell lines and tissues, loss of Fat4 was encountered. Loss of expression by methylation was found in 3/6 established breast cell lines and 3/5 primary breast cancers (Qi et al., 2009). Although this is a limited exercise with reference to human breast cancer, Qi et al. (2009) did go on to show that the loss of Fat4 led to overexpression of YAP, thus linking the loss of Fat4 unequivocally with Hippo signalling. Katoh and Katoh (2006) have claimed that *in silico* analysis Fats 1–4 were differentially expressed in various human neoplasms as well as in normal tissues. How significant Fat signalling is in terms of relevance to tumorigenesis is difficult to determine at present, especially since mutation or deletions of Fat4 are somewhat infrequent (approximately 5%) in gastric tumours as compared with p53 (>70%) for instance (Zang et al., 2012). But we do not know what the status is in breast or other cancers. Genetic prolife of Fat apart, when read in the context of the demonstration that Dachsous can function independently of Fat, Fat deregulation may not seem always relevant for tumorigenesis (Matakatsu and Blair, 2006).

NF2/Merlin as Tumour Suppressor

NF2/Merlin and *Drosophila melanogaster* protein called Expanded are ERM-related cytoskeletal proteins. Merlin has been shown to possess a significant ability of tumour suppression besides its ability to regulate the size of organs. FRMD6 (Willin) is seen as the human homologue of Expanded. Loss of FRMD6 has been reported to produce cellular phenotypic features reminiscent of EMT (Angus et al., 2012). The suppressor effects of the human homologue were seen in breast cancer cell lines where overexpression seemed to have led to the inhibition of cell proliferation colony formation, soft agar colony growth *in vitro* and *in vivo* tumour growth in nude mice. However, these effects occurred independently of Hippo signalling

(Visser-Grieve et al., 2012). NF2 is mutated in many cancers. The presence of inactivating mutations and LOH has suggested that it is a tumour suppressor. Merlin participates in several cellular functions such as intercellular and cell–substratum interaction and adhesion, regulation of cell proliferation and apoptosis. This it does by virtue of functioning as a link between ECM receptors such as CD44, integrins and the actin cytoskeleton and by regulating the function of a number of RTKs and so modulating growth factor signalling, thus integrating ligand binding, cell adhesion and motility and cell proliferation and survival with tumour development. The suppressor function of Merlin seems to derive from its molecular configuration. It assumes a closed conformation as a result of head-to-tail association of the molecule. An open-configured protein results from the phosphorylation of certain C-terminal amino acid residues. In the folded conformation, Merlin exerts its inhibitory activity. But it is a target of several kinases, including PKA, PAK1 and PAK2, and the anti-apoptotic PIKE (phosphoinositide-3-kinase enhancers), which binds and activates Akt. Akt phosphorylates Merlin, abolishes the folded conformation and renders it liable to ubiquitination and proteasome degradation (Stamenkovic and Yu, 2010; Ye, 2007). Subcellular disposition of Merlin is also a possible cause of modulation of function. PAK2-mediated phosphorylation alters its cellular location (Kissil et al., 2002). An interesting observation of relevance is the report by Zhou et al. (2011a) that active Merlin localises Wnt/β-catenin at the cell membrane which renders the proliferative signalling by Wnt ineffective.

Merlin is an upstream regulator of Hippo signalling. Suppression of Merlin leads to enhanced organ growth and eventually to tumorigenesis. It regulates the Hippo kinases MST1 and MST2, activation of Lats and inactivation of YAP. Merlin has also been attributed with the ability to suppress Ras mediation of cell transformation by modulating the disposition or suppressing the functions of downstream effectors of the Ras cascade. It is said to be able to suppress Ras/Rac activation or stop the translocation of Rac to the plasma membrane and in this way contact inhibition of growth (Okada et al., 2005). Equally, it may be as Shaw et al. (2001) found Rac induced the phosphorylation of Merlin and reduced its association with the cytoskeleton. STAT3 and STAT5 signalling is also enhanced in the background of mutated NF2, whilst wild-type NF2 negatively regulates mTORC1 signalling (Stamenkovic and Yu, 2010; Figure 3.1). Merlin has been shown to inhibit the MAP3K kinase MLK3 together with the inactivation of B-Raf/ERK/JNK pathway thus potentially linking its suppressor function with the inhibition of this pathway and Rho GTPase signalling (Zhan et al., 2011).

That Merlin produces changes in the ECM and ECM-mediated function of the cell is established on a reasonable basis. Lallemand et al. (2003) showed some years ago that loss of Merlin led to the deregulation of contact-mediated growth arrest and destabilised cell-to-cell adhesive contacts, which could potentially have led to enhanced cell motility. Merlin might inhibit cell invasion by suppressing the function of FAKs (Poulikakos et al., 2006). More recently, Galcheva-Gargova et al. (2008) transfected NF2 into B16 melanoma cells with metastasising ability and found the transfectants overexpressing NF2 showed reduced metastatic spread in murine models. They also noticed alterations in growth properties in soft agar and *in vitro* motility, and somewhat less convincingly have argued that cell motility and metastasis of the non-transfected parental melanoma cells were inhibited by treatment with cell surface heparan sulphate glycosaminoglycans (HSGAGs) derived from the transfected cells overexpressing NF2. It would be worth noting in this context that HSGAGs occur ubiquitously in or associated with the ECM. Cellular adhesion, migration and proliferation responses to HSGAG demand specific ligand-binding properties of HSGAG, for many ligands such as ECM components and growth factors may be implicated in HSGAG-mediated cellular responses. Hence much further investigation of the molecular specificities of the HSGAGs derived from the transfected cells is necessary before one can firmly deduce the significance of these findings. However, a clear link with aggressive tumour behaviour is endorsed by studies of NF2 expression in breast cancer. NF2 expression is markedly reduced in these tumour tissues and this occurs in parallel with enhanced expression of osteopontin (OPN). Here the loss of Merlin seems to have resulted from induced degradation of the protein by OPN-mediated phosphorylation (Morrow et al., 2011). As discussed earlier OPN was identified as a metastasis-promoter protein some time ago, with the convincing demonstration of its association with metastatic progression in a murine tumour cell model (Oates et al., 1996). OPN expression has generally correlated with poor patient survival (Rudland et al., 2002). A mutual regulatory system might be operating here. Merlin inhibits PI3K/Akt signalling, whilst Morrow et al. (2011) have linked OPN with Akt-mediated phosphorylation and degradation of Merlin. Furthermore, the regulatory link between OPN and Merlin is supported by the fact that CD44, which acts as an OPN receptor and routes its signals downstream, is antagonistic in function to Merlin. Either way OPN negates the function of Merlin and its negative regulation of growth. So prevention of degradation of Merlin by PI3K or by inhibition of OPN could be a valuable approach to respond to the degradation and the insidious effects of loss of Merlin.

Kibra in Tumour Suppression

Kibra was identified, isolated and characterised as a cytoskeletal protein possessing two amino-terminal WW domains, an internal C-2-like domain and a carboxyl-terminal glutamic acid-rich stretch (Kremerskothen et al., 2003). WW-domain proteins participate in cohering together proteins into physiologically important network by specific binding to proline-rich motifs but inter-protein interactions might be more discriminatory since WW domains within the same protein might possess different specificities towards potential interacting proteins (Ingham et al., 2005). Kibra has been implicated in memory dysfunction and genetic alterations of the Kibra gene have been associated with Alzheimer's disease. It has been linked with failure of mammary development. Its notoriety has risen with its recognition and increasing appreciation as an upstream regulator of the Hippo system with a distinctive role of regulating cell proliferation and disease processes. At the molecular level, function of Kibra seems to be regulated by phosphorylation. Buther et al. (2004) reported that Kibra was a target of phosphorylation, for example by PKCζ. Recently phosphorylation of Kibra by mitotic Aurora kinase in conjunction with cell cycle progression has firmly allied Kibra with the regulation of cell proliferation (Xiao et al., 2011). Kibra seems to generate its suppressive effects in collaboration with Merlin with which it forms a complex and this complex regulates Hippo signalling. The formation of this complex is related to the phosphorylation of Kibra, possibly of the Ser539 by Aurora kinase (Xiao et al., 2011; Yu et al., 2010). Although Kibra and Merlin have appeared as two separate upstream regulatory components of the Hippo pathway, they indeed seem to be functionally related. At the practical level, it is of considerable interest that Kibra interacts and forms a complex with DYNLL1 (dynein light chain 1). The Kibra/DYNLL1 complex might be important for the transactivation of ER (Rayala et al., 2006). As discussed in several contexts, ER− and indeed TNBC form a formidable challenge in breast cancer treatment.

One might recall here that DYNLL1 is a participant in several biological functions, especially with regard to apoptosis by abolishing the anti-apoptotic activity of Bcl-2. DYNLL1 might also regulate apoptosis by other means. It is said to interact with Dkk (Dickkopf)-3 (Ochiai et al., 2011). Dkks inhibit Wnt signalling and obviously interaction between Dkk and DYNLL1 might have implications for apoptosis and activation of EMT. Zhang et al. (2008c) found DYNLL1 interacted with RACK1 and BimEL a pro-apoptosis protein of the Bcl-2 family. RACK1 has been postulated to interact with integrin and in this way inhibit cell motility (Buensuceso et al., 2001). DYNLL1 is a component of the dynein complex and is a substrate for PAK1 serine/threonine kinase which participates in cell proliferation, migration and morphogenesis. PAK1 modulates the actin cytoskeleton, and cytoskeletal dynamics are also influenced by the dynein complex and so they are functionally tightly related. As noted before, Merlin is also a substrate for PAK. The obvious phosphorylation-mediated functional linkup between Kibra and Merlin could be theoretically meaningful and important.

Kibra is hypermethylated in leukaemias. Methylation has been encountered in 70% of B-ALL (acute lymphocytic leukaemia), in <20% T-ALL and in around a third of CLL (chronic lymphocytic leukaemia) patients. Of much interest also is the finding that methylation was not seen in remission. The methylated status reflected poor prognosis for CLL patients (Hill et al., 2011; Shinawi et al., 2012).

Kibra methylation may be potentially regarded as relevant to the pathogenetic process since it was associated in ALL with the chromosomal translocation t(12;21)(p13;q22) involving ETV6/RUNX1 commonly associated with childhood ALL.

RASSF Family Suppressor Genes

RASSF family genes are upstream regulators of Hippo signalling. RASSF are so called because they contain the RA (Ras-association domain). Several members of the family are known. The presence of two sub-groups has been acknowledged. RASSF1–RASSF6 (the C-terminal family) carry the RA domain at the C-terminus while the domain occurs at the N-terminus in N-terminal family of RASSF7–RASSF10. The RA domain mediates RASSF binding to GTPases. But they probably differentially interact with Ras proteins *in vitro* (Ikeda et al., 2007). The SARAH domain is also an important motif of RASSFs. SARAH occurs immediately C-terminal to the RA domain in RASSF1–RASSF6. The N-terminal family RASSFs lack the SARAH

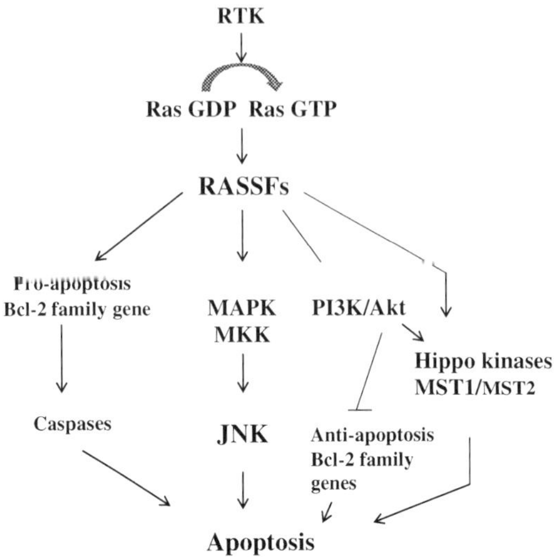

Figure 25.2 The signalling pathways mediated by RASSF proteins in the induction of apoptosis. Four postulated pathways shown include caspase-mediated function of pro-apoptosis Bcl-2 genes or PI3K/Akt-mediated inhibition of anti-apoptosis genes. Akt is also suggested to phosphorylate and activate the Hippo kinases and Hippo also might act independently of Akt-mediated activation. RASSF can also activate MAPK/MKK and pro-apoptotic JNK pathway.

domain (Weyden and Adams, 2007; Richter et al., 2009). The SARAH domain facilitates the heterodimerisation of RASSFs with the Hippo kinases MST1 and MST2. In general, RASSFs are suppressors of tumorigenesis, cell proliferation and regulation of the cell cycle and inducers of apoptosis. Some members may be involved with the stabilisation of cytoskeletal structures (Sherwood et al., 2010). RASSF function is regulated by the phosphorylation of amino acid residues of the RA domain (Richter et al., 2010). However, at present, there is no clarity of distinction in function between the C-terminal and N-terminal RASSFs (but see below). The postulated pathways of RASS signalling are shown in Figure 25.2 which has to be read in conjunction with Figure 25.1 depicting the Hippo signalling and interacting pathways.

RASSF Genes Are Silenced in Tumours

This gene family has received much attention on account of frequent methylation and inactivation of its members in many human neoplasms, for example epithelial cancers, leukaemia and lymphomas and several forms of childhood cancers, so much so they have been widely described as tumour suppressor genes. The C-terminal RASSFs might be tumour suppressors, but the N-terminal RASSFs are not (see below). The degree of methylation might vary and some RASSF genes might be prominently methylated as in neuroblastomas where RASSF7 has been conspicuous in its frequency of methylation (Djos et al., 2012). The inhibition of expression, for example of RASSF1, which was hypermethylated in melanoma cell lines and tumour, was less conspicuous in melanomas as compared with benign nevi. However, there might have been significant contribution from mutation of B-Raf and N-Ras in tumour growth (Reifenberger et al., 2004). In breast cancer also, methylation of RASSF1 has been noticed in a majority of cases but markedly varied in degree, but none in control tissues (Sebova et al., 2012). These authors also suggest that the degree of methylation might be related to tumour size, lymph node involvement and TNM stage. Promoter methylation of RASSF1A was markedly greater in cervical tumours from patients with nodal metastases than in patients who showed no metastatic spread (Malyukova et al., 2004). Reduced RASSF1 expression correlated with tumour grade and nodal involvement in oesophageal squamous cell carcinoma and nasopharyngeal carcinoma (Lo et al., 2007). Similar inverse correlation with nodal metastasis has also been found in melanoma (Yi et al., 2011). In gastric tumours, reduced cytoplasmic expression RASSF6 significantly associated with invasion, regional lymph node metastasis as well as distant metastasis and was reflected in poor disease prognosis. Expression of RASSF6 was lower in metastatic lesions than in corresponding primary tumour (Wen et al., 2011). Also consistent with other reports, methylation was encountered more often in metastatic than in primary tumours. Methylation status in small bowel tumours and matched metastatic lesions were examined for methylation of RASSF1. RASSF1 methylation was encountered more frequently in metastatic tumours than in primary tumours (Zhang et al., 2006b). This is obviously an interesting finding if it were properly established. For, although matched primary and secondary tumours were studied, it is unclear how many metastatic lesions showed increased methylation as compared with the corresponding primary tumours.

A link between epigenetic silencing of RASSF1 and invasion and progression has also been established for endometrial tumours (Liao et al., 2008). In some paediatric tumours, the relationship between methylation of RASSF1 and the disease process was not quite persuasive (Wong et al., 2004). However, Teng et al. (2011) used mesenchymal stem cells derived from bone marrow to show that experimentally methylating RASSF1 and the growth regulatory and tumour suppressor gene HIC (hypermethylated in cancer) 1 transformed them into cancer stem cells. Furthermore, upon implantation into immune compromised murine hosts the stem cells carrying methylated RASSF1 developed into tumours, which upon serial transplantation showed cells with CSC markers. This suggests the expansion of clones carrying the marker into CSC subpopulations. One has to subject these findings to the caveat that the CSC markers may not only display different patterns of presentation but also this diversity or heterogeneity of marker expression might often depend upon the tumour microenvironment. This is an important caveat in this series of experiments since the CSC markers were identified after several serial transplantations. Notwithstanding these reservations, it should be acknowledged that this is indeed a valuable exercise, for cancer stem cells would be akin to adult stem cells and would represent a clone that is amenable to the induction of differentiation, capable of self-renewal and possess proliferative potential. Therefore, they are potential targets for therapeutic intervention routed through RASSF function.

Nucleotide polymorphisms have been identified in RASSF1A and it is possible that besides epigenetic silencing, aberrant RASSF functions could arise from these genetic changes (Gordon et al., 2012). Earlier El-Kalla et al. (2010) reported that nuclear polymorphism of RASSF1A affected its association with microtubules and its suppressor function.

RASSFs Regulate Cell Proliferation and Apoptosis

RASSFs induce the auto-phosphorylation and activation of the pro-apoptosis Hippo kinases, MST1 and MST2. This generalisation may have to be revised with the implication of other signalling pathways in the regulation of the apoptotic process. RASSFs might under certain conditions inhibit Hippo signalling and apoptosis. For instance, Ikeda et al. (2009) found that RASSF6 binds and inhibits MST2 activity and apoptosis, but induced apoptosis when released from binding to MST2 by WW45. Thus RASSF6 seems to possess a direct mode of action independent of Hippo. But then the pro-apoptotic effect of MST2 may be inhibited by other interactions, for example with Raf1. RASSF1A is said to break up this complex leading to downstream signalling via Lats and YAP (Matallanas et al., 2007). Besides, they also seem to promote induction of apoptosis by TNF and related peptides, possibly also involving MST1 function as demonstrated in the case of RASSF5 (Park et al., 2010). Induction of apoptosis can occur by other routes, for example via upregulation of p38 MAPK and inhibition of Akt mediation quite independently of Hippo (Yi et al., 2011). Activation of Bax and induction of apoptosis with or without the mediation of caspases is also a possibility (Ikeda et al., 2007). However, the isoform RASSF1C is said to bring about results that are not compatible with suppressor function (Reeves et al., 2010). It may be that MST2 might be involved in experimental setup in a manner described by Ikeda et al. (2009). On the other hand, this might be the beginning of the emergence of differential functioning of the RASSFs and its isoforms. This might not be a purely theoretical exercise for Reeves et al. (2010) have also gone on to show that RASSF1C not only downregulates caspase expression but also makes tumour cells less sensitive to apoptosis.

RASSF1A might function in collaboration with p53 in regulating apoptosis and tumorigenesis (Tommasi et al., 2011). This is eminently worthy of exploration in human malignancies in the light of the fact that p53 is the most frequently mutated tumour suppressor, especially whether deregulation of p53 function can be countered and neutralised by RASSF signalling. Equally RASSF1A can inhibit cell proliferation by upregulating cyclin-dependent kinase inhibitor $p21^{waf1/cip1}$ via MAPK/Akt route and quite independently of p53 (Thaler et al., 2009).

The N-terminal RASSFs May Be Tumour Promoters

Unlike the C-terminal RASFs, the N-terminal RASSFs are probably not tumour suppressors. RASSF7 is not methylated nor mutated in several tumour cell lines studied, but indeed upregulated in many tumours, for example pancreatic, endometrial and ovarian carcinomas (Brandt et al., 2004; Friess et al., 2003; Logsdon et al., 2003; Tan et al., 2009b). RASSF7 upregulation has been found to occur in response to hypoxia. Its suppression results in suppression of cell proliferation together with abnormalities of spindle formation and abnormal mitotic index and a failure of chromosomal movement to the spindle equatorial plate in mitosis (Recino et al., 2010). Similar abnormalities also accompany other N-terminal RASSFs. Hypoxia/HIF signalling involves many miRNAs and among them is miRNA-210 (Kulshreshtha et al., 2007). HIF1-α activity is involved in the promotion of angiogenesis by certain miRNA, for example miRNA-210. In MCF-7 breast cancer cells hypoxia induced an overexpression of miRNA-210 and this was mediated by HIF-1α/VHL (Camps et al., 2008). Possibly therefore the upregulation of RASSF7 could have repercussions for tumour growth and spread, although there is no evidence for this at present. In any event, the elucidation of the biological function of N-terminal RASSFs is just beginning.

26 HIC1 Suppressor Gene

HIC1 is a tumour suppressor whose expression is suppressed in a wide variety of cancers. The gene encodes a transcriptional repressor with five Krüppel-like C2H2 zinc fingers that facilitate sequence-specific DNA binding and two repressor domains, viz. an N-terminal BTB/POZ (broad complex Tramtrack bric-a-brac/Pox virus and zinc-finger) domain that is characteristic of transcriptional regulators and a central region which engages different co-repressors, e.g. the CtBP (C-terminal EIA binding protein) and NuRD complexes (Deltour et al., 2002; Fleuriel et al., 2009; Pinte et al., 2004; Wales et al, 1995). CtBPs have been strongly implicated in promoting EMT and tumour progression, inhibition of tumour suppressor genes, and suppression of apoptosis (Chinnaddurai, 2009). The NuRD complex has histone deacetylases, histone binding proteins and Mi-2-like proteins and the tumour promoter gene MTA (see discussion about MTA on pp. 77–79).

It has emerged from the recent work that the BTB/POZ (broad complex Tramtrack bric-a-brac/Pox virus and zinc-finger) domain of HIC1 recruits PCR (Polycomb repression complex) 2 (Boulay et al., 2012). The Polycomb group genes form two important multi-protein complexes, namely PRC1 and PRC2. PRC1 is said to stably maintain gene repression, whilst PRC2 contains HDACs and methyltransferses EZH1 or EZH2 (enhancer of Zeste homologue 2), SUZ12, and initiates silencing (Han et al., 2009; Kohler and Villar, 2008). EZH2 is a histone–lysine *N*-methyltransferase which catalyses H3K27me2/3, that is di- or tri-methylation of histone H3 at lysine 27 and with these epigenetic modifications maintain genetic repression. The molecular structure of HIC1 does suggest potentially significant links with the regulation of cell proliferation and growth control, cell survival and DNA damage response and tumorigenesis.

Silencing of HIC1 in Tumours

Gene silencing occurs by one of four mechanisms, namely methylation of the gene promoter, histone deacetylation, histone methylation and ubiquitination. The arginine and lysine residues of histones are the main targets of methylation; histone methylation can repress or activate transcription. Ubiquitination of histones is another mode of regulation of transcription. Ubiquitination of the core histone H2A frequently leads to gene silencing whilst H2B ubiqutination leads to activation.

The methylation of cytosine bases in the DNA can prevent the binding of transcription factors to gene promoters or facilitate the engagement of transcription repressors to methylated DNA. Also methylation can repress transcription by

Therapeutic Strategies in Cancer Biology and Pathology. DOI: http://dx.doi.org/10.1016/B978-0-12-416570-0.00026-3

promoting an inactive chromatin structure. As the name of the gene implies, the expression of HIC1 in tumours is suppressed by methylation. Promoter methylation of HIC1 is seen very frequently in prostate carcinoma specimens where the high frequency of methylation occurred in high Gleason grade tumours (Kilinc et al., 2012). In renal cell tumours, higher levels of methylation was found in primary tumours which had metastasised to regional lymph nodes as compared with those which had not metastasised (Eggers et al., 2012).

HATs and HDACs regulate gene transcription. HATs acetylate the lysine tails of histones leading to an open chromatin state making genes more accessible to transcription. But HDACs reduce histone acetylation and can suppress transcription or deregulate gene expression. Thus modifications of histones at the post-translational level are important regulators of gene expression (Kouzarides, 2007; Li et al., 2007b). Svedlund et al. (2012) have argued the repressive modification of H3 histone (H3K27me2/3, i.e. di- or tri-methylation of histone H3 at lysine 27) rather than DNA methylation might be a major means of HIC1 silencing. Deacetylation is also able to regulate HIC1 activity. HIC1 interacts with the deacetylase SIRT1 (sirtuin 1) and this involves the last four last zinc fingers of HIC1 and a certain defined amino acid sequence of SIRT1 (Dehennaut et al., 2012). This sequence contains the domain called ESA (essential for SIRT1 activity) which serves as a switch to turn on the deacetylase function (Kang et al., 2011a). This in a way confirms that the interaction between HIC1 and SIRT1 has a functional basis.

A Resumé of Apoptosis Pathways

HIC1 is actively interacts with many pathways of apoptosis signalling. So a brief resumé is provided here as prelude. Apoptosis can occur by the intrinsic mitochondrial pathway where it is induced by DNA damage and other stress factors. Caspases are the major participants of the apoptosis process. They are involved in mitochondrial as well as the death receptor pathways. The death receptor pathway invokes the intervention of Bax/Bak/Bcl-2, which can involve regulation by p53. This pathway constituted by TNFR (tumour necrosis factor receptor)/TRADD (TNFR-associated death domain) or Fas receptor (CD95) leads to the activation of caspases via FADD (Fas-associated death domain). TRADD can function via TRAF2 (TNF-associated factor 2) towards cell survival by NF-κB or via JNK signalling towards apoptosis (Sherbet, 2011a). This pathway would activate apoptosis independently of p53. The intrinsic and death receptor pathways meet downstream at the initiation of the so-called executioner caspases (caspase-3, caspase-6 and caspase-7). Apoptosis is negatively regulated by Akt, which in turn is subject to supervision by PTEN. Many of these pathways involve HIC function together with other interacting elements (Figure 26.1) which have been discussed in some detail here.

The Alternative Reading Frame Tumour Suppressor Genes

The alternative reading frame (ARF) locus harbours the tumour suppressors, namely p16ARF ($p16^{INK4a}$, CDKN2A) and p15ARF ($p15^{INK4b}$). The p16ARF locus also

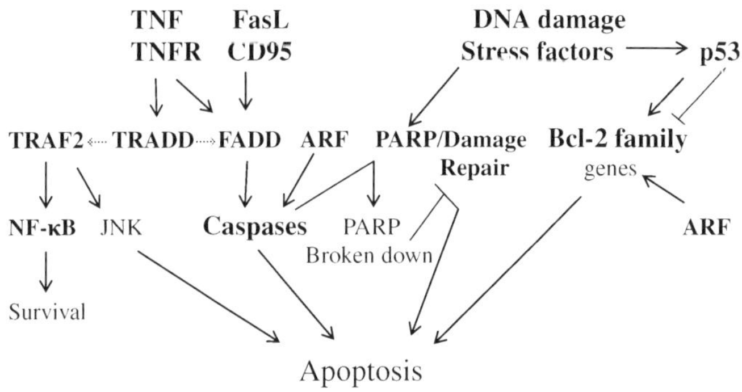

Figure 26.1 A resumé of apoptosis pathways which involve HIC1. The points of involvement of HIC1 or CtBP are not shown here but discussed in the text. ARF locus gene products, cyclin-dependent kinase inhibitors, are closely linked with cell cycle regulation, cell senescence and apoptosis and differentiation. This is the rationale for including them as a part of the discussion of apoptosis signalling.

encodes thee p14ARF (homologue of murine p19ARF). These genes encode cyclin-dependent kinase inhibitors. They are involved with both p53 and Rb signalling pathway. Phosphorylation of Rb by cdk4/6 is regulated by p16ARF, and p19ARF has been implicated in the regulation of mdm2 and p53 degradation (Li et al., 2009c; Serrano, 2000; Sharpless and DePinho, 1999). Furthermore, they are suppressed in genetically reprogrammed and induced pluripotent cells and in embryonic stem cells (Li et al., 2009c). Thus overall the products of the ARF locus are closely linked with cell cycle regulation, cell senescence and apoptosis and differentiation. This is the rationale for including the cyclin-dependent kinase inhibitors as a part of the discussion of apoptosis signalling.

ARFs and Bmi-1 Function

The status of ARFs in cancer stem cells has received much attention in recent years. It is regulated by p53. The expression of ARF is repressed by binding to p53. This inactivation of ARF requires both HDACs and the Polycomb group proteins (Zeng et al., 2011). Bmi-1 (the Polycomb group protein B lymphoma Mo-MLV insertion region 1 homologue) negatively regulates the expression of the pro-apoptotic gene Bim (Jagani et al., 2010). Bmi-1 is overexpressed in many cancers and is essential for stem cell renewal. Hence there is much interest in the fact that Bmi-1 is a regulatory component of the PRC1 complex (see p. 169). The latter has been shown to regulate the transcription of ARFs (Cao et al., 2011). Cell subpopulations capable of self-renewal and tumorigenesis that could be CSCs displayed stem cell markers CD166 and CD44 and were characterised by loss of expression of ARFs (Rath et al., 2011). In mammary epithelium, suppression of the ARF locus restores Bmi-1 and leads stem cell renewal and growth (Pietersen et al., 2008) and by implication of high Bmi-1 expression also to breast cancer tumorigenesis. However in the light of doubts being expressed about the suppression of ARFs by Bmi-1 (Chiba et al., 2008; Xu et al., 2009), one has to look at other lines of evidence for ARF involvement in CSC and tumorigenesis. In glioblastoma multiforme, Bmi-1 is expressed in CD133+ cells (Abdouh et al., 2009), suggesting a close relationship between Bmi-1 and the capacity of stem cells for self-renewal. Indeed Zheng et al. (2008) have implicated Myc/p53/PTEN axis in the regulation of stem cell self-renewal and tumorigenesis of glioblastomas. In the absence of other oncogenic signals, Bml-1 probably acts only partially in an ARF-dependent manner as Dovey et al. (2008) have put it. As stated at the outset, ARFs are involved with both p53 and Rb signalling. With deletion of ARF leading to the restoration of Bmi-1 function, one can conclude that modulation of p32/Rb function and suppression of ARF could achieve the faculty of self-renewal. As alluded to in the introductory chapter, metastatic progression has to be viewed as a stem cell clone with specialised ability of invasion and dissemination. The finding of Mihic-Probst et al. (2007) that Bmi-1 shows greater expression in metastatic lesions of melanomas that in corresponding primary tumours is of prodigiously significant in this context.

ARF Function and p53 Activity

ARFs are often found to function as a tumour suppressor. ARF was shown some time ago to induce apoptosis by recourse to the mitochondrial pathway. ARF activates caspases. ARF also modulates the expression of Bcl-2 family genes and this occurs either dependent upon or independently of p53. Du et al. (2011) transfected a cell line lacking p53 with an inducible p14ARF construct and showed that induction of p14ARF activated p21 and ERK1/2 in the transfectants cells and initiated G1 arrest. ARF is said to downregulate anti-apoptosis Bcl-2 in a p53-dependent manner but upregulate the pro-apoptotsis Bax/Bim via a p53-independent pathway (Nakazawa et al., 2003). Indeed induction of apoptosis by ARF may require different pro-apoptosis genes in genetic environment with or without p53 (Hemmati et al., 2006). In other words, p53-dependent or independent function of ARF is related to which pro-apoptosis genes of the Bcl-2 family are engaged.

ARF Interactions with CtBP

As noted earlier, HIC1 engages the co-repressor CtBP. CtBPs have been strongly implicated in promoting EMT and tumour progression, inhibition of tumour suppressor genes and suppression of apoptosis. Hence the antagonistic interaction between CtBP and ARF has been focused upon by several investigators. A marked inverse correlation of expression of CtBP and ARF has been reported in adenocarcinoma of the colon (Kovi et al., 2010; Straza et al., 2010). The suppression of p19ARF increases *in vitro* invasion by an HCC cell line increases invasion

(Chen et al., 2008c). BRCA tumour suppressor proteins interact with several proteins such as BCCIP (a cofactor for BRCA2 in tumour suppression), the DNA repair proteins Rad50 and Rad51, c-myc, p53 and Rb (Jhanwar-Uniyal, 2003). CtBP and CtIP (CtBP-interacting protein) also interact with BRCA proteins. The biological functions of BRCA1 are mediated by two C-terminal BRCT domains. Proteins containing the LIM domain, that is a tandem repeated zinc-finger structure that mediates interaction between proteins, are known to participate as transcription factors in gene expression and in many cellular functions such as cell adhesion, motility and signal transduction. The BRCT domain binds CtIP and interacts with the LIM domain protein LMO4, thus forming a complex that was suggested to negate the suppressor function of BRCA1 (Sum et al., 2002). Both CtIP and CtBP suppress the transactivation of the p21 promoter by BRCA1 and again its terminal BRCT appears to be involved (Li et al., 1999). There is no new information in this regard or in respect of other tumour suppressors.

Do ARFs Function in Conjunction with 14-3-3σ?

The 14-3-3 family of proteins have been studied in some detail on account of the tumour suppressor function of some members of the family, for example 14-3-3σ. With the association of p53 with ARF function, the natural follow-up in terms of cell cycle checkpoint control is to inquire into whether ARF and 14-3-3σ are connected in terms of their function. The regulation of apoptosis and checkpoint function has prompted investigation into the relevance of the 14-3-3 proteins in carcinogenesis. 14-3-3σ has been cited as a target for interference to make tumour cells susceptible to the action of anticancer agents (Hermeking et al., 2003). This may be a theoretical option, but 14-3-3σ is not known to be expressed to the exclusion of members of the 14-3-3 family which actively promote cell proliferation and EMT. The induction of apoptosis by p14ARF via the intrinsic mitochondrial pathway in relation to the operation and effect of p21 and 14-3-3σ was studied by Hemmati et al. (2008) who reported that p14ARF induced apoptosis irrespective of the presence or absence of 14-3-3σ. In other words, it would be useful to have an overall picture of 14-3-3 representation in the target cells before therapeutic manipulation.

The PARP Pathway

Another important pathway leading to apoptosis is by PARP mediation. This could conceivably function either dependently or independently of p53. DNA damage induces PARP activation which then participates in DNA repair. Caspases can break down PARP, prevent DNA repair and induce apoptosis. On the other hand, there is a view that hyperactivation of PARP could activate AIF and apoptosis (Yu et al., 2002), although lately PARP-independent activation of AIF has been advocated (Kondo et al., 2010).

HIC1 is a Downstream Target of p53

HIC1 targets several genes. Of these some may be directly relevant to the biological processes associated with cell proliferation, apoptosis and transformation. Many investigations have focused on the regulation of apoptosis by HIC1. Wales et al. (1995) not only identified HIC1 as a tumour suppressor gene possessing as a consensus p53 binding site in the 5′ flanking region 4 kB upstream of its transcription site of the gene and also that it is activated by wild-type p53. This was a finding of enormous significance since p53 is mutated in a majority of human cancers leading to a loss of its cell cycle regulatory function. But transfection of HIC1 showed that it can suppress growth independently of the wild-type or mutant status of the recipient cells, thus not to the exclusion of other pathways.

HIC1 Can Function Independently of p53

CtBPs have been strongly implicated in promoting EMT and tumour progression, inhibition of tumour suppressor genes and suppression of apoptosis. The two splice variants of CtBP1 and CtBP2 are widely expressed in tumour tissues and cell lines. It has been reported that the downregulation of CtBPs inhibits tumour cell invasion and promotes apoptosis independently of p53. This is possibly due to transcriptional repression of $p21^{waf1}$, a downstream target of p53 by CtBP. CtBP is able to inhibit the transactivation of the p21 promoter by BRCA1. The integrity of BRCA1 C-terminal repeats is essential for BRCA1 to transactivate the p21 promoter, which suggests that these repeats bind to CtBP to regulate the transcription of p21 (Li et al., 1999). PARP1 also influences p21 expression, but this also requires CtBP (Madison and Lundblad, 2010). With these different modes of functional interaction, it would be needless to reiterate the ability of HIC1 to act independently of p53.

27 The DLC Suppressor Genes

The deleted in liver cancer (DLC) gene, so named because of the original finding that it was deleted in liver cancer, has come to be established as a suppressor of many tumour types. DLC1 is one of a cluster of six genes on chromosome 8 deleted in liver cancer and along with some other genes demonstrated to reflect tumour progression and poor prognosis (Roessler et al., 2012). DLC1 is a member of a family comprising three proteins that function as Rho-GAP (GTPase-activating proteins) by virtue of the GAP domain and so actively involved in the regulation of cell proliferation, modulation of cytoskeletal dynamics and cell migration. DLC1 requires Rho-GAP activity to exert its suppressor function, but this is probably insufficient for complete tumour suppression (Sekimata et al., 1999; Wong et al., 2005; Yam et al., 2006). DLC1 interacts with and potentiates the activity of PLCD1 (phospholipase C1δ) (Homma and Emori, 1995) and PLCD1 is itself a tumour suppressor (see pp. 195–197). Deregulated function of the DLC proteins leads to cell transformation with attendant aberrant cell behaviour. Loss of expression occurs by deletion, promoter methylation or mutations (Liao et al., 2008; Yuan et al., 2003a,b). The suppressor function of DLC1 may be abrogated also by PI3K/Akt-mediated phosphorylation (Ko et al., 2010).

DLC1 and DLC2 show significant homology and share the same domain structure, namely a Rho-GAP domain, a SAM (sterile alpha motif domain related to p73/p63/p53 family SAM domain containing proteins) and a lipid-binding START (StAR-related lipid transfer) domain. Whether DLC2 has similar tumour suppressor ability as DLC1 is yet undetermined. Both the genes are suppressed in a number of human tumours. Early work by Ching et al. (2003) showed that DLC2 is downregulated in human HCC, probably at a lower frequency than DLC1 is reported to be. But DLC2 does possess GAP activity specific for small GTPases RhoA and Cdc42. Furthermore, the GAP domain inhibited the formation of actin stress fibres by Rho mediation and suppressed Ras-mediated cell transformation. Holeiter et al. (2008) found that whilst loss of DLC1 enhanced migration, suppression of DLC2 had no effect on cell migration. According to Yau et al. (2009), DLC2 deficiency did not make host mice prone to chemically induced tumorigenesis. Holeiter et al. (2008) believe that the differences in their suppressor function might be related to the subcellular distribution. DLC3 is a third member of the family with tumour suppressor function. It is expressed in normal tissues but markedly downregulated in several forms of cancer (Durkin et al., 2007).

Therapeutic Strategies in Cancer Biology and Pathology. DOI: http://dx.doi.org/10.1016/B978-0-12-416570-0.00027-5

Effects of Re-expression of DLC1

The re-expression of DLC1 alters cell morphology, cell proliferation and migration and apparently also restores the non-metastatic phenotype. With the unequivocal demonstration that DLC genes are silenced by promoter hypermethylation, demethylation has provided a viable means to obtaining re-expression of the genes. Flavones have been tested and shown to restore the expression of DLC1 with attendant inhibitory effects on cell motility (Herzog et al., 2004; Ullmannova and Popescu, 2007). The expression of numerous genes is altered two- to threefold by flavone (Herzog et al., 2004) and it would be an onerous responsibility to attribute a major influential role to DLC1, a proverbial needle in a haystack exercise. Ullmannova and Popescu (2007) again found that several genes were upregulated and DLC1 among them. Flavone enhanced the expression of DLC1 two- to sixfold in breast cancer cell lines and around fivefold in HT-29 colon carcinoma cell lines. In parallel they noticed the induction of apoptosis and cell cycle arrest at G2–M accompanied by an increase in $p21^{waf/cip1}$, which could have arrested the cell cycle. The G2–M arrest could also been an effect of downregulation of Stathmin. Ullmannova and Popescu (2007) found that it was downregulated in response to flavone. Stathmin can sequester p53 which functions as a G2–M checkpoint regulator (Cajone and Sherbet, 1999) and its downregulation would restore the normal p53 checkpoint function at G2–M transition.

Gene or cDNA transfection is another favourite ploy employed in such studies. DLC1 expression is not detected in Raji and Daudi cell lines of Burkitt's lymphoma. Upon re-expression, however, these cells displayed inhibition of growth and cell migration, and induction of apoptosis (Feng et al., 2011b). Transfection of DLC1 cDNA into HCC cells and into metastatic cell lines induced apoptosis, and inhibited growth, invasion and tumorigenesis (Goodison et al., 2005; Zhou et al., 2004). In the study with HCC, restoration of DLC1 reduced tumour size but did not prevent tumour formation. Inexplicably there seems to have been no attempt there to look at pulmonary metastases. In a subsequent study, this group of workers showed that re-expression of DLC1 did indeed reduce pulmonary metastasis *in vivo*, as assayed by an unduly sophisticated method (Goodison et al., 2005). Kim et al. (2007) employed the HCC cell line SNU-368 and transfected them with DLC1. The gene transfer resulted in the inhibition of cell proliferation and increase in sub-G1 population. Forced expression of DLC1 also inhibited cell migration which was apparently linked to the function of membrane adhesive proteins, such as FAK and paxillin.

Apart from this, DLC1 can influence intercellular signalling involving E-cadherin, a well-established suppressor of cell motility and a suppressor of aggressive behaviour of tumours, has been shown to involve small GTPases. E-cadherin is believed to activate them and they then regulate the effects of E-cadherin on intercellular adhesion (Watanabe et al., 2009). Tripathi et al. (2013) have shown recently that DLC1 induces the expression of E-cadherin by virtue of its Rho-GAP function and suppression invasion. E-cadherin is also known to activate other routes in regulating intercellular adhesion. In intestinal epithelial cells, it can downregulate

MEK/ERK signalling via the PI3K/Akt cascade (Laprise et al., 2004). But then MEK/ERK signalling might itself involve the Rho system. Pullikuth and Catlin (2010) showed ERK functions through the agency of p190A Rho-GAP in activating fibronectin-mediated focal adhesion and spreading of cells. The expressions of E-cadherin and fibronectin are closely linked in EMT. For, EMT induced by TGF-β is associated with increased fibronectin expression and loss of E-cadherin (Doherty et al., 2006) and E-cadherin is known to inhibit fibronectin expression (Solanas et al., 2008).

DLC1 induced enhancement of E-cadherin suppresses the migratory ability of cells. The suppressor function of DLCs is due to their Rho-GAP activity negatively regulating RhoA, RhoB, RhoC and to a limited extent cdc42 (Healy et al., 2008; Tripathi et al., 2013). In NSCLC cells, E-cadherin expression has been linked with a reduction in the activity of Rho family proteins, RhoA and Cdc42. Reduction in RhoA required DLC1 and p190 Rho-GAP and was associated with an increase in a membrane-associated p190 Rho-GAP/p120 Ras-GAP complex in which p190 Rho-GAP is activated by binding to p120 Ras-GAP. Extracellular signals seem to elicit responses from Rho GTPases in the form of actin cytoskeletal rearrangements and p190 Rho-GAP regulates Rho which mediates cell adhesive interactions.

Tensin appears to have some bearing on DLC function. As one might recall that tensin binds to actin filaments via a PTB (phosphotyrosine-binding) domain to the cytoplasmic tail of integrins thus establishing a link between integrin receptors and the actin cytoskeleton (Lo, 2004), which might thus influence cell motility. Besides, loss of tensin has been suggested to lead to tumorigenesis (Martuszewska et al., 2009). Both DLC1 and DLC3 bind to human tensin1, the binding being mediated by the SH2 (Src homology 2) and PTB domains of tensin proteins and specific sequences of the DLC protein. DLC1 is said to compete with other cytoskeleton modulators such as integrins for binding to tensin (Qian et al., 2007b). This has raised the possibility that DLC1 might have recourse to a mode of function independent of its Rho function. Equally DLCs might be recruited to membrane receptors in this way to alter cell motility. Stress fibres play an important role in cell adhesion and motility by constant modulation of the actin cytoskeleton and we know that DLC overexpression leads to the dis-assembly of stress fibres (Kim et al., 2007).

DLC Expression in Metastatic Spread

There is much evidence that loss of expression of the DLC genes whether experimentally contrived or occurring naturally does affect cell migration and metastases *in vivo* tumour models. However we have only a limited amount of information about how this expression relates to the aggressive behaviour of human tumours and whether the degree of expression of either DLC gene might offer information about the potential for metastasis. Ullmannova and Popescu (2006) investigated the expression of DLC genes in many human tumours and noticed reduced expression in between 60% and 90% of tumour specimens. They were underexpressed in samples of primary ductal breast cancers from patients with regional lymph node metastases.

However, no samples from patients with localised cancers seem to have been tested. Nor were the clinical stage or histological grade reckoned in the inverse correlation. Guan et al. (2006) have published some interesting data on DLC1 downregulation and promoter methylation encountered in prostate cancer. Benign hyperplasia of the prostate showed rather high frequency of methylation. Without reading too much into it, one might suggest that perhaps the downregulation of DLC1 in hyperplasia could presage progression to carcinoma. Perhaps the progression of colon cancer might be a suitable model for investigating if indeed DLC1 expression status reflects progression from early and incipient stages of neoplasia.

28 The LKB1 (STK11) Suppressor Gene

The LKB1 gene (STK11) located at chromosome 19p13.3 has been focused upon vigorously in recent years and some have described it to be a novel suppressor gene. It was identified and recognised for its tumour suppressor properties many years ago. LOH and both germ line and somatic mutations of the gene were linked with the Peutz–Jeghers syndrome (PJS) and enhanced risk of developing cancers associated with the syndrome. Germ line mutation of the suppressor gene PTEN has also been associated with PJS. The incidence of mutations of LKB1 in sporadic tumour is not remarkable.

LKB1 is a serine/threonine kinase that is known directly to target AMPKs (AMP-activated protein kinase). Several AMPKs have been identified of which the majority are activated by LKB1 by phosphorylation of a conserved threonine residue in their T-loop region (Bright et al., 2009). AMPK functions as an energy sensing device and maintains energy balance by lipid and glucose metabolism, by integrating the generation of ATP and inhibiting cellular processes that consume ATP (Hardie et al., 2012; Wang et al., 2012b). The rise of LKB1 to prominence has flowed from its association with cell cycle regulation, microtubule dynamics, cell polarity and differentiation. Not unexpectedly LKB1 regulates metabolic functions as well. Mutations of the gene suppress or reduce the kinase activity.

The enzymatic activity and subcellular localisation of LKB1 requires it to form complexes with STRAD (STE20-related kinase adaptor) and MO25 (mouse protein 25, the scaffolding calcium-binding protein also known as CAB39) (Boudeau et al., 2004; Hawley et al., 2003) and LKB1 is constitutively activated in complex with these proteins. Complex formation with STRAD is essential for the G1 suppressor function of LKB1 (Baas et al., 2003) and polarisation of cells. Specific LOH of STRAD has been encountered, but no somatic mutations in a small number of gastric adenocarcinomas (De Leng et al., 2005). It would seem therefore that inactivation of the gene might not be a solitary requisite in the pathogenesis of gastric carcinomas, but its importance might be linked with activation of LKB1. This might also partly account for the not so striking a correlation between loss of LKB1 and tumorigenesis.

Expression of LKB1 and Tumorigenesis

Tumorigenesis is likely to arise from mutations, both germ line and somatic, of the LKB1 gene. Genetic alterations as homozygous deletions or LOH are frequently

Therapeutic Strategies in Cancer Biology and Pathology. DOI: http://dx.doi.org/10.1016/B978-0-12-416570-0.00028-7

associated with tumorigenesis, so also alterations in gene expression consequent upon promoter methylation. As mentioned before, the incidence of mutations of LKB1 in sporadic tumour is not remarkable. Somatic mutations were recorded in around a third of lung adenocarcinomas (Ji et al., 2007; Weir et al., 2007), but Suzuki et al. (2012) found mutations in only 14% of samples. A small proportion of NSCLC carry altered LKB1 (Strazisar et al., 2009), but interestingly they add the rider that changes occurred mainly in advanced stage tumours. Okuda et al. (2011) found LKB1 mutated only in 5/174 patients. Mutations have been reported in four of seven sporadic non-PJS mucinous bronchioalveolar carcinomas. Allelic loss and mutations occurred in 70% of cases, but percentages do not reflect much when the patient numbers are small (Osoegawa et al., 2011). Mutations were even less frequent in cervical cancer (Forbes et al., 2011; Wingo et al., 2009) and melanoma (Forbes et al., 2011; Guldberg et al., 1999; Rowan et al., 1999). A recent study of acinar cell carcinoma of the pancreas has revealed no mutations, deletions or promoter hypermethylation of LKB1, nor alterations in its expression (De Wilde et al., 2011), but a caveat to be attached to this finding is that only five sporadic carcinomas were studied here.

A familiar approach to the loss of expression by methylation of LKB1 as a cause of tumorigenesis has led to early reports of hypermethylation of the gene in many forms of human cancer. Some early investigations reported hypermethylation of the gene in a small number of tumour cell lines and also methylation was not encountered too frequently in tumour tissues. It ought to be mentioned that small numbers of tumour samples were used in this study. Nonetheless, hypermethylation also occurred in tumours from PJS patients (Esteller et al., 2000). Loss of expression of LKB1 is not remarkable in invasive ductal carcinoma of the breast. However, a proportion of *in situ* and invasive carcinomas with loss of LKB1 expression have turned out to be high grade tumours; also in some of the invasive tumours LKB1 was methylated (Fenton et al., 2006). No significant methylation of the promoter was found in sporadic colon adenocarcinomas, but LOH showed some association with advanced stage tumours (Trojan et al., 2000). Overall, the evidence is not overwhelming that promoter methylation is a mechanism for the silencing of LKB1 in tumours. However, inactivation of LKB1 by homozygous deletion or LOH at the LKB1 locus has been noted in a vast majority of NSCLC (Gill et al., 2011).

Other molecular and mechanistic modes than heritable and somatic mutations and epigenetic modifications of modulation of LKB1 can be envisaged. Regulation of transcription of the gene is an alternative mode of modulating gene expression. LKB1 possesses four cis-acting elements in its promoter, which are able to bind to transcription factors such as Sp1, NFY (nuclear transcription factor Y) and FOXO3 (forkhead box O) and FOXO4. LKB1 promoter activity is greatly enhanced when these transcription factors are overexpressed (Luetzner et al., 2012).

NF-κB and Sp1 activation seems to be closely integrated. Sp1 can activate or repress genetic transcription dependent upon the signalling ligand. It is involved in biological processes of cell differentiation, proliferation and apoptosis and immune responses. Sp1 regulates both RelA (NF-κB p65 subunit) and p50 (NF-κB1) subunits of NF-κB. RelA binds to the zinc finger region of Sp1 in a specific interaction. Significantly, there are indications that NF-κB can induces Sp1 and they co-operate to promote cell proliferation and survival. Sp1 is highly expressed in many tumours and has been linked with the regulation of cell-cycle-related genes, growth factor signalling, apoptosis and it is also said to regulate tumour suppressor genes (Culver et al., 2011; Li and Davie, 2010; Sankpal et al., 2011). Sp1 is markedly associated with induction of cell proliferation and inhibition of apoptosis. Many cell cycle and apoptosis-related genes such as p53, cyclins and IκB are degraded by the ubiquitin–proteasome pathway. NFY induces proteasome gene expression and knockdown of the same inhibits the proteasome pathway and proteasome genes (Xu et al., 2012).

The FOXO family transcription factors regulate many biological processes such as apoptosis and cell cycle arrest. They inhibit cell proliferation and promote apoptosis via activation of pro-apoptosis genes of the Bcl family or by enhancing the expression of FasL death receptor ligands and TRAIL or by inducing cell cycle arrest. In this way FOXO transcription factors function as tumour suppressors. In fact, loss of FOXO function with the accompaniment of increased cell survival has been reported in many human cancers.

Akt kinase inhibits FOXO and promotes cell survival. FOXO1 is downregulated in Hodgkin's (but not in non-Hodgkin's lymphoma) (Xie et al., 2012). FOXO3 is inactivated in mantle cell lymphoma cells (Obrador-Hevia et al., 2012). It is worth noting that the VEGF gene is a target of and repressed by FOXO3a. VEGF and FOXO3a show inverse correlation in breast cancer tissues (Karadedou et al., 2012). Loss of expression would be conducive to the progression of aggressive disease. But the function of FOXO3 has been controversial. For, it has been attributed with promotion of invasion. Furthermore FOXO3a might activate the NF-κB survival pathway (Li et al., 2012e). Largely this is incongruous in some ways, for LKB1 transcription is enhanced by these transcription factors. As noted, many genes are targeted by these transcription factors and one has to guard against overinterpretation of their part in LKB1 expression. Equally LKB1 might activate Akt signalling and promote cell survival. Akt phosphorylated and repressed FOXO3a in LKB1 wild-type but not null cells. Also depletion of LKB1 in wild-type cells reduced FOXO3 phosphorylation. This suggests LKB1-activated Akt led to suppression of FOXO3a and cell survival (Zhong et al., 2008). Nevertheless, as discussed below, expression of LKB1 does indeed suppress angiogenesis as well as invasion.

LKB1 Suppresses Invasion and Metastasis

Genetic alterations as homozygous deletions or LOH and loss of expression by epigenetic mechanisms of LKB1 are not infrequently reported in many tumours. In fact, loss of LKB1 expression might be conducive to invasion and metastasis. LKB1 (−/+) female mice develop endometrial adenocarcinomas highly invasive into the myometrium. LKB1 (L/L homozygous floxed allele) mice were injected intrauterine with Ad-Cre recombinase to excise the floxed alleles to achieve endometrium-specific deletion of LKB1. This produced highly invasive endometrial adenocarcinomas. In human endometrial cancers, loss of LKB1 correlated with tumour progression more aggressive disease (Contreras et al., 2008).

Transfection of LKB1 into MDA-MB-435 breast cancer cells lacking the gene inhibited invasion *in vitro* and *in vivo* led to inhibition of tumour growth and reduced tumour associated microvessel density. Suppression of angiogenesis seemed to result from suppression of angiogenic agents such as MMPs, VEGF and bFGF (Zhuang et al., 2006). In contrast, LKB1 might activate TGF-β signalling, and in the absence of LKB1 TGF-β signalling is suppressed. This implicates LKB1 as an important element in TGF-β induced angiogenesis (Londesborough et al., 2008). There is an implied suggestion here that the angiogenic effects of LKB1 can be uncoupled from its ability to suppress cell proliferation. Londesborough et al. (2008) suggest that changes in cell proliferation or apoptosis do not have a bearing on the effects of loss of LKB1 on angiogenesis. It would not be out of place to state here that TGF-β can signal by Smad-dependent as well as independent non-canonical pathways, which might generate different outcomes in terms of cell proliferation, apoptosis and angiogenesis (Sherbet, 2011a).

Signalling Systems in Cross Talk with LKB1

In common with several tumour promoter and suppressor genes discussed here, LKB1 seems to collaborate with and its expression modulated by interacting pathways. LKB1 might also be regulated in its function by oncogenes or signalling systems by modulating those that are operational in tumour development, invasion and spread. Looking directly at LKB1 as suppressor of cell proliferation, the Ras signalling system comes readily to mind. Activated Ras transduces its signals via three distinct routs, namely the ERK1/2 pathway, the PI3K pathway and the Ral/GDS system. The PI3K/Akt pathway signals can be carried along several sub-routes. Akt targets many genes, among them are the apoptosis inducing BAD gene, caspase 9 and FasL as well as NF-κB (Figure 28.1). Of interest also is the activation of mTOR signalling with its implications for EMT (see Figures 3.1 and P2.1).

Of particular interest is the demonstration by Liu et al. (2012e) that loss of LKB1 expression and K-Ras activation resulted in the formation of melanomas. Furthermore somatic deletion of p53 and K-Ras activation also increased tumorigenesis. Also inactivation of LKB1 led to the expansion of subpopulation of tumour cells such as CD24+ cells that could be representing CSCs and participating in metastatic spread.

Little collaborative signalling is evident in adenocarcinoma of the lung. Not only was the incidence of genetic changes low, even in those cases where mutations did occur, simultaneous changes of LKB1 or with EGFR were noted only in a small number of patients (Suzuki et al., 2012). Okuda et al. (2011) found no concomitant mutation of LKB1 with Ras or EGFR in 5 that carried LKB1 mutation of 174 cases examined. LKB1 is reported to be mutated simultaneously with K-Ras, EGFR and p53 in non-PJS mucinous bronchioloalveolar carcinomas; with K-Ras in two tumours, with EGFR in two and with p53 in one tumour (Osoegawa et al., 2011). LKB1 haploinsufficiency in conjunction with K-Ras has been associated with murine and human pancreatic ductal carcinoma. Depressed levels of LKB1

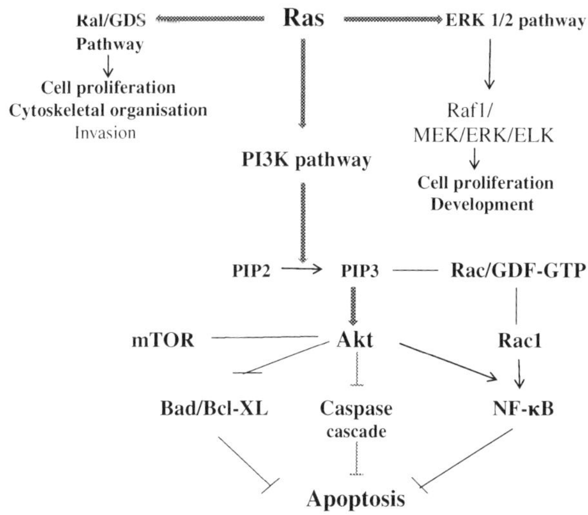

Figure 28.1 The three major pathways of Ras signalling, with particular focus on Akt signalling affecting apoptosis and cell population size. Akt inhibits FasL-mediated apoptosis. This is not shown here. The ERK and Ral/GDS (guanine nucleotide dissociation stimulator) pathways transduce Ras signal to promote cell proliferation, modulate cytoskeletal organisation and induce cell motility and invasion. Signalling along these two pathways can synergise in their phenotypic effect and lead to the generation of an aggressive phenotype.

expression in human tumours correlated with reduced levels of p21 together with loss of growth arrest (Morton et al., 2010). Overall these findings are underwhelming and unremarkable *per se*. The association of EGFR on LKB1 is rather tenuous but deserves investigation in other tumours, as also the putative links with Ras and p53. One would do well to recall there that EGFR expression might be co-ordinately regulated by suppressor genes such as nm23, and conversely nm23 can negatively regulate EGF/EGFR and Ras-mediated activation of the ERK pathway and cellular proliferation.

LKB1 in Cross Talk with Oestrogens and ER/HER2

Aside from its conventional roles of regulating the cell cycle and cell proliferation and apoptosis, p53 seems to be associated with the regulation of lipid metabolism (Goldstein and Rotter, 2012). With the pervasive infiltration of conventional signalling systems operating growth signal transduction into the LKB1 suppressor function and given that LKB1 also regulate glucose and lipid metabolism and further that p53 is a target of ERα, it would be appropriate to look at ER/HER2 in the backdrop of the LKB1 scenario, with its inherent contradictions and varied nuances.

ERα as a tumour promoter has marked influence on the expression of other suppressor genes. It downregulates NDRG1 and their expressions are inversely correlated in breast cancer samples. ERα seems able to inhibit LKB1 in breast cancer cells. Oestradiol inhibits LKB1 expression, and siRNA-mediated suppression of ERα upregulates LKB1 transcription. ER binds to the LKB1 promoter on which several oestrogen response elements have been detected (Brown et al., 2011; Linher-Melville et al., 2012). So one would expect ERα would positively correlate with EGFR. Nonetheless any postulated relationship between LKB1 and EGFR is indeed not robust. The lack of strong evidence of concomitant genetic changes in LKB1 and EGFR may also be interpreted in this way and suggest they may be suppressed as parallel events but independently of each other. One has to exercise some caution in interpreting the effects involving ERα or indeed ERα and ERβ.

In murine and human adipocytes, 17β-oestradiol/ERα upregulated LKB1 transcription and phosphorylation of AMPK, but androgens produced an opposite effect (McInnes et al., 2012). However, so far as breast cancer cell lines are concerned, oestradiol reduces the expression of LKB1. 5-Aminoimidazole-4-carboxamide 1-D-ribonucleoside (AICAR)-mediated activation of AMPK exerts anti-proliferative effects on breast cancer cell lines MCF-7, MDA-MB-231 and T47D, but the degree of responses differed considerably. T47D (ER+/mutant p53+) were most sensitive to the effects of AICAR and MCF-7 (ER+/wild-type p53+) and MDA-MB-231 (ER−/mutant p53+) are the least sensitive. Of these T47D showed cell cycle arrest but other two cell lines showed induction of apoptosis. As discussed previously, AMPK activation has been attributed to the activation of p53 and p21, but here in MDA-MB-231, the apoptotic effect seems to occur independently of p53. In MCF-7 cells apoptosis is not known to be dependent upon p53 and probably also the case upon AICAR treatment. But a comparison between T47D and MDA-MB-231 would suggest ER expression might be relevant to the inhibition of cell growth. Phenformin (a biguanide related to metformin) also exerted similar effects with the same sensitivity pattern (El-Masry et al., 2012). However, the situation is more complex than one can imagine. For instance, metformin suppresses HER2 (Vazques-Martin et al., 2012). MDA-MB-231 cells do indeed overexpress HER2, so the effects noticed might have occurred independently of ER expression.

Nath-Sain and Marignani (2009) showed previously that binding occurred between LKB1 and ERα in the nucleus and suggested that LKB1 is thus targeted to ER-responsive genes. More debatably, LKB1 transactivated ERα, which is incongruous with the propensity of tumour promotion with ERα. But in the event, it is possible to take the view that ERα and LKB1 could form a mutually regulatory loop. A potential outcome could be alteration of the balance of the tumour-promoting ERα and inhibitory ERβ. It has become apparent from studies concerning the suppressor effects of nm23-H1 and nm23-H2 isoforms that ERα and ERβ can outweigh each other's effects on promotion of cell growth and migration. The phenotypic outcome can be a consequence of the function of the equilibrium between ERα and ERβ being tilted in favour of either receptor.

It is needless to say that this would remain in the realm of speculation until we know if LKB1 exerts any effects on ERβ. Nonetheless, one would not fail to notice that there is a fundamental contradiction here and therefore any suggestions of a therapeutic spin off would be premature. Oestrogens function via ERα to induce proliferation. But then many genes, besides ERα, that influence tumour growth and progression would need to be reckoned with. Among them are S100A4, 14-3-3σ and 14-3-3ζ genes. MiRNAs such as miRNA-451 can influence the expression of 14-3-3ζ as well as EGFR and HER2. There is evidence that EGFR might be involved with Hippo signalling to modulate cell proliferation (see references provided in relevant sections).

At the experimental level, there is much credible information about the involvement of p53 in LKB1 function. That LKB1 induced the expression of p21 in a p53-dependent fashion and led to growth arrest was demonstrated some years ago (Tiainen et al., 2002). LKB1 has been shown to be functionally linked to the p53/p21 axis and indeed shown to activate the same. LKB1 is said to interact with p53 and possibly phosphorylate it. LKB1 is targeted to the p21/WAF1 promoter by the mediation of p53 (Zeng and Berger, 2006). Yao et al. (2008) have identified a p53 and Sp1 binding site in the LKB1 promoter, which seems to be essential for the function of p53 as a transcription factor. Inhibition of endogenous LKB1 promotes G1–S transition together with increase in Rb phosphorylation and this was accompanied by reduction in p53 and p16 expression (Liang et al., 2010a). Induction of p21 by LKB1 probably can occur also independently of p53 (Setogawa et al., 2006).

LKB1 Suppresses EMT

The significance of the loss of LKB1 in tumorigenesis is dramatically underscored by the putative attribution to it of the ability to inactivate EMT. Suppression of LKB1 has had marked effects of enhanced cell motility in lung adenocarcinoma cells and this was accompanied by activation of EMT by ZEB1 mediation (Roy et al., 2010). Using murine bladder epithelia as a model, Shorning et al. (2011) have demonstrated that LKB1 with PTEN, but not individually, suppressed EMT. This seemed to involve suppression of mTOR signalling. When LKB1 and PTEN were deleted at the same time, mTOR signalling was upregulated and eventually led to the activation of EMT and tumorigenesis. As discussed at another location, PI3K/Akt signalling can and does function through mTOR/Rac and influence both cell proliferation and migration. However it is unclear how a link between LKB1 and mTOR can be causally interpreted. Several suppressor proteins are able to suppress mTOR signalling, KAI and 14-3-3 proteins, for instance. Akt suppresses the function of LKB1 by 14-3-3ζ mediation. Akt phosphorylates LKB1 at a specific serine residue and promotes 14-3-3ζ binding to the suppressor interfering in this way with its binding to STRAD and blocking its suppressor function. Inhibition of 4-3-3ζ using specific siRNAs promoted translocation of LKB1 from the nucleus to the cytoplasm (Liu et al., 2012b). On the other hand, it has been shown that 14-3-3ζ prevents LKB1 from phosphorylating its AMPK substrate and this effectively inhibits the suppressor effects of LKB1 (Bai et al., 2012).

MiRNAs might be another candidate that have to be reckoned with. Many miRNAs, Let-7d, miRNA-34c, miRNA-194 and miRNA-200c, inhibit EMT (see pp. 18–21). Therefore is it is of some interest to note here that activation of EMT by

LKB1 suppression also suppressed miRNA-200a/c expression (Roy et al., 2010) and inversely correlated with that of ZEB1 in cell clones with suppressed LKB1.

Another path to suppression of EMT by PTEN is via upregulation of PTEN leading to the upregulation NDRG1 (see pp. 111–112); the latter would eventually inhibit EMT. This builds up a complex scenario of LKB1 interaction with PTEN which is known to be an LKB1 interacting protein and a substrate of the kinase (Mehenni et al., 2005).

LKB1 and Stem Cell Survival and Pluripotency

The suppression of EMT by LKB1 has inevitably led to investigations of its effects on formation and maintenance of CSCs subpopulations in tumours and acquisition of EMT. Indeed, apart from the conventional features associated with tumour suppressors, LKB1 has to be accorded with prominence because there are indications of its potential involvement in stem cell perpetuation. It may be recalled here that cell clones carrying CD44+CD24+ESA+ markers are regarded as putative stem and these clones might be expanded and the tumour enriched in respect of this subpopulation that is capable of not only initiating the tumour development and possibly also related to progression of tumours to an aggressive state (see pp. 35, 214–215). Liu et al. (2012e) found that forced inactivation of LKB1 activity in the presence of activated oncogenic Ras, probably also in the backdrop of loss of p53 led to the expansion of CD24+ cell population which could represent CSCs. Activation of AMPK, a direct target of LKB1, can prevent the transcriptional activation of Oct4 (Vazquez-Martin et al., 2012). Metformin, also an AMPK activator, inhibited Oct4 expression. Although reduced Oct4 transcription would suggest AMPK-initiated regulation of stem cell programming, no stem cell marker expression was assessed here.

The opposite view has been expounded by other reports. Reduced LKB1 expression has been correlated with upregulated expression of genes associated with pluripotency in human embryonic stem cells. The transcription factors Oct4, Nanog and Sox2 are said to be expressed at higher levels in the absence of LKB1 expression and this is accompanied by the downregulation of differentiation markers, for example Runx1, AFP, GATA, Brachyury, Sox17 and Nestin (Lai et al., 2012). It would be interesting to see if miRNAs operate in this. Let-7miRNAs suppress stem cell pluripotency and oppose the effects of the miRNA-290 cluster that promote pluripotency. This functional disparity is due to the ability of miRNA-290 to maintain the expression of Lin28 which inhibits the maturation of Let-7. The pluripotency factors Oct4, Sox2, REX1 and Nanog induce miRNAs that promote pluripotency and cell survival, whilst Let-7s inhibit pluripotency (Li and He, 2012). Notwithstanding the disagreements, it is quite obvious from the limited resources at hand that deregulation of LKB1 signalling can lead to defects in differentiation, stem cell perpetuation and activation of EMT encountered in tumorigenesis, both PJS-associated and not associated with PJS. The extensive interactions of LKB1 with proliferation, cell adhesion linked features of motility and invasion-related signalling pathways together with cross talk with those operating in differentiating system lend themselves to positive interpretation of LKB1 in stem cell programming and perpetuation (Udd and Makela, 2011). Given that EMT activation, cell proliferation, apoptosis and other cell features such as cell polarity, cell adhesion and motility involve pathways with which LKB1 can and does cross talk, the apparent contradictions could arise from the different signalling pathways with which LKB1 might be operating. One has to differentiate stem cells into three types, namely embryonic stem cell, adult reprogrammed stem cells and CSCs. Due to epigenetic relationships, CSCs might resemble embryonic stem cells. CSCs share stem cell features with adult stem cells but differ with regard to homoeostatic regulation, survival and perpetuation. These differences would indeed flow from the different pathways operating in their generation and are bound to impinge upon the function of modulators of stem cell biology, such as LKB1 among others.

LKB1, Cytoskeletal Dynamics and Cell Motility and Invasion

Among the features that are most conducive cancer cell migration, invasion and metastasis is the deregulation of cytoskeletal dynamics. Some tumour promoter genes are known to destabilise microtubule dynamics, whilst suppressor genes promote microtubule stability. In fact it has been argued that destabilising cytoskeletal dynamics might enhance cell invasion leading to metastatic spread. Contrary to conventional wisdom, LKB1 phosphorylates and activates the AMPK-related MARK2 (microtubule affinity regulating kinase) and leads to Tau protein phosphorylation resulting in the suppression of tubulin polymerisation (Kojima et al., 2007). This has been confirmed in differentiating systems where LKB1 destabilises microtubule dynamics (Mian et al., 2012).

Equally, epithelial structural integrity is based on adherens junctions involving cadherins and catenin and desmosomes. Desmosomes are membrane plaques which form links via connecting filaments and provide adhesion between cells and in turn desmosomes are linked with intermediate filaments of the cell cytoskeletons of adjacent cells. It is of interest therefore that loss of LKB1 results in the deterioration of desmosome-mediated intercellular adhesion and loss of structural integrity of epithelia (Partanen et al., 2012). It might be recalled that the development of carcinomas is characterised by a lack of desmosomes. Partanen et al. (2012) have gone on to show that loss of LKB1 *per se* may not lead to tumorigenesis but tumours do develop in the presence of oncogenes. Generally consistent with this the introduction of LKB1 into LKB1-deficient HeLa cells has resulted in the activation of Rho GTPase and promotion of actin filament assembly with focal adhesion (Xu et al., 2010).

Therapeutic Assessment of LKB1

The loss of LKB1 in many tumour types and the effects of it loss on invasion has instilled investigations into potential benefits of reactivating its expression as a modality to control tumour growth and progression. Liposome-mediated transfection of LKB1 (Lps-LKB1) into NSCLC has provided some encouraging results in that intratumoral injection of Lps-LKB1 together with low-dose cisplatin inhibited the growth of A549 tumour xenografts. Intravenous injection of Lps-LKB1 plus cisplatin also reduced pre-existing metastatic lung nodules. There was also increase in survival of animals that received Lps-LKB1 plus cisplatin (Ou et al., 2012).

It might be possible to take advantage of the ability of LKB1 to suppress pluripotency. Exploration of reactivation of LKB1 with miRNA such as Let-7 that can suppress stem cell pluripotency is eminently worthy of consideration, especially in the light of the expression of Lin28 a suppressor of Let-7, correlating strongly with aggressive behaviour of tumours. Suppression of miRNA-199a-3p is said to enhance LKB1 expression. Also activation of FXR (Farnesoid X receptor) reduces levels of this miRNA in HepG2 cells (Lee et al., 2012). The non-steroidal synthetic FXR ligand GW4064 (3-[2-[2-chloro-4-[[3-(2,6-dichlorophenyl)-5-(1-methylethyl)-4-isoxazolyl]methoxy]phenyl]ethenyl]benzoic acid) might provide a useful tool for preliminary laboratory investigations. GW4064 markedly reduces cell viability as tested on human HCC cell lines. It also inhibited cell viability synergistically with acyclic retinoid (ACR), a synthetic retinoid targeting RXRα. The effect was by the induction of apoptosis, and again there was synergy between the two agents. This combination also induced G_0/G_1 arrest (Ohno et al., 2012). Interestingly, FXR and its agonists might exert additive effect by other means too. GW4064 can elevate plasma glucocorticoid levels (Hoekstra et al., 2012), possibly a beneficial side effect in the regulation of tumour growth. Also FXR is able to upregulate the expression of p62, which induces autophagy and activate NF-κB towards cell survival and balance this with induction of apoptosis by activation of caspase-8 (Williams et al., 2012). To contrast with its effects on miRNA-199a-3p, GW4064 upregulates miRNA-29a in quiescent hepatic stellate cells and notably miRNA-29a downregulates the expression of ECM proteins (Li et al., 2011a).

On the down side, some problems have been identified which might restrict the use of this agonist. GW4064 has poor water solubility making it difficult to test bioavailability. A vehicle containing Tween80, HPMC (hydroxypropyl methylcellulose), Cremophor EL (a formulation vehicle used for drugs with poor solubility), polyethylene glycol 400/ethyl alcohol and Capmul MCM (a co-surfactant), was formulated to overcome the problem of limited solubility (Chiang et al., 2011). Nevertheless, assessing the efficacy of GW4064 in conjunction with the other agents that can amplify efficacy might be rewarding. Also many novel and improved analogues have been synthesised.

A possible novel approach would be to take advantage of the several interacting and cross talking signalling pathways and press this advantage home by engaging them into upregulation of LKB1 and reinstating its suppressor function. As discussed earlier, oestradiol reduces LKB1 promoter activity (Brown et al., 2011). Downregulation of ER/PR and deployment of anti-oestrogens might possibly be of benefit in the restoration of LKB1 expression where transcription is repressed.

AMPK as a Therapeutic Target

LKB1 directly targets AMPKs and is associated with cell cycle regulation, microtubule dynamics, cell polarity and differentiation, and not unexpectedly attributed with activation of AMPKs and regulation of metabolic functions. Given the tumour suppressor function of LKB1, AMPK provides a metabolic link with pathogenesis of cancer besides acting as an active nexus with signalling systems related to apoptosis and cell cycle regulator genes. Upon activation by LKB1, AMPKs liaise with many tumour suppressor genes and interact with several signalling systems such as p53, BRCA, Wnt/β-catenin and mTOR, among others. Therefore AMPKs have been viewed as a legitimate therapeutic target and have been the focus of much attention and investigation. Many agents that activate AMPKs have been identified and some studied in great detail. Several erudite reviews have been written on the subject, but a restricted purview is provided here.

AMPK functions through the FOXO family transcription factors which participate in the regulation of cell proliferation, apoptosis. FOXO transcription factor isoforms FOXO1, FOXO3 and FOXO4 are phosphorylated by PI3K/Akt and sequestered to the cytoplasm and prevented from inducing cell cycle arrest and apoptosis. FOXO-dependent induction of cell cycle arrest and apoptosis are suppressed in response to growth factors. AMPK also phosphorylates FOXO but at regulatory sites different from those phosphorylated by Akt. FOXO activation by AMPK leads to cell cycle arrest and apoptosis (Salih and Brunet, 2008). They are functionally downregulated in many forms of cancer and might thus lead to cell survival and tumorigenesis. Some isoforms may be associated with angiogenesis, so by implication in tumour progression and prognosis. An overall assessment would be that FOXOs are tumour suppressors, but the inevitable diversity of the transcription function of individual FOXOs is reflected in their being ascribed a role in the maintenance of CSCs such as leukaemic stem cells.

AICAR (5-aminoimidazole-4-carboxamide 1-D-ribonucleoside)

The AICAR is an AMPK activator. Inhibition of AMPK has led to the downregulation of p53 and p21 and attendant promotion of cell proliferation. Loss of AMPK function also leads to enhanced cell motility, whereas in contrast AMPK activation by AICAR produced the opposite effect (Zhou et al., 2009). Oestrogen/ER and p53 background signalling might be relevant in AICAR-mediated activation of AMPK (El-Masry et al., 2012). Activation of p53 by AICAR seems to engage the ATM and mTOR pathways. Inhibitors of ATM kinase suppress the activation and so do also mTOR inhibitors (Zajkowicz and Rusin, 2011). Obviously ATM pathway is operational in this experimental system. As discussed already, the ATM pathway regulates many signalling systems that culminate in DNA repair, apoptosis or cell survival. It involves the downstream checkpoint kinases Chk1 and Chk2, and p53 and BRCA1 as well as c-Abl and NF-κB. The checkpoint kinases in turn target cdc25/cdc2 (Figure 14.2). ATM can directly activate p53 and via cdk1 and lead to G2/M arrest. ATM can also use the p53/Chk2/cdc5 to produce the same effect. In other words, cell cycle arrest can occur totally independently of AMPK.

AICAR and phenformin can both activate AMPK and lead to growth inhibition when Raf/MEK/ERK signalling operates in the background (Petti et al., 2012). Contraindicated is that AMPK activation might be inhibited by MAPK, but this has given rise to the possibility of more effective AICAR mediation of growth inhibition when combined with MAPK inhibition. A useful combination especially in the management of TNBCs might be of metformin to inhibit HER2 and activate AMPKs. AICAR is said to be less efficient than metformin in inhibiting HER2 (Vazquez-Martin et al., 2012). HER2 has an inhibitory effect on AMPK and inhibitors of HER2/EGFR have been found able to activate AMPK as Spector et al. (2007) showed some time ago. So the inhibitory effects of AICAR and metformin on cell proliferation might be further contributed to and compounded by HER2 inhibition and the resultant restraint of its inhibition of AMPK. Resveratrol is another possibility. It has been found to activate AMPK irrespective of ER status (Lin et al., 2010). AICAR has been shown to inhibit the mTOR pathway (Saha et al., 2011) and so mTOR inhibitors might be another valid option for exploration.

Metformin Activates AMPK

Metformin is a biguanide which activates AMPKs and activation of the latter might involve binding of metformin to the γ-subunit of AMPK (Zhang et al., 2012e). Metformin is said to increase AMP/ATP ratio and this activates AMPK.

Metformin inhibits cell proliferation and exerts marked anti-inflammatory effects. It has also been found to be an effective anti-cancer agent. The pathways of metformin function are being elucidated, but there are clear indications that it can arrest cell cycle progression and induce apoptosis. A major effector is AMPK, but possibly these effects might occur independently of AMPK. Brown et al. (2010) reported that not only this activation but also found that LKB1 promoter activity is enhanced, which in turn might potentiate AMPK activation. The inhibitory effect does seem to be dependent upon LKB1.

AMPK activation has been achieved by gene transfer in cells not expressing LKB1 and there is some agreement that AMPK might function by the inhibition of mTOR signalling and might also downregulate the ERK pathway (Klubo-Gwiezdzinska et al., 2012; Xiao et al., 2012). Metformin activates AMPK and inhibits mTOR signalling but it can also directly target mTOR quite independently of AMPK (see Dowling et al., 2011). It activated AMPK and inhibited mTOR independently of Akt in lymphoma tissues from a large number of patients with B- and T-cell lymphoma. Inhibition of AMPK resulted in upregulation of mTOR and its downstream targets. Proliferation of established lymphoma cell lines was markedly inhibited by metformin. The growth of Daudi and Jurkat tumour xenografts was also blocked by intraperitoneal administration of metformin. A heightened growth inhibitory response was seen when administered with doxorubicin and mTOR inhibitor. The combination treatment was more effective in inducing autophagy than when administered alone (Shi et al., 2012). In ovarian cancer cells, Akt inhibition does compound the inhibitory effects of metformin (Li et al., 2012b). Paclitaxel is a mitotic inhibitor capable of promoting tubulin polymerisation to microtubules and stabilising them. It blocks cell cycle progression in G2–M. It is an approved therapeutic agent in the management of NSCLC, ovarian and breast cancer, and Kaposi sarcoma. Upon combination with metformin, paclitaxel seems to add to the loss of cells by apoptosis induced by metformin via AMPK activation and suppression of mTOR survival pathway with cell cycle arrest at G2–M thus enhancing the overall growth inhibition (Rocha et al., 2011). But Rocha et al. (2011) state that paclitaxel on its own could activate AMPK and inhibit mTOR and its downstream targets. It may be that there is no indisputably synergistic effect here, but one cannot exclude the possibility of AMPK inducing acetylation and stabilisation of p53 which would induce apoptosis. AMPK also inactivates the p53 deacetylase, SIRT1 (Lee et al., 2012), thus consolidating the acetylation, stability and activation of p53. One should also mention here that the mTOR suppressor temsirolimus and the microtubule destabiliser vinblastine generate marked anti-tumour effects with attendant suppression of anti-apoptosis and pro-survival proteins survivin, Bcl-2, and Mcl-1 (Zhou et al., 2012b).

Metformin inhibited insulin-induced proliferation of ovarian theca-interstitial cells with the upregulation of the cell cycle regulatory proteins, cyclin D3 and cdk4. In this system, metformin is seen to function via activation of AMPK and inhibiting ERK1/2 activation and the effects were blocked by the inhibition of AMPK by compound C ((6-[4-(2-piperidin-1-ylethoxy)phenyl]-3-pyridin-4-ylpyrazolo[1,5-a] pyrimidine) (Will et al., 2012).

The apoptotic pathway has also been invoked in relation to LKB1/AMPK repressor function. AMPK could be inducing apoptosis by the mitochondrial caspase cascade and AIF. Also postulated is the participation of PI3K/Akt pathway and MAPK/38-ERK signalling (Kim and Choi, 2012; Liu et al., 2012a). As shown in Figure 28.2, AMPK might also be able to induce apoptosis by inhibiting mTOR signalling and also NF-κB. Independently, AMPK can induce cell cycle arrest via upregulation of cyclins and cdks. Inhibition of NF-κB survival pathway has also not been excluded. Metformin suppresses its activation, and inhibition of both AMPK and PTEN is able to restore NF-κB. Thus metformin might be a highly effective growth suppressor using the vast versatility of AMPK functions drawing together many signalling systems to suppress cell survival and induce apoptosis.

It is of much interest to note that metformin inhibits cell proliferation irrespective of steroid or growths factor receptor status. It is inhibitory of ERα+, HER2+ as well as TNBC cells (Liu et al., 2009, 2012a). Liu et al. (2009) have described TNBC-specific inhibition of cell proliferation partly by cell cycle arrest and induction of apoptosis. In parallel they could see the induction of phosphorylated AMPK and reduced levels and activation of EGFR and also suppression of MAPK, src kinases, cyclins D1 and E and finally enhancement of PARP cleavage indicating induction of apoptosis. But then Alimova et al. (2009) have come up with all-inclusive signalling process. Metformin was effective against ER+ and ER−, HER2 expressing breast

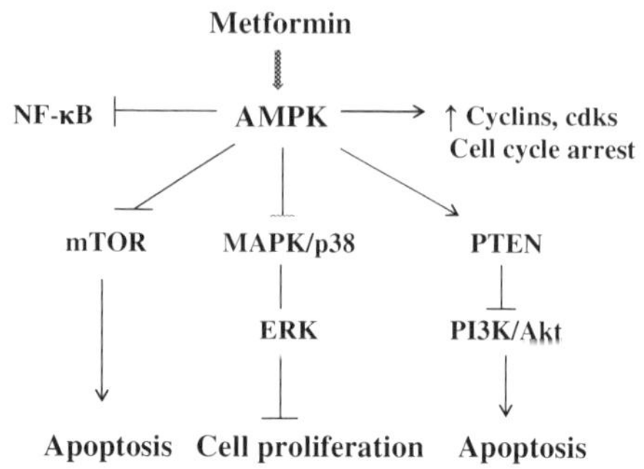

Figure 28.2 Postulated signalling systems operating the metformin/AMPK-mediated induction of apoptosis and inhibition of cell proliferation.

cancer and normal cells. Cell proliferation was due to G1–S transition, but without induction of apoptosis. They also encountered MAPK, Akt and mTOR suppression in cells irrespective of their steroid receptor and HER2 level of expression.

Significant from the point of view is the apparent inhibitory effects of metformin on cancer metastasis. Rattan et al. (2011) injected A2780 ovarian cancer cells intraperitoneally into nude mice and studied tumour growth and dissemination. They found that activation of AMPK and inhibition of mTOR signalling by metformin suppressed angiogenesis as indicated by CD31 expression. Reduced microvessel density occurred with reduction in VEGF expression. Furthermore although tumours occurred on the surface of liver, spleen and kidney, they were not invaded by the tumour. Animals not treated with metformin showed deposition of tumour in the lungs and this was reduced in the treated animals. Unfortunately details of how immunohistochemistry results were assessed and graded have not been provided. Nonetheless, this is a highly creditable effort and if confirmed would be invaluable. Quite obviously the authors are trying to portray a situation analogous to human ovarian cancer where tumour spread occurs by peritoneal adhesion of cells released from the primary lesion. So notwithstanding the demonstration that without metformin treatment the tumours showed metastatic spread to the lungs, it is essential that cell surface determinants such as CA125 (MUC16), mesothelin and others which promote peritoneal implantation should have been assessed. Some adhesion molecules such as intercellular adhesion molecule (ICAM) might be influenced by metformin. Germeyer et al. (2011) have noticed that metformin increased and insulin decreased, but not beyond reasonable doubt, the expression of ICAM gene. But earlier Bilgir et al. (2009) reported a marked reduction in soluble ICAM in patients with polycystic ovary syndrome. In TNBC cells, metformin reduced CD24 expression (Vazquez-Martin et al., 2011).

Does Metformin Selectively Destroy CSCs?

LKB1/AMPK signalling has also been implicated in the suppression of CSCs. Hirsch et al. (2009) reported selective killing of CSCs ($CD44^{high}/CD24^{low}$) by metformin, in contrast with doxorubicin which was not able to target CSCs preferentially in this way but did kill non-CSC cancer cells.

They also isolated CSC populations by flow cytometric cell sorting. CSCs were highly susceptible to killing by metformin, whilst non-CSC tumour cells were virtually unaffected. But intriguingly tumour growth *in vivo* was suppressed only by a combination of metformin and doxorubicin. Song et al. (2012a) have claimed that they have noted a preferential deletion of high CD44/low CD24 expressing subpopulation of MCF-7 breast cancer cells presumed to be CSCs, with reductions in size of populations ranging from around 3% to 1%. They have also argued that metformin-activated AMPK with concomitant reduction in levels of phosphorylated mTOR and its downstream effectors S6K1 and 4EBP1. The suppression of the mTOR pathway by metformin is well documented. But here mTOR suppression would occur as a general effect in the population and so somewhat difficult to link it up with the perceived reduction in the CSCs. Experimentally inactivating LKB1 might collaborate with activated oncogenic factors to promote the survival of CD24+ cells suggestive of CSC population expansion. The CD24+ cell population also showed increased metastatic behaviour in comparison with CD24- cells (Liu et al., 2012e).

In general research circles CD44, CD133, CD24 and ESA are employed in the identification of CSCs. Among these the status of CD24 is somewhat uncertain, but it has been associated with pancreatic cancer and invasive breast cancer (Baumann et al., 2005; Li et al., 2007b). As discussed above, CD24+ do represent a tumorigenic fraction and indeed can and should be regarded as a marker of CSCs. Cells expressing CD44+ phenotype are associated with initiation of many forms of cancer, for example the breast, colon, pancreas and the prostate. ESA is the favoured phenotype of breast and pancreatic cancer CSCs. CD133 has been linked with glioblastoma, and colon and prostate cancers (Al-Hajj, 2003; Collins, 2005; Gudjonsson, 2002; O'Brien, 2007; Singh et al., 2004). It is needless to say that great care has to be exercised in employing CSC markers. Recently ALDH (aldehyde dehydrogenase) has been touted as a CSC marker. The ALDH isoform ALDH1A3 has been suggested to identify breast CSCs (Marcato et al., 2011). In the light of this, it is interesting to note that metformin has been shown to inhibit the growth *in vitro* of ovarian ALDH-expressing cells (Shank et al., 2012).

Application of Metformin in Cancer Management

Numerous clinical trials are in progress on the efficacy of metformin as a single agent or in combination with standard chemotherapy as an anti-cancer agent by evaluating biological and molecular markers related to tumour growth. Markers of cell proliferation and apoptosis, which are not indisputable and appropriate as markers of metastasis, are frequently employed, but one needs trials to assess clinical endpoints. Many tumour types are being studied.

Possible effects of CSCs are also being evaluated. Metformin has been credited with a preventive and protective role against cancer, but this view has not had wide acceptance. At present there is no incontestable evidence that metformin reduces cancer risk. Pre-surgical administration of the drug has had little effect on cell proliferation markers in a randomised breast cancer study; possibly the status of insulin resistance might have had some bearing on this outcome (Bonanni et al., 2012). Metformin has often been described as well tolerated, but one should recognise that this might be because low doses employed for controlling diabetes had been administered (Pollak, 2012). There is significant evidence that metformin sensitises tumour cells to cytotoxic therapy. Indeed one sees considerable scope for extending the efficacy of metformin by taking into account the hormonal and growth factor receptor status. Combination with inhibitors of growth factor receptors and inhibitors of signalling pathways identified as being activated by AMPK seems eminently worthy of pre-clinical assessment. Of immense value would be targeting CSCs, which by available accounts are targeted by metformin/AMPK activation. Perhaps, sesquiterpene lactone parthenolide, putatively attributed with the ability to target CSCs, might be worthy of pre-clinical combination with metformin. Parthelonide is a naturally occurring inhibitor of NF-κB signalling. It has been shown to inhibit the IKK kinase complex which is the core element of the NF-κB cascade (Bork et al., 1997; Hehner et al., 1999; Kwok et al., 2001). It is also said to promote ubiquitination and degradation of mdm2 and activate p53 (Gopal et al., 2009), two important pathways among those influenced by AMPK. Several studies indicate that parthelonide might target CSCs, albeit this is subject to caveats regarding the choice of markers and using radiation resistance for identifying CSCs often with somewhat convoluted argumentation.

29 PLCD1 Suppresses Tumorigenesis

The PLCD1 (phospholipase Cδ1), a member of the phospholipase C family, which together with other members PLCγ, PLCζ, PLCη and PLCε, converts phosphatidylinositol 4,5-bisphospate to diacylglycerol (DAG) and inositol 1,3,5-trisphosphate (IP3), both possessing second messenger function. IP3 regulates intracellular Ca^{2+} levels by mobilising calcium from intracellular endoplasmic reticulum-associated stores and probably also by stimulating Ca^{2+} influx into cells. The transduction of signals imparted to the cell by the binding of extracellular ligands to their respective membrane receptors generates repetitive calcium spikes or oscillations and these transient increases of intracellular calcium are mediated by IP3. Such calcium oscillations occur in many physiological phenomena. IP3R are required for IP3-mediated mobilisation of calcium from intracellular stores, but not for calcium influx. A variety of biological response modifiers transduce their signals via cytoskeletal structure and IP3 signalling has obvious links with cytoskeletal dynamics.

Deregulation of IP3 and downstream calcium signalling has serious consequences for biological responses in developmental systems as well in pathogenesis (Sherbet, 2001). DAG promotes RasGRP (Ras guanyl nucleotide-releasing protein)-mediated Ras-GDP to Ras-GTP exchange activity, and this leads to the activation of the Ras/MEK/ERK signalling cascade to generate phenotypic effects of cell proliferation and differentiation. By another route DAG can activate Rac-GTP signalling to influence cytoskeletal dynamics, cell migration, adhesion and invasion (Griner and Kazanietz, 2007). With such important signalling connections and its loss in many tumour types, PLCD1 has earned the epithet of tumour suppressor.

Loss of PLCD1 Expression in Tumours and its Biological Outcome

The tumour suppressor function of PLCD1 is reinforced by the perceived loss of its expression in several tumour types, such as oesophageal squamous cell carcinomas, gastric and colon cancers. But there is some uncertainty in relation to the status of its expression in breast cancer. The PLCD1 gene is located on chromosome 3p22.3 and this happens to be one of the regions frequently deleted in oesophageal squamous cell carcinoma. Loss of expression was encountered in 26/50 carcinomas and 4/9 tumour cell lines. This resulted from promoter methylation or loss of one or both alleles or loss of an allele plus methylation of the second (Fu et al., 2007). Of particular interest is the correlation noticed between loss of expression and dissemination of tumour to

Therapeutic Strategies in Cancer Biology and Pathology. DOI: http://dx.doi.org/10.1016/B978-0-12-416570-0.00029-9

the lymph nodes. Loss of expression was far greater in primary lesions with metastasis than in tumours that had not metastasised. In colon cancers too, PLCD1 expression was reduced in a majority of tumour samples, but PLCγ1 showed a marked increase (Nomoto et al., 1995). Similarly in gastric tumours, loss of PLCD1 by methylation was noticed in 16/19 gastric cancer cell lines and 61/98 primary cancers; methylation was associated with higher tumour grade (Hu et al., 2009). Epigenetic silencing of the gene was found in bone marrow mononuclear cells of CML patients but not in normal adult marrow mononuclear cells (Song et al., 2012c). Promoter methylation was encountered in just over half of breast cancer specimens and in 7/9 cell lines tested (Xiang et al., 2010).

Among biological abilities that PLCD1 seems to suppress are cell adhesion, invasion and cell proliferation, which are consistent with its putative tumour suppressor function. Forced expression of PLCD1 in non-expressing cells reduced MMP7 expression and brought about alterations in cytoskeletal dynamics apparently by inactivating cofilin (Hu et al., 2009). Cofilin, a plasma membrane-associated protein, promotes actin-depolymerisation, but primarily it is an F-actin binding protein. Its activity and subcellular localisation is regulated by phosphorylation (Bamburg, 1999). Phosphorylated inactive cofilin has been shown to inhibit cell migration (Popow-Wozniak et al., 2012). Equally other means which would affect migration can be visualised. The suppressor protein DLC1 interacts with and potentiates the activity of PLCD1 (Homma and Emori, 1995). As discussed before DLC1 is a member of a family comprising three proteins that function as Rho GTPase-activating proteins by virtue of the GAP domain and so actively involved in the regulation of cell proliferation, modulation of cytoskeletal dynamics and cell migration. Kawai et al. (2004) determined the cellular distribution of p122/RhoGAP, a GTPase-activating protein of Rho small GTPase family, which activates PLCD1. P122 co-localised with vinculin an important link to the actin cytoskeleton. Thus much confirmatory evidence supports the role of PLCD1 in cell adhesion phenomena and in cell migration.

PLCD1 is said to restrain cell cycle progression at G1–S transition checkpoint by upregulating p21 expression. Suppression of cell proliferation signalling via Akt is also a mechanism mooted for the suppressor function of PLCD1 (Fu et al., 2007). In a couple of cell lines, cell cycle arrest at G1–S transition has been reported. FACS (fluorescence activated cell sorter) analysis has shown accumulation of cells in G0/G1 and a corresponding decrease in S-phase cells (Kaproth-Joslin et al., 2008). G1 arrest has been shown to occur in a CML cell line with ectopic expression of PLCD1 (Song et al., 2012c). Although there is consensus about G1 arrest, G2–M arrest is said to have occurred in breast cancer cell lines (Xiang et al., 2010).

The mode of function and nature of the effects of PLCD1 on cell proliferation need to be explored further since views contradicting cell proliferation suppression have also been expressed. It has been claimed that C6 glioma cells siRNA-mediated inhibition of PLCD1 actually decreased cell proliferation and FACS data have suggested tardy S-phase progression and a resultant reduction in G2/M cells (Stallings et al., 2008). Earlier, Leung et al. (2004) had claimed that overexpression of PLCD4 by transfection upregulated the expression of EGFR and erbB2 expression and led to the activation of ERK1/2 proliferative signalling in MCF-7 cells. But MCF-7 cells do express EGFR at low levels. Intriguingly they find that the PLCD1 effects on growth are not reversible. This is possibly a consequence of other interacting factors, such as DLC1 which is known to interact with PLCD1 which influences some cellular features such as cell migration partly by an EGFR-MEK-dependent mechanism (Shih et al., 2012). PLCD4 seems to play an important part in liver regeneration and its disruption does interfere with cell proliferation. Possibly it co-operates with PKC isoforms (Akutagawa et al., 2006). Many tissues do show differential expression of PLC isoenzymes and also isoforms of PLCD itself. The differences in the mode of action of PLCD1 and PLCD4 are still unclear. For all intents and purposes, the PLCD isoforms have similar function, but they might differ with regard to membrane targeting due to differences in membrane binding properties (Ananthanarayanan et al., 2002) and these can quite conceivably translate into functional disparities and distinctions.

30 Inhibitor of DNA Binding Proteins in Tumours

Inhibitor of DNA binding proteins (IDs1–4) are dominant negative bHLH transcription factors. These ID bHLH proteins lack the basic DNA binding domain and so have no DNA binding ability but they do heterodimerise with and negatively regulate the activity of other bHLH transcription factors, by sequestering and preventing DNA binding by the target transcription factor.

As transcription factors IDs are involved in the regulation of several signalling pathways that determine many biological functions including cell proliferation, differentiation, neurogenesis, myogenesis and differentiation (see Morgenstern and Atchley, 1999; Norton et al., 1998). Salient biological properties are summarised in Table 30.1.

Table 30.1 ID bHLH Proteins and Their Biological Properties

ID	Synonyms	Chromosome Location	Biological Properties
ID1	bHLHb24	20q11	EGR→ID1 cell proliferation; inhibits NF-κB and apoptosis; CSC maintenance; invasion; angiogenesis; Parallel↑EGFR; VEGF?
ID2	bHLHb26	2p25	EGR2→ID2; inhibits RANKL/ NF-κB→ inhibits apoptosis; CSC maintenance; angiogenesis; Parallel↑EGFR; VEGF?
ID3	HEIR-1, bHLHb25	1p36.13-p36.12	CSC maintenance Angiogenesis; Parallel↑EGFR; VEGF?
ID4	bHLHb27	6p22.3	Tumour and growth suppression; stem cell programme; Cdc42→ID4↓ ID4→BRCA1↓; angiogenesis?

Source: Information based on Human Gene Nomenclature Committee and references cited in the text.

Therapeutic Strategies in Cancer Biology and Pathology. DOI: http://dx.doi.org/10.1016/B978-0-12-416570-0.00030-5

Is ID4 a Tumour Suppressor?

A large body of evidence has built up supporting the tumour suppressor function of IDs. ID4 is downregulated in many tumours but not without exceptions. The loss of expression is occasioned by epigenetic modification. So ID4 is deemed a tumour suppressor protein. The gene is silent in a vast majority of AML and CLL patients, and its association with DAPK1 (death-associated protein kinase 1) in CLL has underscored the relevance of ID4 in the pathogenesis of CLL. DAPK is involved in apoptosis and autophagy and its loss of expression occurs in many tumours. It may be recalled that DAPK1 shows epigenetic modulation and reduced expression in sporadic CLL (Raval et al., 2007). Hypermethylation has been linked with metastatic spread (Martinez-Glez, 2007). In fact, AML and CML differed in respect of frequency of methylation of ID4, SFRP1 and SHP1, methylation being higher in AML patients than CML patients. But the suppressor gene DLC1 and cadherin, among others, displayed no differences (Uhm et al., 2009). This possibly indicates sharing of functions between ID4 and different suppressor entities in AML and CML and could relate to the characteristics of the acute and chronic conditions. Myelodysplastic syndrome (MDS) which carries a 1/3 risk of transformation into AML has shown ID4 promoter methylation. Frequency of methylation was higher in advanced MDS subtypes and in the higher risk groups assessed by the international prognostic scoring system. Also ID4 methylation correlated with poor prognosis and reflected more rapid progression in patients with ID4 methylation as compared with those without methylation (Wang et al., 2010a).

The frequency of methylation and the degree of methylation are the criteria for defining the state of methylation and sometimes implied or interpreted as equivalent measures. The extent or degree of methylation is of overriding importance for it is suggestive of more aggressive disease and correlated in prostate cancer with tumour grade and reflected potential earlier recurrence (Sharma et al., 2012; Vinarskaja et al., 2012). Umetani et al. (2005) reported some time ago that ID4 methylation was found in 16/24 node-positive and 7/36 node-negative T1 primary breast cancers. Another study has also drawn attention to the frequent loss of ID4 expression by promoter methylation in 117/170 of primary breast cancers. ID4 promoter methylation reflected poor prognosis in terms of recurrence-free survival and further it was associated with tumour spread to the regional lymph nodes. ID4 loss did not correlate with tumour size, grade or ER/PR status (Noetzel et al., 2008). This is intriguing. But the authors had made no attempt to look at the expression of cell proliferation markers. Furthermore, inclusion of ductal carcinomas *in situ* (DCIS) of different grades would have shown whether ID4 loss affected the invasive process. But earlier, de Candia et al. (2006) noted a striking inverse relationship between ER and ID4 expression in normal breast epithelia. Virtually all DCIS and invasive carcinomas that showed downregulation of ID4 mRNA were ER+ and 3/31 ID4+ invasive carcinomas were ER−. A point to be stressed here is that the debate might be inconclusive when attribution is restricted to global ER rather than ERα and ERβ, which have divergent biological effects.

In colorectal cancer, no methylation was detected in normal epithelia and adenomas, but primary carcinomas and liver metastases showed hypermethylation in 49/92 and 19/26 cases, respectively (Umetani et al., 2004). Thus, to date the indications are that hypermethylation of the gene seems to reflect tumour progression. However, methylation status of colon cancer did not correlate with prognosis (Tanaka et al., 2011). But Vinarskaja et al. (2012) have added a note of caution in assessing the degree of methylation. For, they found methylation-specific PCR and pyrosequencing techniques were not always consistent with each other. The degree of gene silencing might relate to the gene copies methylated or the degree of methylation (Matzke et al., 1994). But one cannot be certain or assume that the degree of methylation relates pro rata to gene silencing.

Cdc42 has recently been ascribed a role in the methylation of ID4. The latter was found to be downregulated by Cdc42, which is overexpressed in a majority of colorectal adenocarcinomas (Del Pulgar et al., 2008). The mechanism involved has not been elucidated. However, the implications of this in terms of accentuating the effects of loss of ID4 suppressor function can be visualised immediately because Cdc42 is a Rho-GTPase which transduces signals arising from the activation of a variety of membrane receptors into induction of apoptosis, regulation of the cell cycle, modulation of cytoskeletal organisation leading to cell migration (Figure 30.1).

As discussed earlier, DLC1 possesses a Rho-GAP domain and GAP activity specific for small GTPases RhoA and Cdc42. The suppressor function of DLCs is said to derive from their Rho-GAP activity negatively regulating RhoA, RhoB, RhoC and to a limited extent Cdc42. The KAI1 suppressor gene might selectively inhibit Rho-GTPases/Rho/Rac/cdc42 signalling cell motility and also modulate signalling

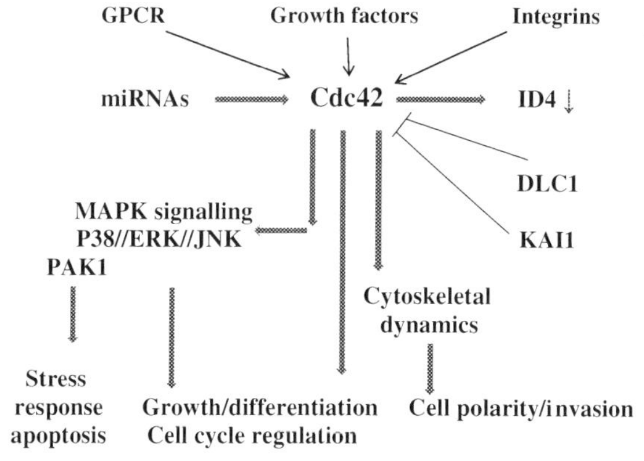

Figure 30.1 The Cdc42/Rho-GTPases/Rho/Rac signalling pathway regulating apoptosis, cell cycle regulation, growth differentiation and cell invasion. The operation of this pathway is modulated by some miRNAs and the suppressor genes DLC1 and KAI1. It would be worth mentioning here that ID4 aids the development of the mammary gland by suppressing MAPK/p38 activity, but this signalling is active in ID4−/− cells (Dong et al., 2011).

involved with both cell proliferation and survival. Therefore besides downregulating ID4 expression, there are other pathways by which Cdc42 can produce its phenotypic effects. Equally TNBCs showed far greater ID expression than did non-TNBCs. ID4 was found in more than 50% of tumour cells in 76/101 TNBCs, whilst the comparable figure was 6/113 non-TNBCs. Here it is argued that this is due to the fact that ID4 inhibits and downregulates BRCA1 suppressor (Wen et al., 2012). ID4 is overexpressed in basal-like breast cancer of which a majority tend to be triple negative. BRCA1 methylation is not remarkably higher, but mRNA levels were greatly reduced in cancer than in control, which suggests the possibility of ID4 being involved in the diminished transcription of BRCA1 (Turner et al., 2007). This postulate runs counter to the suppressor function of ID4.

MiRNA-335 has been drawn into the picture and has been implicated in regulating ID4, ERα, IGF1R and Sp1 known to induce BRCA1 expression and by this means regulate BRCA1. Not only miRNA-335 is downregulated in breast cancer, but its expression correlated positively with BRCA1. The miRNA upregulates BRCA1 expression (Heyn et al., 2011). This is in agreement with the view that miRNA-335 is an inhibitor of metastasis. Compatibly, miRNA-335 reduces ERα level but when the endogenous miRNA is suppressed using antisense oligonucleotides, ID4 expression increases, possibly a compensatory effect indicating a regulatory loop involving miRNA-335, BRCA1 and ERα. But BRCA1 expression is said to inversely relate to both ERα and ID4 (Roldan et al., 2006), which insinuates a notion of co-regulatory link involving the miRNA, BRCA1 and ID4 but not ERα. DAPK1 promoter hypermethylation and the presence of miRNA-335 would be expected to be just as effective in determining potential aggressive behaviour of tumours.

Induction of miRNA-335 in ER− MDA-MB-231 and BT474 breast cancer cells does indeed inhibit cell migration and metastasis as assayed by the less than ideal assay of lung colonisation upon tail vein injection even in the absence of ER. This in fact seems to occur by a different route, namely by inhibiting Sox4, an activator of EMT (Zhang et al., 2012c). MiRNA-206 also downregulates ERα but not ERβ, possibly as a result of binding to two sites in the 3′-untranslated region of ERα mRNA (Adams et al., 2007). Several miRNAs are associated with ER+/node-negative and one in ER−/node-negative breast cancers and TNBCs (Foekens et al., 2008). The sheer numerical aspect of miRNA expression is a complicating factor that cannot be easily addressed whilst setting up experimental protocols (see Table 2.1 and Figures 3.4, 3.5 and 3.6).

Equally, it has been argued not infrequently that ID4 is upregulated in some forms of cancer, promoted cell proliferation and as discussed below also linked with the propagation and maintenance of CSCs. As pointed out earlier, ID4 has been seen to block the suppressors BRCA1 gene transcription *in vitro* and could downregulate its expression *in vivo* (Wen et al., 2012). Indeed, Ren et al. (2012) regard ID4 as an oncogene, amplified in 32% of high-grade serous ovarian cancers and overexpressed in most primary ovarian cancers but not in normal ovary. Overexpression also occurred in a number of tumour cell lines of endometrial cancer, breast cancer and glioblastoma origin.

ID1, ID2 and ID3 Expression in Tumours

As opposed to ID4, IDs1–3 appear to be associated with more aggressive tumour and also linked with invasion and angiogenesis as discussed below. ID1 expression has often been noted to relate with aggressive tumours. At the experimental level, transfection of ID1 into oesophageal carcinoma cells increased cell proliferation by the activation of PI3K/Akt pathway, reduced apoptosis and enhanced metastatic spread as assayed by experimental metastasis assay (Li et al., 2009a). But this cannot be directly linked with the increased angiogenesis also noted in this study.

Correlation with tumour size, grade and stage has been frequently noticed. ID1 and ID3 expression corresponded with the degree of differentiation in gastric adenocarcinomas, with higher expression in poorly differentiated tumours than in well-differentiated counterparts. Their expression was also related to clinical stage. However, expression levels did not relate unequivocally to metastasis. ID1 levels correlated with lymph node metastasis, but not with distant metastasis. But this is at odds with the finding that high expression levels were associated with poor prognosis in gastric cancer. High ID1 and ID3 expression was associated with aggressive disease (Yang et al., 2011c). ID1 expression was related to size and to lymph node-positive status of oral carcinomas where it is also related to tumour recurrence (Dong et al., 2010). Nuclear ID1 expression has been detected in squamous and non-squamous cell carcinoma of the lung and ID+ samples also expressed src kinase and MMP-9. Src kinase and MMP expression correlated with each other and with tumour grade (Rothschild et al., 2011). Src kinase is a major regulator of IDs. In many tumour cell lines, ID1 expression is downregulated when src kinase is inhibited. Furthermore, src and ID inhibition decreased cell invasion (Gautschi et al., 2008). Consistent with this is that ID1 expression has been allied with the depth of myometrial invasion in endometrial cancers. So also in ovarian cancer, ID1 expression was related to clinical stage. The survival rate was lower in patients with high ID1 expression than in those who had lower ID1 (Maw et al., 2009, 2010). ID1 and ID2 expressions have been found to be overexpressed and ID3 and ID4 expressions were reduced in prostate cancers compared to nodular hyperplasia (Yuen et al., 2006). But in some cases, ID3 may be expressed highly with aggressive disease.

The significance of the patterns of expression, namely distribution of the IDs between the cytoplasm and nucleus in relation to prognosis has been focused upon, but no firm conclusions can be based on the currently available evidence, often based on immunohistochemical assessment with the attendant pitfalls. For instance, there are reports about the suppressor ID4 presence in the nucleus correlating with disease parameters, but the accent appears on the cytoplasmic presence of ID1 and ID2, and not infrequently ID3 with disease state. More investigations are warranted to rationalise the significance of the distribution pattern IDs. For instance, one needs to know if the subcellular distribution pattern is a consequence of the importation of regulatory molecules into nucleus or a result of their export to the cytoplasm. In other words, one has to look at the ratios of distribution than mere detection in the cellular compartments before any rationalisation can be attempted and significance of patterns of distribution interpreted in relation to the disease process.

IDs and Tumour Angiogenesis

Aside from the overexpression reported in some tumours, more evidence has emerged how it can support tumour growth and even metastatic spread. Overexpression of some IDs promotes metastatic spread. ID1, ID2 and ID3 were shown some time ago to be able to induce angiogenesis. Indeed many of the cases where ID expression levels correlated with clinical stage and secondary dissemination also displayed greater degree of angiogenesis. Breast tumour cells either failed to grow or grow into poorly vascularised and necrotic tumours when injected into animals with altered genetic background of ID1+/− or ID3−/− (Lyden et al., 1999; Benezra, 2001). In NSCLC, ID1 expression has been found to occur in parallel with src kinase and MMP-9 (Rothschild et al., 2011). This is of interest since MMPs are associated with tumour invasion and induction of angiogenesis. In endometrial cancers, ID1 expression was associated with increased microvessel density (Maw et al., 2010). Ling et al. (2005) transfected the ID1 gene into prostate cancer cells and showed the enforced expression of the gene-induced transcription of the VEGF gene and secretion of VEGF. Prostate cancer cell xenografts also showed enhanced VEGF expression in parallel with increased ID1 expression. *In vitro* assays indicated that HUVEC cells differentiated into tubule formation when grown in media conditioned with ID1 transfected cells. However, it is surprising that when xenografts were at hand, no attempt was made to investigate or measure microvascular density.

A parallel upregulation of EGFR and VEGF with ID1 has been reported in colorectal cancers, where these features also correlated with tumour stage (Meteoglu et al., 2009), but a causal relationship between these is not visible. However, it is worth reiterating here that EGF and HB-EGF do indeed possess angiogenic properties (Sherbet, 2011a). Whilst clinical stage does reflect progression, one ought to recognise that colorectal cancer progression is characterised by the acquisition of definable molecular features and these could have provided deeper insight into the parallel upregulation of ID1, EGFR and VEGF.

In contrast with the tumour suppressor function discussed earlier, glioblastoma tissues have been found to overexpress ID4 and cell lines derived from these tumours expressed ID4 10-fold higher than normal astrocytes. Highly vascularised tumours formed when the high ID4 cells were xenografted. These tumours displayed enhanced growth, but *in vitro* the high expresser cell lines did not show enhanced proliferation or alteration in apoptosis. So the enhanced growth of the xenografts seems to be a natural increase aided by higher vascularisation (Kuzontkoski et al., 2010). Unfortunately angiogenesis was determined by histological assessment and no markers of angiogenesis were employed and no attempt was made to elucidate potential molecular mechanisms involved in the induction of angiogenesis. Furthermore despite so much investment in terms of animal model, it is rather surprising that the enhancement in angiogenesis was not followed up with the metastatic spread of the xenografts.

A new link of much potential has been forged between angiogenic effect of ID4 and p53 mutation. ID4 is demonstrably overexpressed in the presence of mutant p53. A gain of function mutant p53 together with the transcription factors p65 and E2F1 assemble at regions of ID4 promoter and transcriptionally enhances its expression, ID4 now stabilises certain pro-angiogenic factors thus bringing together mutant p53 function and induction of angiogenesis (Fontemaggi et al., 2009). Given that IDs can act as inducers of angiogenesis, this study provides a tangible link between ID4 function in conjunction with mutated p53. There are suggestions that IDs might link up with Rb protein, another cell cycle regulator.

IDs being small proteins would be able to freely translocate between the cytoplasmic and nuclear compartment. Nonetheless, how their subcellular distribution is altered in response to growth factors and other signalling modulators is not fully understood. It is needless to emphasise here that reconciliation is required of this angiogenesis promoting ability with the tumour suppressor capability attributed to ID4. ID4 is known to form complexes with and sequester other bHLH transcription factors in the cytoplasm and prevent their biological function. Induction of overexpression of cytoplasmic ID4 by BMP promotes the translocation of the transcription factors OLIG1 and OLIG2 and they are sequestered by ID4 (Samanta and Kessler, 2004). Many genetic programmes are activated in a temporal sequence during tumour development and progression. It may be that IDs display a stage-dependent and differential subcellular distribution that could account for the tumour suppressor function of ID4 early in the development of a tumour and induction of angiogenesis occurring as subsequent event and also explain the tumour and proliferation promoting abilities of the other IDs. This is in the realm of speculation, but it needs to be explored in the light of IDs participating in stem cell propagation and maintenance as discussed below.

IDs in Cancer Stem Cell Propagation and Maintenance

An important question being asked is whether IDs participate in stem cell perpetuation and maintenance or induction of their differentiation. The answer when a definitive one is available is of immense therapeutic value. ID2 and ID4 have been connected with the differentiation of CSCs. In glioblastoma, cultures containing CSCs have shown overexpression of ID2 and ID4. Wu et al. (2012c) have explored a possible link between these. Induced overexpression of the IDs seemed to promote differentiation of the CSCs as indicted by reduced expression of CSC markers.

Contrary to this, ID1, ID2 and ID3 have been found to enhance neural stem cell proliferation and maintenance. Nestin and the transcription factor Sox2 commonly linked with proliferating neural stem cell were employed as neural stem cell markers. Forced expression of the IDs generated more Ki-67+ and Sox2 expressing cells (Jung et al., 2010). Williams et al. (2011) found that deubiquitination of ID proteins promoted their stability and stem cell features in osteosarcoma. In some human osteosarcomas USP (ubiquitin-specific peptidase), a deubiquitinating enzyme and IDs displayed parallel upregulation. Experimental suppression of USP1 led to destabilisation of ID proteins and induced cell cycle arrest. The opposite effects were observed with forced expression of USP1. The reason for this is that USP1 is phosphorylated by Cdks and thus prevented from degradation; USP1 deubiquitintes and stabilises IDs, so promoting cell proliferation. When USP1 is suppressed, IDs are ubiquitinated and destabilised leading to cell cycle arrest and apoptosis (Cotto-Rios et al., 2011). Ubiquitin/proteasome-mediated degradation of ID1 seems to lead to TNF-α-induced apoptosis in prostate cancer cells (Ling et al., 2006). But equally this can occur by other routes, for example by ubiquitination of NF-κB which TNF-α is known to activate. USP11 can negatively regulate TNF-α-induced activation of NF-κB (Sun et al., 2010). The Pax transcription factors are differentiation markers and are arguably valid for CSC markers. Pax7 has been shown to induce ID2 and ID3 and block premature differentiation of quiescent satellite myogenic stem cells (Kumar et al., 2009).

Some evidence has emerged relating to ID4. The latter has also been attributed with gliomas stem cell promotion by enhancing Sox2 expression but by an indirect route, namely by negating the suppressor effects of miRNA-9*, the 3′ transcript of miRNA-9 (Jeon et al., 2011). Earlier Jeon et al. (2008) had shown that ID4 might be functioning via cyclin E and notch signalling. Whilst cell proliferation is an integral part of CSC propagation, one ought to make a distinction between purely proliferating subpopulations and CSCs. Cell proliferation is not obligatory in CSCs and markers such as Nestin may be expressed with no proliferation marker (Ernst and Christie, 2006). The difficulties arise with markers for proliferation signalling employed with or without standard CSC markers.

Dependent upon the biological model, various markers such as Wnt proteins, SFRP, the negative regulators of Wnt, and the transcription factors Oct have been employed as haematopoietic CSC makers, and Pax and Sox transcription factors as differentiation markers. In the absence of standard protocols, it is difficult to assert the intrinsic ability of the IDs in the promotion and maintenance of CSCs or the promotion of their differentiation.

One should also bear in mind the potential effects of other suppressor genes on the CSC markers. LKB1 regulates many of the markers, such as Oct, Sox 2 and Nestin. Oct4 and Sox2 tend to be overexpressed in the absence of LKB1 expression, together with suppression of differentiation markers such as Nestin.

Do ID Proteins Activate Other Signalling Systems and Promote Cell Proliferation?

In common with other bHLH proteins, one can expect that they would be involved in activation of some and repression of other genes. Therefore hardly surprising it is that ID4 with notable exceptions functions as a suppressor of tumour growth, invasion and metastasis. Also subject to confirmation (see below), forced expression of ID1 promotes apoptosis and reduces cell proliferation provisionally attributed to S-phase arrest and a prolonged G2–M. Attendant upon these are changes in the expression of cell cycle regulatory p53 pathway, enhancing the expression p21, p27 and their regulator p53 (Carey et al., 2009). Given that loss of p21 and p27 might lead to tumour progression, the effects on these proteins by ID4 are compatible with the suppressor function of the latter. The Rb protein has also been included in the purview of ID function. IDs1–3 are shown to be often associated with promotion of cell proliferation. Again as with ID4, these functions are an outcome of activation of many signalling systems. As noted on many occasions, activation of NF-κB signalling leads to the inhibition of apoptosis and promotion of cell population expansion. In many cancer cell lines, ID1 has induced activation of NF-κB by directly interacting with the p65 subunit of the NF-κB complex (Peng et al., 2012). Investigating the expression of ID1 and p65 in NPC (nasopharyngeal carcinoma), Sun et al. (2012) have concluded that enhanced ID1 and NF-κB/p65 correlated with poor prognosis. They also inhibited ID1 and NF-κB/p65 using specific shRNAs in an NPC cell line and found that the treatment had reduced MMP-9 levels and *in vitro* migration of the carcinoma cells.

ID1 has been found to directly interact with the zinc finger transcription factors EGR1 which is induced by growth factors. Exposure of neuroblastoma cell line, SK-N-MC to FGF-2 induced ID1 expression, via activation ERK1/2 pathway and activation of EGR1. The latter interacted with ID1 promoter and led eventually to cell proliferation (Passiatore et al., 2011). EGR2 can function directly by inhibiting RANKL/NF-κB and suppress apoptosis, but it can also interact with ID2 promoter and induce its expression (Kim et al., 2012c). Some evidence has been adduced for the participation of the TGFβ/Smad canonical pathway in ID function.

Are IDs Suitable Therapeutic Targets?

Many aspects of the function of IDs have interesting features that are exploitable to render IDs potential targets for therapeutic management. Some efforts have been made in this regard using conventional technology such as RNA interference to suppress ID genes that can promote tumour growth and spread. Many pathways of the regulation of IDs that operate in producing changes in the growth and behaviour of tumours have been suggested as possible targets. Elements in the regulation of apoptotic events, such as NF-κB function, the effects on the stability of IDs from ubiquitination and deubiquitination have been implicated. Certain miRNAs have been conspicuously cited as candidates to modulate IDs, but the sheer multiplicity of the modes of their function does not lend themselves as the most attractive elements. Among other notable features is the promotion of angiogenesis by ID1, ID2 and ID3. IDs may be upregulated in parallel with EGFR and VEGF but at present there is no visible causal link. Although mechanistic studies of how IDs promote these effects are still needed, fundamental findings are in place. Also of much significance is that ID4 on occasions does downregulate the suppressor gene BRCA1. In general ID expression is higher in TNBCs than non-TNBC, which could be a result of downregulation of the BRCA1. So in breast cancer subsets where BRCA1 is silenced, clarification of this inverse relationship would be worthy of therapeutic pursuit. The postulated regulatory loop consisting of ER and ID4, BRCA1 and certain MiRNAs also holds much yet undefined promise. Inducing re-expression of ID4 by targeted demethylation might merit some approach. We know from the work of Noetzel et al.

(2008) that ID4 re-expression is possible. Do note that the operative emphasis is on targeted demethylation. Targeted methylation of specified genes is a possibility albeit its practical value in a therapeutic setting is a long way away. Li et al. (2007c) have demonstrated that methylation can be targeted to promoters of specific genes and the selective methylation can lead to its repression. As the authors state this could indeed offer a powerful means of regulating the inappropriate expression of genes. Whether this approach will work in the natural genetic environment is an open question. Finally, the demonstration of complex operation of IDs in the propagation and maintenance of cancer stem cells has provided extra impetus to these efforts.

31 PDCD4 (Programmed Cell Death 4)

Another important tumour suppressor gene which functions as a cell proliferation regulator and which is downregulated or inactivated in expression in many tumours is PDCD4 (programmed cell death 4). PDCD4 can regulate both genetic transcription and the process of translation of the message. The protein is phosphorylated and regulated by Akt at two Akt/S6K phosphorylation sites, one N-terminal and one C-terminal. Phosphorylation induces nuclear translocation of PDCD4, but PDCD4 is redistributed in the cytoplasm if its phosphorylation is inhibited. Nuclear accumulation of c-Jun reflects target gene activity in terms of induction of apoptosis. Although c-Jun phosphorylation might not be an absolute requirement for its translocation to the nucleus, PI3K/Akt-mediated phosphorylation does regulate nuclear translocation and AP-1-mediated transcription activity of PDCD4. Given the distribution of c-Jun in the nucleus and JNK-1 in the cytoplasm, it is postulated that the basis of functional differentiation arising from intracellular distribution is that whereas nuclear PDCD4 inhibits c-Jun by directly interacting with it, in the cytoplasm PDCD4 inhibits JNK-1 and prevents it from phosphorylating c-Jun (Palamarchuk et al., 2005). PDCD4 interacts with the eukaryotic translation initiation factors eIF4AI and eIF4AII and co-localises with eIF4A in the cytoplasm. Regulation of translation is mediated by interaction with initiation factors eIF4A and eIF4G via the two tandem highly conserved helical MA3 domains each consisting of about 130 amino acid residues (Waters et al., 2007; Yang et al., 2003a).

PDCD4 shows loss of expression in a number tumour cell lines; reduction in protein correlated with that of the mRNA, although not invariably. Loss of PDCD4 expression was described in lung tumour cell lines and in a large number of tumour samples. The loss occurred in a majority of the samples and associated with high tumour grade and stage and corresponded with poor prognosis (Chen et al., 2003b). Downregulation of PDCD4 mRNA and protein has been encountered in 2/3 gastrointestinal stromal tumours and correlated with proliferation Ki-67 index, which more than halved in high PDCD4 expressing tumours (Ding et al., 2012). The authors do claim that loss of expression is greater in the so-called risk group, but they do not define what they mean by risk groups, how the risk factor is determined or assessed.

The suppressor function of the gene might be achieved via the operation several signalling pathways. The transcription factor myc regulates several pathways leading to cell proliferation, cell cycle regulation and apoptosis. Myc induces Ras/MAPK signalling. Inhibition of the activation of MAPK suppresses the oncogenic action of c-myc (Gramling and Eischen, 2012). The suppressor function of PDCD4 might flow from its ability to inhibit MAPK signalling. Suppression of PCDC4 seems to result in the upregulation of MAP4K1, which is also increased by c-myc overexpression (Wang et al., 2012a). The anti-proliferative, pro-apoptosis function and suppression of

Therapeutic Strategies in Cancer Biology and Pathology. DOI: http://dx.doi.org/10.1016/B978-0-12-416570-0.00031-7

invasion might involve PI3K/Akt signalling with regulation by PTEN. Overexpression of PDCD4 is associated with enhanced PTEN expression and reduced Akt (Wei et al., 2012a). As discussed later in this section, both PI3K/Akt/mTOR and MAPK/ERK pathways are probably operational in the inhibition of angiogenesis by PDCD4.

PDCD4 is targeted and regulated by miRNA-21, as noted earlier in connection with the discussion of cell proliferation and metastasis. It is downregulated or inactivated in expression in many tumours. Early studies with miRNA-21 has generated much interest because it was found to be markedly upregulated and expressed in relation to grade in breast cancer (Iorio et al., 2005; Volinia et al., 2006). MiRNA-21 is anti-apoptotic (Carletti et al., 2010; Chan et al., 2005) and induces cell proliferation (Asangani et al., 2008; Roldo et al., 2006; Si et al., 2007). MiRNA-21 is said to be able to inhibit the tumour suppressor PDCD4 (Asangani et al., 2008; Frankel et al., 2008; Li et al., 2010d). The anti-apoptotic effect might also result from its ability to inhibit PTEN, which negatively regulates Akt and Fas/FasL signalling (Yang et al., 2011a; Zhang et al., 2012a). In fact, Zhang et al. (2012a) found that miRNA-21 is overexpressed in gastric carcinoma tissues and suppression of its expression levels correlates with upregulated expression of PTEN.

MiRNA-21 seems to regulate PDCD4 levels as evidenced by the fact that inhibition of miRNA-21 lead to approximately 3.5-fold upregulation of PDCD4 expression (Frankel et al., 2008). Suppression of miRNA-21 has led to reduction of invasion of MDA-MB-231 *in vitro* and reduced lung metastases in animal models (Zhu et al., 2008b). Interestingly these authors also reported that the expression of miRNA-21 inversely correlated with that of PDCD4 and Maspin (a putative metastasis inhibitor) in breast cancer specimens. Unfortunately this is a study limited to breast cancer specimens and corresponding normal tissues. No clinical information on the specimens is available. Glioblastomas also show an inverse relationship. Downregulation of miRNA-21 correlated with enhanced expression of PDCD4, together with reduced cell proliferation and increased apoptosis (Gaur et al., 2011). Experimentally altering the levels of miRNA-21 of malignant peripheral nerve sheath tumour cells by transfection the cells with miRNA-21 inhibitor suppressed cell proliferation with induction of apoptosis. This was accompanied by enhanced PDCD4 expression (Itani et al., 2012). Horiuchi et al. (2012) investigated a large series of colorectal cancer patients. Levels of PDCD4 were negatively related to Dukes grade. Low PDCD4 levels correlated with poor prognosis in each of Dukes' stages B, C and D. Furthermore, PDCD4 mRNA expression was negatively related to miRNA-21 (Horiuchi et al., 2012). This is a clear example of progression-related downregulation of PDCD4 with parallel enhanced miRNA-21 expression. Nonetheless, dissenting voices are heard. For instance, Torres et al. (2011) have reported marked upregulation of miRNA-21 in endometrial cancers, but Maspin also showed considerably increased expression. PDCD4 expression was not altered. None of them showed any correlation with FIGO stage. Other possibilities also need to be reckoned with. As discussed in another location, miRNA-21 can regulate the PI3K/Akt/mTOR pathway albeit operating via PDCD4 (see below). It can also promote proliferation and invasion by downregulating p12 (CDK2-AP1), which suppresses cell proliferation by inhibiting cdk2 and G1–S transition. MiRNA-21 has also been associated with the induction of EMT by TGF-β.

Activation of PI3K signalling subserves many cellular functions. An important downstream effector of PI3K signalling is mTOR, which participates in the regulation of mitogenic signals, among others (Wullschleger et al., 2006). Akt links and co-ordinates signals accrued from activated receptors with downstream effectors to generate phenotypic effects of proliferation and cell motility. Thus the PI3K/Akt/mTOR pathway is pre-eminently involved in the regulation of cell proliferation, growth, apoptosis and cell motility. A link with this signalling system is indicated by the ability of miRNA-21 to regulate the PI3K/Akt/mTOR pathway by operating via PDCD4. This signalling is in turn subject to the function of the tumour suppressor PTEN, which can negatively regulate PI3K/Akt/mTOR signalling. MiRNA-21 is able to inhibit apoptosis and it seems to achieve this by regulating the expression of PTEN (Chan et al., 2005; Meng et al., 2007).

MiRNA-21-mediated inhibition of PDCD4 can occur by other means. Inhibition of MAPK signalling pathway might be involved here (Yang et al., 2006). From their study of gastric tumours, Li et al. (2011c) have provided some indirect evidence for this. They found miRNA-223 is able to downregulate EPB41L3 expression by directly targeting its 3′-untranslated regions. As further subsidiary evidence they state that overexpression of miRNA-223 in primary gastric carcinomas correlated with poor metastasis-free survival.

In this context, one should cite the recent findings of Imanaka et al. (2011). They found miRNA-141 to be highly expressed in cisplatin-resistant cells, and further that enforced expression of miRNA-141 in cisplatin-sensitive cells increased cell viability. They found that this miRNA-141 targeted and downregulated YAP1 expression. One would recall that YAP1 (the 65 kDa YAP transcriptional co-activator, see also Hippo signalling pathway, pp. 155–160) is an important mediator of p73-mediated apoptosis. YAP1 binds to p73, regulates its transcriptional activity and p73-dependent apoptosis. Akt phosphorylates YAP to inhibit its binding to p73. YAP is said to be able to stabilise p73 in response to DNA damage and induce p73-dependent apoptosis. By virtue of its ability to induce apoptotic cell loss, YAP has been labelled as a tumour suppressor (Di Agostino et al., 2010). This pathway can conceivably also link up with PTEN/Akt signalling in apoptosis.

The PI3K/Akt/mTOR and MAPK/ERK pathways have been suggested to be involved in another putative function of PDCD4. Suppression of the latter using shRNAs has been shown to increase angiopoietin-2 (Ang-2) mRNA and protein (Krug et al., 2012). These authors have argued without much conviction that PI3K/Akt route rather than MAPK/ERK might be operating in their experimental setup. Surprisingly the bHLH transcription factor HIF-1 and its transcriptional target VEGF were not affected in any way. Indeed signalling via both PI3K/Akt/mTOR and MAPK/ERK pathways might operate in the induction of angiogenesis.

Can PDCD4 Be Manipulated for Therapy?

The loss of PDCD4 in human tumours with attendant effects on tumour growth, invasion and progression inevitably raises the question whether PDCD4 loss can be prevented and or its expression restored for therapy. Promoter methylation is a means by which the gene is silenced. PRMTs (protein arginine methyltransferases) modify arginine residues and so epigenetically modify and regulate chromatin structure and transcriptionally activate or repress the expression of target genes. Promoter methylation of PDCD4 was reported in breast cancer cell lines some time ago (Wen et al., 2007) and subsequently in gliomas, where methylation occurred in around half of tumour issues tested. Methylation corresponded with loss of PDCD4 (Gao et al., 2009). But Powers et al. (2011) have found that PRMT5, a member of the PRMT family, methylates a site in the N-terminal region of PDCD4. When co-expressed, PDCD4 and PRMT5 enhanced breast cancer growth as orthotopic grafts. They suggest that the situation is analogous to a proportion of breast cancers expressing high PDCD4 have poor prognosis, with high PRMT5 expression functioning as a co-factor. The role of miRNA-21 in transcriptionally regulating PDCD4 expression was discussed earlier. Inhibition of miRNA-21 to enhance PDCD4 expression could suppress the aggressive behaviour of tumours. Prevention of ubiquitin-mediated degradation of PDCD4 is another potential route. PDCD4 is phosphorylated by both Akt and p70(S6K) and targeted for ubiquitination and proteasomal degradation (Dorrello et al., 2006; Schmid et al., 2011). Erioflorin (NSC144151), which inhibits proliferation of many tumour cell lines, is said to inhibit this process by interacting with E3-ubiquitin ligase (Blees et al., 2012). So several approaches, conventional as well as novel, are available to restore PDCD4 suppressor function and prevent its degradation. Some are worthy of further exploration.

Epilogue

The evolution of an appropriate therapeutic strategy that enables the identification of effective disease control without profound side effects and collateral damage to normal biological function is a formidable challenge. In order that this can be achieved, identification of genetic determinants that induce the neoplastic phenotype and a detailed study of the signalling pathways activated and operated by these determinants are pre-eminent requirements. With this in view one can order and organise the means of approach as identifiable modules into (i) molecular targets related to tumour growth and cell proliferation, apoptosis and angiogenesis; targets differentially expressed *in situ* and invasive and metastatic disease, such as metastasis promoter and suppressor genes, (ii) cell signalling pathways, e.g., modulation of hormonal and growth factor signalling, and (iii) motility-related factors including remodelling of the cell membrane and the extracellular matrix and cytoskeletal dynamics (Sherbet, 2009). Within this conceptual framework, no single essential element can be identified that can counteract the many facets of cancer development and progression. This is largely due to the fact that stages of tumour development are perceptibly dependent upon more than one signalling, possibly interactive, pathway to generate the biological phenotypic qualities. Besides identifying the major modifiers of biological response such as the tumour suppressor and promoter genes and the signalling pathways that they activate, it is essential to elucidate the interaction of and cross talk between multi-functional signalling systems, so that these interactions could be employed to promote and drive collaborative and concerted signalling to a beneficial outcome. The focus naturally would be upon detection of the primary signal initiators and the links in the flow of information and the meeting or intersecting points of the signalling pathways.

Virtually all the categories of genetic foci listed above have provided potential targets for manipulation; the conventional tumour promoter and suppressor genes, inducers of growth, growth factors/receptors and the signalling systems that they activate and modulators of transcription and translation are among notable ones. All of them have been invariably and unerringly labelled tumour/metastasis suppressors or promoters. It is here the conventional description of tumour promoters and suppressors inevitably breaks down. All these modulate tumour growth and metastasis in one direction or the opposite direction. However, some targets unmistakably emerge to be of greater benefit than others. Notable among metastasis suppressor being targeted for manipulation is nm23. The latter is unregulated by medroxyprogesterone acetate, whose deployment as an anti-metastatic and anti-angiogenic agent is being clinically evaluated. But several other tumour suppressor and promoter genes, some with demonstrably wide ranging biological effects, have had lukewarm reception.

Among these NDRG1, RKIP and LKB1 are worthy of attention. The potentialities of LKB1 are underscored by the findings that it directly targets AMPKs. Furthermore, metformin, phenformin and AICAR activate AMPKs. This opens up routes for targeting CSCs and tumour cell proliferation. So there are clear and vibrant views to make a robust case for its clinical assessment. Despite the diversity of biological effects flowing from the inappropriate expression of S100A4 and interactions of S100A4 signalling with several modifiers of biological response, there are no attempts to-date to assess the merits of subjugating S100A4 expression as a possible therapeutic approach. S100A4 function is allied with growth factor and steroid hormone receptors and osteopontin which is itself subject to regulation by Wnt, and NF-κB among others, so here is an ample provision of options available for therapeutic intervention.

One might legitimately ask what the possible approaches might open up genetic manipulation. Restoration of down regulated suppressors would not seem a practical proposition. Most downregulation is by methylation or alteration of acetylation status; so targeted alterations of the status of methylation or acetylation is not a practicable route. Nonetheless, it ought to be recognised that methylation can be targeted to promoters of specific genes and the selective methylation can lead to its repression. This could indeed offer a powerful means of regulating the inappropriate expression genes. Whether this approach will work in the natural genetic environment is an open question.

MicroRNAs have a powerful influence on the expression and signalling by many genes that inhibit or promote tumorigenesis and progression and they might yet turn out to be a major force. But the sheer magnitude of numbers and their ability to modulate signal flow of opposing directions makes them an unenviable target for manipulation. Nonetheless, miRNAs could make a perceptible difference in treatment modality, to witness those shown to up regulate growth factor receptors such as HER2 and make tumour cells amenable to treatment, or paradoxically promote proliferation of tumour cells and render them susceptible to cytotoxic therapy. Indeed, many miRNAs can modulate chemo and radiation sensitivity of tumours.

The process of EMT has overwhelming implications for tumour initiation, growth and progression and has quite deservedly received significant attention directed towards inhibiting EMT activation or setting in motion the reverse process of mesenchyme epithelial transition. The role of CSC in tumour initiation and progression possibly goes hand in hand with EMT in that they share many signalling pathways such as the Hedgehog, Wnt and miRNAs and so amenable to the induction of differentiation and might be seen as potential targets for therapeutic intervention. However, before therapeutic strategies can be developed an intrinsic problem has to be resolved. This relates to the identification of stem cell markers. Often a circuitous argument is employed, of imputing a marker as specific for CSCs and in parallel implicitly recognising cells that express them as stem cells. Nonetheless, the perceived temporal acquisition of stem cell features with attendant resistance to chemotherapy makes the objective of targeting CSC and EMT eminently worthy of evaluation and exploration. The basic strategy has to find a marker that can be targeted in a CSC clone that carries the conventional panoply of CD44+, CD24+,

ESA+ and CD133+ together with the presence or absence notably of p53, cadherins, and vimentin among others. Although some of these are looked for in potential EMT activation, insufficient cognizance is taken of their significance whilst determining the treatment. Besides being a regulator of cell cycle progression, p53 can up regulate the expression of miRNAs that can contribute positively to p53-mediated inhibition of cell proliferation and inhibition of EMT activation. Some miRNAs seem to be able to inhibit pancreatic CSCs. A combined inhibition of Hh and mTOR with conventional chemotherapy has also been effective in blocking pancreatic CSC propagation. Also imperative is the identification of agents that might selectively target and destroy CSCs. Targeting the SMO component of Hh signalling by using small molecule inhibitors has had profound impact on the growth of some forms of cancer and on CSC inhibition, so much so that some inhibitors are in clinical trials and their efficacy in combination with standard chemotherapy is also being assessed.

Salinomycin appears to inhibit Wnt signalling with associated consequent effects on cell proliferation, apoptosis and migration and is regarded as an inhibitor of CSCs. Metformin seems to inhibit the proliferation of ALDH expressing ovarian CSCs and it is of interest that ALDH isoform ALDH1A3 is expressed in breast CSCs. Metformin inhibits CSCs identifiable with high CD44 and low CD24 expression and also it is able to suppress mTOR signalling. One can amply justify the employment of metformin being able to sensitive cells to cytotoxic agents. As stated earlier, the efficacy of metformin can be amplified with the hormonal and growth factor status of cancers such as breast cancer and deployment of their inhibitors. Much of the progress made in the identification of suitable means of targeting CSCs and cancer cells often seems serendipitous. There are overwhelming opportunities to investigate the exploitation of the growth factors other than the conventional ones in targeted therapy.

Several signalling components seem to be able to play tumour promoter or inhibitor roles. Several important elements have been identified so far, but from the practical point of view these are difficult to target specifically. On the other hand, modulators of signalling function do seem to occupy a prominent position in the context of the questions being addressed here. The 14-3-3 family proteins to modulate cell signalling systems, cell proliferation, cell cycle regulation and apoptosis, cytoskeletal dynamics, functions that are crucial to tumour development and progression. They target and bind components of signalling systems and in this way modulate information flow leading to the positive or negative regulation of cell motility and invasion, cell proliferation and apoptosis, EMT and metastasis. By virtue of this ability, the 14-3-3 proteins have been allied with tumour progression or suppression. Of interest also is that some isoforms might determine response to chemotherapy and associated with radiation and drug resistance. However, efforts towards taking advantage of their biological action using inhibitors are still at an incipient stage, possibly due to the lack consensus concerning the effects of the individual members. Small molecules capable of disrupting 14-3-3 interaction with their targets are being designed and tested for their biological efficacy.

Science and serendipity are inseparable but serendipity is also linked with intensity of effort. The massive literature that identifies and describes tumour promoters and suppressors and their modes of function bears overwhelming testimony to this intensity of effort and dedication to the cause of advancing human wellbeing. Humility is the hallmark of science but as the quoted Thirukkural says science can legitimately be proud of its achievements arising from the intensity of the pursuit. This alone by accident or rational design might yield fruit.

References

Aarumugam, T., Simeone, D.M., Schmidt, A.M., Logsdon, C.D., 2004. S100P stimulates cell proliferation and survival via receptor for activated glycation end products (RAGE). J. Biol. Chem. 279, 5059–5065.

Abdelrahim, M., Smith III, R., Burghardt, R., Safe, S., 2004. Role of Sp proteins in regulation of vascular endothelial growth factor expression and paroliferation of pancreatic cancer cells. Cancer Res. 64, 6740–6749.

Abdouh, M., Facchino, S., Chatoo, W., Balasingam, V., Ferreira, J., Bernier, G., 2009. BMI1 sustains human glioblastoma multiforme stem cell renewal. J. Neurosci. 29, 8884–8896.

Abe, H., Kamai, T., Shirataki, H., Oyama, T., Arai, K., Yoshida, K., 2008a. High expression of RAN GTPase is associataed with localinvasionand metastasis of human clear cell renal carcinoma. Int. J. Cancer 122, 2391–2397.

Abe, M., Sugiura, T., Takahashi, M., Ishii, K., Shimoda, M., Shirasuna, K., 2008b. A novel function of CD82/KAI-1 on E-cadherin-mediated homophilic cellular adhesion of cancer cells. Cancer Lett. 266, 163–170.

Abrami, L., Kunz, B., Deuquet, J., Bafico, A., Davidson, G., van der Goot, F.G., 2008. Functional interactions between anthrax toxin receptors and the WNT signalling protein LRP6. Cell. Microbiol. 10, 2509–2519.

Abu Ghazaleh, H., Chow, R.S., Choo, S.L., Pham, D., Olesen, J.D., Wong, R.X., et al., 2010. 14-3-3 mediated regulation of the tumor suppressor protein, RASSF1A. Apoptosis 15, 117–127.

Adams, B.D., Furneaux, H., White, B.A., 2007. The micro-ribonucleic acid (miRNA) miR-206 targets the human estrogen receptor-alpha (ER alpha) and represses ER alpha messenger RNA and protein expression in breast cancer cell lines. Mol. Endocrinol. 21, 1132–1147.

Adams, C.J., Graham, A.L., Jansson, M., Coutts, A.S., Edelmann, M., Smith, L., et al., 2008. ATM and Chk2 kinase target the p53 cofactor Strap. EMBO Rep. 9, 1222–1229.

Aguilera, C., Fernandez-Majada, V., Ingles-Esteve, J., Rodilla, V., Bigas, A., et al., 2006. Efficient nuclear export of p65-IkappaBalpha complexes requires 14-3-3 proteins. J. Cell. Sci. 119, 3695–3704.

Aka, J.A., Zerradi, M., Houle, F., Huot, J., Lin, S.X., 2012. 17beta-hydroxysteroid dehydrogenase type 1 modulates breast cancer protein profile and impacts cell migration. Breast Cancer Res. 14, R92.

Akahira, J.I., Suzuki, F., Suzuki, T., Miura, I., Kamogawa, N., Miki, Y., et al., 2007. Decreased expression of RIZ1 and its clinicopathological significance in epithelial ovarian carcinoma: correlation with epigenetic inactivation by aberrant DNA methylation. Pathol. Int. 57, 725–733.

Akiba, J., Murakami, Y., Noda, M., Watari, K., Ogasawara, S., Yoshida, T., et al., 2011. N-myc downstream regulated gene1/Cap43 overexpression suppresses tumor growth by hepatic cancer cells through cell cycle arrest at the G(0)/G(1) phase. Cancer Lett. 310, 25–34.

Akiyama, Y., Maesawa, C., Ogasawara, S., Terashima, M., Masuda, T., 2003. Cell-type-specific repression of the maspin gene is disrupted frequently by demethylation at the promoter region in gastric intestinal metaplasia and cancer cells. Am. J. Pathol. 163, 1911–1919.

Akutagawa, A., Fukami, K., Banno, Y., Takenawa, T., Kannagi, R., Yokoyama, Y., et al., 2006. Disruption of phospholipase C delta 4 gene modulates the liver regeneration in cooperation with nuclear protein kinase C. J. Biochem. 140, 619–625.

Al-Alwan, M., Olabi, S., Ghebeh, H., Barhoush, E., Tulbah, A., Al-Tweigeri, T., et al., 2011. Fascin is a key regulator of breast cancer invasion that acts via the modification of metastasis-associated molecules. PLoS ONE 6, e27339.

Al-Hajj, M., Wicha, S.M., Benito-Hernandez, A., Morrison, S.J., Clarke, M.F., 2003. Prospective identification of tumorigenic breast cancer cells. Proc. Natl. Acad. Sci. U.S.A. 100, 3983–3988.

Alimova, I.N., Liu, B., Fan, Z., Edgerton, S.M., Dillon, T., Lind, S.E., et al., 2009. Metformin inhibits breast cancer cell growth, colony formation and induces cell cycle arrest in vitro. Cell Cycle 8, 909–915.

Al-Mulla, F., Hagan, S., Al-Ali, W., Jacob, S.P., Behbehani, A.I., Bitar, M.S., et al., 2008. Raf kinase inhibitor protein: mechanism of loss of expression and association with genomic instability. J. Clin. Pathol. 61, 524–529.

Al-Mulla, F., Bitar, M.S., Taqi, Z., Rath, O., Kolch, W., 2011. RAF kinase inhibitory protein (RKIP) modulates cell cycle kinetics and motility. Mol. Biosyst. 7, 928–941.

Alvarez, C.E., 2008. On the origins of arrestin and rhodopsin. BMC Evol. Biol. 8, 222.

Ambros, V., Lee, R.C., Lavanway, A., Williams, P.T., Jewell, D., 2003. MicroRNAs and other tiny endogenous RNAs in C. elegans. Curr. Biol. 13, 807–818.

Ananthanarayanan, B., Das, S., Rhee, S.G., Murray, D., Cho, W., 2002. Membrane targeting of C2 domains of phospholipase C-delta isoforms. J. Biol. Chem. 277, 3568–3575.

Andersen, K., Nesland, J.M., Holm, R., Florenes, V.A., Fodstad, O., Maelandsmo, G.M., 2004. Expression of S100A4 combined with reduced E-cadherin expression predicts patient outcome in malignant melanoma. Mod. Pathol. 17, 990–997.

Andersen, S.D., Keijzers, G., Rampakakis, E., Engels, K., Luhn, P., El-Shemerly, M., et al., 2012. 14-3-3 checkpoint regulatory proteins interact specifically with DNA repair protein human exonuclease 1 (hEXO1) via a semi-conserved motif. DNA Repair 11, 267–277.

Ang, E.S.M., Pavlos, N.J., Chai, L.Y., Qi, M., Cheng, T.S., Steer, J.H., et al., 2009. Caffeic acid phenethyl ester, an active component of honeybee Propolis attenuates osteoclastogenesis and bone resorption via the suppression of RANKL-induced NF-kappa B and NFAT ctivity. J. Cell. Physiol. 221, 642–649.

Angeles, C., Moreno-Bueno, G., Romero-Perez, L., Van De Vijver, K., Biscuola, M., Angeles, L.G.M., et al., 2011. Micro-RNA signature of the epithelial–mesenchymal transition in endometrial carcinosarcoma. J. Pathol. 223, 72–80.

Angus, L., Moleirinho, S., Herron, L., Sinha, A., Zhang, X., Niestrata, M., et al., 2012. Willin/FRMD6 expression activates the Hippo signaling pathway kinases in mammals and antagonizes oncogenic YAP. Oncogene 31, 238–250.

Antognelli, C., Del Buono, C., Ludovini, V., Gori, S., Talesa, V.N., Crino, L., et al., 2009. CYP17, GSTP1, PON1 and GLO1 gene polymorphisms as risk factors for breast cancer: an Italian case-control study. BMC Cancer 9, 115.

Aprelikova, O., Pace, A.J., Fang, B., Koller, B.H., Liu, E.T., 2001. BRCA1 is a selective coactivator of 14-3-3 sigma gene transcription in mouse embryonic stem cells. J. Biol. Chem. 276, 25647–25650.

Apte, U., Zeng, G., Muller, P., Tan, X., Micsenyi, A., Cieply, B., et al., 2006. Activation of Wnt/beta-catenin pathway during hepatocyte growth factor-induced hepatomegaly in mice. Hepatology 44, 992–1002.

Arao, T., Yanagihara, K., Takigahira, M., Takeda, M., Koizumi, F., Shiratori, Y., et al., 2006. ZD6474 inhibits tumor growth and intraperitoneal dissemination in a highly metastatic orthotopic gastric cancer model. Int. J. Cancer 118, 483–489.

Arrendale, A., Kim, K., Choi, J.Y., Li, W., Geahlen, R.L., Borch, R.F., 2012. Synthesis of a phosphoserine mimetic prodrug with potent 14-3-3 protein inhibitory activity. Chem. Biol. 19, 764–771.

Asangani, I.A., Rasheed, S.A., Nikolova, D.A., Leupold, J.H., Colburn, N.H., Post, S., et al., 2008. MicroRNA-21 (miR-21) post-transcriptionally down regulates tumor suppressor Pdcd4 and stimulates invasion, intravasation and metastasis in colorectal cancer. Oncogene 27, 2128–2136.

Asaoka, Y., Kanai, F., Ichimura, T., Tateishi, K., Tanaka, Y., Ohta, M., et al., 2010. Identification of a suppressive mechanism for Hedgehog signaling through a novel interaction of Gli with 14-3-3. J. Biol. Chem. 285, 4185–4194.

Asdaghi, N., Kilani, R.T., Hosseini-Tabatabaei, A., Odemuyiwa, S.O., Hackett, T.L., Knight, D.A., et al., 2012. Extracellular 14-3-3 from human lung epithelial cells enhances MMP-1 expression. Mol. Cell. Biochem. 360, 261–270.

Aslan, J.E., You, H., Williamson, D.M., Endig, J., Youker, R.T., Thomas, L., et al., 2009. Akt and 14-3-3 control a PACS-2 homeostatic switch that integrates membrane traffic with TRAIL-induced apoptosis. Mol. Cell. 34, 497–509.

Avizienyte, E., Fincham, V.J., Brunton, V.G., Frame, M.C., 2004. Src SH3/2 domain-mediated peripheral accumulation of Src and phospho-myosin is linked to deregulation of E-cadherin and the epithelial–mesenchymal transition. Mol. Biol. Cell. 15, 2794–2803.

Avraamides, C.J., Garmy-Susini, B., Varner, J.A., 2008. Integrins in angiogenesis and lymphangiogenesis. Nat. Rev. Cancer 8, 604–617.

Aylon, Y., Michael, D., Shueli, A., Yabuta, N., Nojima, H., Oren, M., 2006. A positive feedback loop between the p53 and Lats2 tumor suppressors prevents tetraploidization. Genes Dev. 20, 2687–2700.

Azuma, K., Kawahara, A., Hattori, S., Taira, T., Tsurutani, J., Watari, K., 2012. NDRG1/Cap43/Drg-1 may predict tumor angiogenesis and poor outcome in patients with lung cancer. J. Thorac. Oncol. 7, 779–789.

Bahra, M., Kamphues, C., Boas-Knoop, S., Lippert, S., Esendik, U., Schuller, U., et al., 2012. Combination of hedgehog signaling blockage and chemotherapy leads to tumor reduction in pancreatic adenocarcinomas. Pancreas 41, 222–229.

Baas, A.F., Boudeau, J., Sapkota, G.P., Smit, L., Medema, R., Morrice, N.A., et al., 2003. Activation of the tumour suppressor kinase LKB1 by the STE20-like pseudokinase STRAD. EMBO J. 22, 3062–3072.

Bai, Y., Zhou, T., Fu, H., Sun, H., Huang, B., 2012. 14-3-3 interacts with LKB1 via recognizing phosphorylated threonine 336 residue and suppresses LKB1 kinase function. FEBS Lett. 586, 1111–1119.

Bair III, W.B., Cabello, C.M., Uchida, K., Bause, A.S., Wondrak, G.T., 2010. GLO1 overexpression in human malignant melanoma. Melanoma Res. 20, 85–96.

Bamburg, J.R., 1999. Proteins of the ADF/cofilin family: essential regulators of actin dynamics. Ann. Rev. Cell. Dev. Biol. 15, 185–230.

Bandyopadhyay, S., Pai, S.K., Hirota, S., Hosobe, S., Tsukada, T., Miura, K., et al., 2004. PTEN up-regulates the tumor metastasis suppressor gene Drg-1 in prostate and breast cancer. Cancer Res. 64, 7655–7660.

Bao, B., Wang, Z., Ali, S., Kong, D., Li, Y., Ahmad, A., et al., 2011a. Notch-1 induces epithelial–mesenchymal transition consistent with cancer stem cell phenotype in pancreatic cancer cells. Cancer Lett. 307, 26–36.

Bao, W., Fu, H.J., Xie, Q.S., Wang, L., Zhang, R., Guo, Z.Y., et al., 2011b. HER2 interacts with CD44 to up-regulate CXCR4 via epigenetic silencing of microRNA-139 in gastric cancer cells. Gastroenterol 141, 2076–2100.

Bar, N., Dikstein, R., 2010. MiR-22 forms a regulatory loop in PTEN/AKT pathway and modulates signaling kinetics. PLoS ONE 5, e10859.

Barbareschi, M., Cuorvo, L.V., Girlando, S., Bragantini, E., Eccher, C., Leonardi, E., et al., 2012. PI3KCA mutations and/or PTEN loss in Her2-positive breast carcinomas treated with trastuzumab are not related to resistance to anti-Her2 therapy. Virchows Archiv. 461, 129–139.

Baritaki, S., Chapman, A., Yeung, K., Spandidos, D.A., Palladino, M., Bonavida, B., 2009. Inhibition of epithelial to mesenchymal transition in metastatic prostate cancer cells by the novel proteasome inhibitor, NPI-0052: pivotal roles of Snail repression and RKIP induction. Oncogene 28, 3573–3585.

Barkham, T., Carlsson, B., Nilsson, Y., Enmark, E., Gustafsson, J., Nilsson, S., 1998. Differential response of estrogen receptor alpha and estrogen receptor beta to partial estrogen agonists and antagonists. Mol. Pharmacol. 54, 105–112.

Bartel, D.P., 2009. MicroRNAs: target recognition and regulatory functions. Cell 136, 215–233.

Bass, R., Wagstaff, L., Ravenhill, L., Ellis, V., 2009. Binding of extracellular maspin to beta 1 integrins inhibits vascular smooth muscle cell migration. J. Biol. Chem. 284, 27712–27720.

Baumann, P., Cremers, N., Kroese, F., Orend, G., Chiquet-Ehrismann, R., Uede, T., et al., 2005. CD24 expression causes the acquisition of multiple cellular properties associated with tumor growth and metastasis. Cancer Res. 65, 10783–10793.

Beach, S., Tang, H., Park, S., Dhillon, A.S., Keller, E.T., Kolch, W., et al., 2008. Snail is a repressor of RKIP transcription in metastatic prostate cancer cells. Oncogene 27, 2243–2248.

Belshaw, N.J., Elliott, G.O., Foxall, R.J., Dainty, J.R., Pal, N., Coupe, A., et al. 2008. Profiling CpG island field methylation in both morphologically normal and neoplastic human colonic mucosa. Br. J. Cancer 99, 136–142.

Benezra, R., 2001. Role of Id proteins in embryonic and tumor angiogenesis. Trends Cardiovasc Med. 11, 237–241.

Berardi, R., Santinelli, A., Onofri, A., Brunelli, A., Pierantoni, C., Pisa, E., et al., 2012. Maspin expression is a favorable prognostic factor in non-small cell lung cancer. Anal. Quant. Cytol. Histol. 34, 72–78.

Bergamaschi, A., Katzenellenbogen, B.S., 2012. Tamoxifen downregulation of miR-451 increases 14-3-3 zeta and promotes breast cancer cell survival and endocrine resistance. Oncogene 31, 39–47.

Berge, G., Pettersen, S., Grotterod, I., Bettum, I.J., Boye, K., Maelandsmo, G.M., 2011. Osteopontin – an important downstream effector of S100A4-mediated invasion and metastasis. Int. J. Cancer 129, 780–790.

Berger, N., Ben Bassat, H., Klein, B.Y., Laskov, R., 2007. Cytotoxicity of NF-kappa B inhibitors Bay 11-7085 and caffeic acid phenethyl ester to Ramos and other human B-lymphoma cell lines. Exp. Hematol. 35, 1495–1509.

Berman, D.M., Karhadkar, S.S., Hallahan, A.R., Pritchard, J.I., Eberhart, C.G., Watkins, D.N., et al., 2002. Medulloblastoma growth inhibition by hedgehog pathway blockade. Science 297, 1559–1561.

Bernstein, E., Caudy, A.A., Hammond, S.M., Hannon, G.J., 2001. Role for a bidentate ribonuclease in the initiation step of RNA interference. Nature 409, 363–366.

Beshir, A.B., Ren, G., Magpusao, A.N., Barone, L.M., Yeung, K.C., Fenteany, G., 2010. Raf kinase inhibitor protein suppresses nuclear factor-kappa B-dependent cancer cell invasion through negative regulation of matrix metalloproteinase expression. Cancer Lett. 299, 137–149.

Bhati, R., Patterson, C., Livasy, C.A., Fan, C., Ketelsen, D., Hu, Z., et al., 2008. Molecular characterization of human breast tumor vascular cells. Am. J. Pathol. 172, 1381–1390.

Bian, H.B., Pan, X., Yang, J.S., Wang, Z.X., De, W., 2011. Upregulation of microRNA-451 increases cisplatin sensitivity of non-small cell lung cancer cell line (A549). J. Exp. Clin. Cancer Res. 30, 20.

Bianchini, L., Saada, E., Gjernes, E., Marty, M., Haudebourg, J., Birtwisle-Peyrottes, I., et al., 2011. Let-7 microRNA and HMGA2 Levels of expression are not inversely linked in adipocytic tumors: analysis of 56 lipomas and liposarcomas with molecular cytogenetic data. Genes Chromosomes Cancer 50, 442–455.

Bilban, M., Ghaffari-Tabrizi, N., Hintermann, E., Bauer, S., Molzer, S., Zoratti, C., et al., 2004. Kisspeptin-10, a KiSS-1/metastin-derived decapeptide, is a physiological invasion inhibitor of primary human trophoblasts. J. Cell. Sci. 117, 1319–1328.

Bilgir, O., Kebapcilar, L., Taner, C., Bilgir, F., Kebapcilar, A., Bozkaya, G., et al., 2009. The effect of ethinylestradiol (EE)/cyproterone acetate (CA) and EE/CA plus Metformin treatment on adhesion molecules in cases with polycystic ovary syndrome (PCOS). Int. Med. 48, 1193–1199.

Birner, P., Jesch, B., Schultheis, A., Schoppmann, S.F., 2012. RAF-kinase inhibitor protein (RKIP) downregulation in esophageal cancer and its metastases. Clin. Exp. Metastasis 29, 551–559.

Blees, J.S., Bokesch, H.R., Rubsamen, D., Schulz, K., Milke, L., Bajer, M.M., et al., 2012. Erioflorin stabilizes the tumor suppressor Pdcd4 by inhibiting its interaction with the E3-ligase beta-TrCP1. PLoS ONE 7, e46567.

Boelte, K.C., Gordy, L.E., Joyce, S., Thompson, M.A., Li, Yang., Lin, P.C., 2011. Rgs2 mediates pro-angiogenic function of myeloid derived suppressor cells in the tumor microenvironment via upregulation of MCP-1. PLoS ONE 6, e18534.

Bonanni, B., Puntoni, M., Cazzaniga, M., Pruneri, G., Serrano, D., Guerrieri-Gonzaga, A., et al. 2012. Dual effect of metformin on breast cancer proliferation in a randomized presurgical trial. J. Clin. Oncol. 30, 2593–2600.

Bonavida, B., Baritaki, S., 2011. Dual role of NO donors in the reversal of tumor cell resistance and EMT: downregulation of the NF-kappa B/Snail/YY1/RKIP circuitry. Nitric Oxide Biol. Chem. 24, 1–7.

Bordonaro, M., Tewari, S., Atamna, W., Lazarova, D.L., 2011. The notch ligand Delta-like 1 integrates inputs from TGFbeta/activin and Wnt pathways. Exp. Cell Res. 317, 1368–1381.

Borel, F., Han, R., Visser, A., Petry, H., van Deventer, S.J., Jansen, P.L., et al., 2011. ATP-Binding cassette transporter genes up-regulation in untreated hepatocellular carcinoma is mediated by cellular microRNAs. Hepatology 55, 821–832.

Bork, P.M., Schmitz, M.L., Kuhnt, M., Escher, C., Heinrich, M., 1997. Sesquiterpene lactone containing Mexican Indian medicinal plants and pure sesquiterpene lactones as potent inhibitors of transcription factor NF-kappaB. FEBS Lett. 402, 85–90.

Borkosky, S.S., Gunduz, M., Nagatsuka, H., Beder, L.B., Gunduz, E., Ali, M.A.S., et al., 2009. Frequent deletion of ING2 locus at 4q35.1 associates with advanced tumor stage in head and neck squamous cell carcinoma. J. Cancer Res. Clin. Oncol. 135, 703–713.

Borkosky, S.S., Gunduz, M., Beder, L., Tsujigiwa, H., Tamamura, R., Gunduz, E., et al., 2010. Allelic loss of the ING gene family loci is a frequent event in ameloblastoma. Oncol. Res. 18, 509–518.

Boudeau, J., Scott, J.W., Resta, N., Deak, M., Kieloch, A., Komander, D., et al., 2004. Analysis of the LKB1–STRAD–MO25 complex. J. Cell. Sci. 117, 6365–6375.

Boulay, G., Dubuissez, M., Van Rechem, C., Forget, A., Helin, K., Ayrault, O., et al., 2012. Hypermethylated in cancer 1 (HIC1) recruits Polycomb repressive complex 2 (PRC2) to a subset of its target genes through interaction with human Polycomb-like (hPCL) proteins. J. Biol. Chem. 287, 10509–10524.

Bouquet, F., Pal, A., Pilones, K.A., Demaria, S., Hann, B., Akhurst, R.J., et al., 2011. TGF beta 1 inhibition increases the radiosensitivity of breast cancer cells in vitro and promotes tumor control by radiation in vivo. Clin. Cancer Res. 17, 6754–6765.

Brabletz, S., Bajdak, K., Meidhof, S., Burk, U., Niedermann, G., Firat, E., et al., 2011. The ZEB1/miR-200 feedback loop controls Notch signalling in cancer cells. EMBO J. 30, 770–782.

Bradley, K.A., Mogridge, J., Mourez, M., Collier, R.J., Young, J.A.T., 2001. Identification of the cellular receptor for anthrax toxin. Nature 414, 225–229.

Bradley, K.A., Young, J.A.T., 2003. Anthrax toxin receptor proteins. Biochem. Pharmacol. 65, 309–314.

Brandt, R., Grutzmann, R., Bauer, A., Jesnowski, R., Ringel, J., Lohr, M., et al., 2004. DNA microarray analysis of pancreatic malignancies. Pancreatology 4, 587–597.

Brasch, J., Harrison, O.J., Ahlsen, G., Carnally, S.M., Henderson, R.M., Honig, B., et al. 2011. Structure and binding mechanism of vascular endothelial cadherin: a divergent classical cadherin. J. Mol. Biol. 408, 57–73.

Bright, N.J., Thornton, C., Carling, D., 2009. The regulation and function of mammalian AMPK-related kinases. Acta Physiol. 196, 15–26.

Bromberg-White, J.L., Boguslawski, E., Duesbery, N.S., 2009. Perturbation of mouse retinal vascular morphogenesis by Anthrax lethal toxin. PLoS ONE 4, e6956.

Brown, K.A., Hunger, N.I., Docanto, M., Simpson, E.R., 2010. Metformin inhibits aromatase expression in human breast adipose stromal cells via stimulation of AMP-activated protein kinase. Breast Cancer Res. Treat. 123, 591–596.

Brown, K.A., McInnes, K.J., Takagi, K., Ono, K., Hunger, N.I., Wang, L., et al., 2011. LKB1 expression is inhibited by estradiol-17 beta in MCF-7 cells. J. Steroid Biochem. Mol. Biol. 127, 439–443.

Bruning, A., Juckstock, J., Blankenstein, T., Makovitzky, J., Kunze, S., Mylonas, I., 2010. The metastasis-associated gene MTA3 is down regulated in advanced endometrioid adenocarcinomas. Histol. Histopathol. 25, 1447–1456.

Bryant, H.E., Petermann, E., Schultz, N., Jemth, A.S., Loseva, O., Issaeva, N., et al., 2009. PARP is activated at stalled forks to mediate Mre11 dependent replication restart and recombination. EMBO J. 28, 2601–2615.

Buensuceso, C.S., Woodside, D., Huff, J.L., Plopper, G.E., O'Toole, T, 2001. The WD protein Rack1 mediates protein kinase C and integrin-dependent cell migration. J. Cell. Sci. 114, 1691–1698.

Buonamici, S., Williams, J., Morrissey, M., Wang, A., Guo, R., Vattay, A., et al., 2010. Interfering with resistance to smoothened antagonists by inhibition of the PI3K pathway in medulloblastoma. Sci. Transl. Med. 2, 51ra70.

Burk, U., Schubert, J., Wellner, U., Schmalhofer, O., Vincan, E., Spaderna, S., et al., 2008. Areciprocal repression between ZEB1 and members of the miR-200 family promotes EMT and invasion in cancer cells. EMBO Rep. 9, 582–589.

Burks, P.J., Isaacs, H.V., Pownall, M.E., 2009. FGF signalling modulates transcriptional repression by Xenopus Groucho-related 4. Biol. Cell 101, 301–308.

Buther, K., Plaas, C., Barnekow, A., Kremerskothen, J., 2004. KIBRA is a novel substrate for protein kinase Cζ. Biochem. Biophys. Res. Commun. 317, 703–707.

Cabello, C.M., Qiao, S., Bair III, W.B., Davis, A.L., Lesson, J.L., Wondrak, G.T., 2010. GLO1 Downregulation in human renal cell carcinoma enables targeted chemotherapeutic intervention using methylglyoxal as cytotoxic heat shock inducer. Free Radic. Biol. Med. 49, S73–S74.

Cadenas, C., Franckenstein, D., Schmidt, M., Gehrmann, M., Hermes, M., Geppert, B., et al., 2010. Role of thioredoxin reductase 1 and thioredoxin interacting protein in prognosis of breast cancer. Breast Cancer Res. 12, R44.

Cai, C., Ashktorab, H., Pang, X., Zhao, Y., Sha, W., Liu, Y., et al., 2012. MicroRNA-211 expression promotes colorectal cancer cell growth in vitro and in vivo by targeting tumor suppressor CHD5. PLoS ONE 7, e29750.

Cajone, F., Sherbet, G.V., 1999. Stathmin is involved in S100A4-mediated regulation of cell cycle progression. Clin. Exp. Metastasis 17, 865–871.

Caldwell, G.M., Jones, C., Gensberg, K., Jan, S., Hardy, R.G., Byrd, P., et al., 2004. The Wnt antagonist sFRP1 in colorectal tumorigenesis. Cancer Res. 64, 883–888.

Camps, C., Buffa, F.M., Colella, S., Moore, J., Sotiriou, C., Sheldon, H., et al., 2008. hsa-miR-210 is induced by hypoxia and is an independent prognostic factor in breast cancer. Clin. Cancer Res. 14, 1340–1348.

Cao, L., Bombard, J., Cintron, K., Sheedy, J., Weetall, M.L., Davis, T.W., 2011. BMI1 as a novel target for drug discovery in cancer. J. Cell. Biochem. 112, 2729–2741.

Caplen, N.J., 2004. Gene therapy progress and prospects. Down regulating gene expression: the impact of RNA interference. Gene Ther. 11, 1241–1248.

Carey, J.P.W., Asirvatham, A.J., Galm, O., Ghogomu, T.A., Chaudhary J. 2009. Inhibitor of differentiation 4 (Id4) is a potential tumor suppressor in prostate cancer. BMC Cancer 9, Art. no. 173.

Carletti, M.Z., Fiedler, S.D., Christenson, L.K., 2010. MicroRNA 21 blocks apoptosis in mouse periovulatory granulosa cells. Biol. Reprod. 83, 286–295.

Carmell, M.A., Hannon, G.J., 2004. RNase III enzymes and the initiation of gene silencing. Nat. Struct. Mol. Biol. 11, 214–218.

Carney, J.P., Maser, R.S., Olivares, H., Davis, E.M., Le Beau, M., Yates III, J.R., et al., 1998. The hMre11/hRad50 protein complex and Nijmegen breakage syndrome: linkage of double-strand break repair to the cellular DNA damage response. Cell 93, 477–486.

Carson-Walter, E.B., Nanda, A., Vogelstein, B., Kinzler, K.W., Croix, B.S., 2001. Cell surface tumor endothelial markers are conserved in mice and humans. Cancer Res. 61, 6649–6655.

Casas, E., Kim, J., Bendesky, A., Ohno-Machado, L., Wolfe, C.J., Yang, J., 2011. Snail2 is an essential mediator of Twist1-induced epithelial mesenchymal transition and metastasis. Cancer Res. 71, 245–254.

Castano, J.P., Martinez-Fuentes, A.J., Gutierrez-Pascual, E., Vaudry, H., Tena-Sempere, M., Malagon, M.M., 2009. Intracellular signaling pathways activated by kisspeptins through GPR54: do multiple signals underlie function diversity? Peptides 30, 10–15.

Cecchinato, V., Chiaramonte, R., Nizzardo, M., Cristofaro, B., Basile, A., Sherbet, G.V., et al., 2007. Resveratrol-induced apoptosis in human T-cell acute lymphoblastic leukaemia MOLT-4 cells. Biochem. Pharmacol. 74, 1568–1574.

Cella, N., Contreras, A., Latha, K., Rosen, J.M., Zhang, M., 2006. Maspin is physically associated with β1 integrin regulating cell adhesion in mammary epithelial cells. FASEB J. 20, 1510–1512.

Cengiz, B., Gunduz, E., Gunduz, M., Beder, L.B., Tamamura, R., Bagci, C., et al., 2010. Tumor-specific mutation and downregulation of ING5 detected in oral squamous cell carcinoma. Int. J. Cancer 127, 2088–2094.

Ceppi, P., Mudduluru, G., Kumarswamy, R., Rapa, I., Scagliotti, G.V., Papotti, M., et al., 2010. Loss of miR-200c expression induces an aggressive, invasive, and chemoresistant phenotype in non-small cell lung cancer. Mol. Cancer Res. 8, 1207–1216.

Chamorro-Jorganes, A., Araldi, E., Penalva, L.O.F., Sandhu, D., Fernandez-Hernando, C., Suarez, Y., 2011. MicroRNA-16 and microRNA-424 regulate cell-autonomous angiogenic functions in endothelial cells via targeting vascular endothelial growth factor receptor-2 and fibroblast growth factor receptor-1. Artherioscl. Thromb. Vasc. Biol. 31, 2595.

Chan, J.A., Krichevsky, A.M., Kosik, K.S., 2005. MicroRNA-21 is an antiapoptotic factor in human glioblastomacells. Cancer Res. 65, 6029–6033.

Chan, K.T., Lung, M.L., 2004. Mutant p53 expression enhanced drug resistance in a hepatocellular carcinoma cell line. Cancer Chemother. Pharmacol. 53, 519–526.

Chan, S.Y.Y., To, K.F., Leung, S.F., Yip, W.W.L., Mak, M.K.F., Chung, G.T.Y., et al., 2010. 14-3-3 sigma expression as a prognostic marker in undifferentiated nasopharyngeal carcinoma. Oncol. Rep. 24, 949–955.

Chang, C.J., Hsu, C.C., Chang, C.H., Tsai, L.L., Chang, Y.C., Lu, S.W., et al., 2011a. Let-7d functions as novel regulator of epithelial–mesenchymal transition and chemoresistant property in oral cancer. Oncol. Rep. 26, 1003–1010.

Chang, C.J., Chao, C.H., Xia, W., Yang, J.Y., Xiong, Y., Li, C.W., et al., 2011b. p53 regulates epithelial–mesenchymal transition and stem cell properties through modulating miRNAs. Nat. Cell. Biol. 13, 317–323.

Chang, H., Li, Q., Moraes, R.C., Lewis, M.T., Hamel, P.A., 2010. Activation of Erk by sonic hedgehog independent of canonical hedgehog signalling. Int. J. Biochem. Cell. Biol. 42, 1462–1471.

Chang, J.T., Wang, H.M., Chang, K.W., Chen, W.H., Wen, M.C., Hsu, Y.M., et al., 2005. Identification of differentially expressed genes in oral squamous cell carcinoma (OSCC): overexpression of NPM, CDK1 and NDRG1 and underexpression of CHES1. Int. J. Cancer 114, 942–949.

Chang, S., Wang, R.H., Akagi, K., Kim, K.A., Martin, B.K., Cavallone, L., et al., 2011c. Tumor suppressor BRCA1 epigenetically controls oncogenic microRNA-155. Nat. Med. 17 1275–U308

Chang, T.C., Liu, C.C., Hsing, E.W., Liang, S.M., Chi, Y.H., Sung, L.Y., et al., 2012a. 14-3-3 sigma regulates beta-catenin-mediated mouse embryonic stem cell proliferation by sequestering GSK-3 beta. PLoS ONE 7, e40193.

Chang, T.H., Tsai, M.F., Su, K.Y., Wu, S.G., Huang, C.P., Yu, S.L., et al., 2011d. Slug confers resistance to the epidermal growth factor receptor tyrosine kinase inhibitor. Am. J. Resp. Crit. Care Med. 183, 1071–1079.

Chang, Z.G., Wei, J.M., Qin, C.F., Hao, K., Tian, X.D., Xie, K., et al., 2012b. Suppression of the epidermal growth factor receptor inhibits epithelial–mesenchymal transition in human pancreatic cancer PANC-1 cells. Dig. Dis. Sci. 57, 1181–1189.

Chapman, E.J., Carrington, J.C., 2007. Specialization and evolution of endogenous small RNA pathways. Nat. Rev. Genet. 8, 884–896.

Chatterjee, D., Bai, Y., Wang, Z., Beach, S., Mott, S., Roy, R., et al., 2004. RKIP sensitizes prostate and breast cancer cells to drug-induced apoptosis. J. Biol. Chem. 279, 17515–17523.

Chauhan, D., Singh, A., Brahmandam, M., Podar, K., Hideshirna, T., Richardson, P., et al., 2008. Combination of proteasome inhibitors bortezomib and NPI-0052 trigger in vivo synergistic cytotoxicity in multiple myeloma. Blood 111, 1654–1664.

Che, N., Zhao, X.L., Sun, T., Zhao, X.M., Gu, Q., Dong, X.Y., et al., 2011. The role of Twist1 in hepatocellular carcinoma angiogenesis: a clinical study. Hum. Pathol. 42, 840–847.

Chen, J., Li, S., Yang, Z., Lu, G., Hu, H., 2008a. Correlation between NDRG1 and PTEN expression in endometrial carcinoma. Cancer Sci. 99, 706–710.

Chen, J., Hui, S.T., Couto, F.M., Mungrue, I.N., Davis, D.B., Attie, A.D., et al., 2008b. Thioredoxin-interacting protein deficiency induces Akt/Bcl-xL signaling and pancreatic beta-cell mass and protects against diabetes. FASEB J. 22, 3581–3594.

Chen, J.K., Taipale, J., Cooper, M.K., Beachy, P.A., 2002. Inhibition of Hedgehog signaling by direct binding of cyclopamine to Smoothened. Genes Develop. 16, 2743–2748.

Chen, L., Zheng, J., Zhang, Y., Yang, L., Wang, J., Ni, J., et al., 2011a. Tumor-specific expression of microRNA-26a suppresses human hepatocellular carcinoma growth via cyclin-dependent and -independent pathways. Mol. Ther. 19, 1521–1528.

Chen, Q., Chen, X., Zhang, M., Fan, Q., Luo, S., Cao, X., 2011b. miR-137 is frequently down-regulated in gastric cancer and is a negative regulator of cdc42. Dig. Dis. Sci. 56, 2009–2016.

Chen, R.X., Xia, Y.H., Xue, T.C., Ye, S.L., 2011c. Osteopontin promotes hepatocellular carcinoma invasion by up-regulating MMP-2 and uPA expression. Mol. Biol. Rep. 38, 3671–3677.

Chen, R.X., Xia, Y.H., Xue, T.C., Ye, S.L., 2012a. Suppression of microRNA-96 expression inhibits the invasion of hepatocellular carcinoma cells. Mol. Med. Rep. 5, 800–804.

Chen, S.L., Wu, Y.S., Shieh, H.Y., Yen, C.C., Shen, J.J., Lin, K.H., 2003a. p53 is a regulator of the metastasis suppressor gene Nm23-H1. Mol. Carcinog. 36, 204–214.

Chen, Y., Knosel, T., Kristiansen, G., Pietas, A., Garber, M.E., Matsuhashi, S., et al., 2003b. Loss of PDCD4 expression in human lung cancer correlates with tumour progression and prognosis. J. Pathol. 200, 640–646.

Chen, Y., Li, S., Tong, C., Zhao, Y., Wang, B., Liu, Y., et al., 2010. G protein-coupled receptor kinase 2 promotes high-level Hedgehog signaling by regulating the active state of Smo through kinase-dependent and kinase-independent mechanisms in Drosophila. Genes Develop. 24, 2054–2067.

Chen, Y., Sasai, N., Ma, G., Yue, T., Jia, J., Briscoe, J., et al., 2011d. Sonic Hedgehog dependent phosphorylation by CK1alpha and GRK2 is required for ciliary accumulation and activation of smoothened. PLoS Biol. 9, e1001083.

Chen, Y.L., Xu, S.W., Eastwood, M., Black, C.M., Denton, C.P., Leask, A., et al., 2006. Contribution of activin receptor-like kinase 5 (transforming growth factor beta receptor type I) signaling to the fibrotic phenotype of scleroderma fibroblasts. Arthritis Rheum. 54, 1309–1316.

Chen, Y.W., Paliwal, S., Draheim, K., Grossman, S.R., Lewis, B.C., 2008c. p19(Arf) inhibits the invasion of hepatocellular carcinoma cells by binding to C-terminal binding protein. Cancer Res. 68, 476–482.

Chen, Z., Zhang, D., Yue, F., Zheng, M., Kovacevic, Z., Richardson, D.R., 2012b. The iron chelators Dp44mT and DFO inhibit TGF-beta-induced epithelial–mesenchymal transition via up-regulation of N-myc downstream-regulated gene 1 (NDRG1). J. Biol. Chem. 287, 17016–17028.

Cheng, J.C., Auersperg, N., Leung, P.C.K., 2012a. EGF-induced EMT and invasiveness in serous borderline ovarian tumor cells: a possible step in the transition to low-grade serous carcinoma cells? PLoS ONE 7, e34071.

Cheng, J.C., Chang, H.M., Leung, P.C.K., 2012b. Epidermal growth factor induces human oviductal epithelial cell invasion by down-regulating E-cadherin expression. J. Clin. Endocrinol. Metab. 97, E1380–E1389.

Cheng, S., Maier, D., Neubueser, D., Hipfner, D.R., 2010. Regulation of smoothened by Drosophila G-protein-coupled receptor kinases. Develop. Biol. 337, 99–109.

Cheng, W.L., Tsai, M.M., Tsai, C.Y., Huang, Y.H., Chen, C.Y., Chi, H.C., et al., 2012c. Glyoxalase-I is a novel prognosis factor associated with gastric cancer progression. PLoS ONE 7, e34352.

Chiang, P.C., Thompson, D.C., Ghosh, S., Heitmeier, M.R., 2011. A formulation-enabled preclinical efficacy assessment of a farnesoid X receptor agonist, GW4064, in hamsters and Cynomolgus monkeys. J. Pharmaceut. Sci. 100, 4722–4733.

Chiba, T., Miyagi, S., Saraya, A., Aoki, R., Seki, A., Morita, Y., et al., 2008. The polycomb gene product BMI1 contributes to the maintenance of tumor-initiating side population cells in hepatocellular carcinoma. Cancer Res. 68, 7742–7749.

Ching, Y.P., Wong, C.M., Chan, S.F., Leung, T.H.Y., Ng, D.C.H., Jin, D.Y., et al., 2003. Deleted in liver cancer (DLC) 2 encodes a RhoGAP protein with growth suppressor function and is underexpressed in hepatocellular carcinoma. J. Biol. Chem. 278, 10824–10830.

Chinnadurai, G., 2009. The transcriptional corepressor CtBP: a foe of multiple tumor suppressors. Cancer Res. 69, 731–734.

Chishti, A.H., Kim, A.C., Marfatia, S.M., Lutchman, M., Hanspal, M., Jindal, H., et al., 1998. The FERM domain: a unique module involved in the linkage of cytoplasmic proteins to the membrane. Trends Biochem. Sci. 23, 281–282.

Cho, J.H., Kim, H.S., Park, C.S., Kim, J.K., Jung, K.Y., Shin, B.K., et al., 2007. Maspin expression in early oral tongue cancer and its relation to expression of mutant-type p53 and vascular endothelial growth factor (VEGF). Oral. Oncol. 43, 272–277.

Cho, S.H., Park, Y.S., Kim, H.J., Kim, C.H., Lim, S.W., Huh, J.W., et al., 2012. CD44 enhances the epithelial–mesenchymal transition in association with colon cancer invasion. Int. J. Oncol. 41, 211–218.

Choi, U.J., Jee, B.K., Lim, Y., Lee, K.H., 2009. KAI1/CD82 decreases Rac1 expression and cell proliferation through PI3K/Akt/mTOR pathway in H1299 lung carcinoma cells. Cell. Biochem. Funct. 27, 40–47.

Chopra, A.P., Boone, S.A., Liang, X., Duesbery, N.S., 2003. Anthrax lethal factor proteolysis and inactivation of MAPK kinase. J. Biol. Chem. 278, 9402–9406.

Chow, A., Hao, Y., Yang, X., 2010. Molecular characterization of human homologs of yeast MOB1. Int. J. Cancer 126, 2079–2089.

Christgen, M., Bruchhardt, H., Ballmaier, M., Krech, T., Laenger, F., Kreipe, H., et al., 2008. KAI1/CD82 is a novel target of estrogen receptor-mediated gene repression and down regulated in primary human breast cancer. Int. J. Cancer 123, 2239–2246.

Chua, M.S., Sun, H., Cheung, S.T., Mason, V., Higgins, J., Ross, D.T., et al., 2007. Overexpression of NDRG1 is an indicator of poor prognosis in hepatocellular carcinoma. Mod. Pathol. 20, 76–83.

Chun, S.G., Zhou, W., Yee, N.S., 2009. Combined targeting of histone deacetylases and hedgehog signaling enhances cytoxicity in pancreatic cancer. Cancer Biol. Ther. 8, 1328–1339.

Chung, L.C., Tsui, K.H., Feng, T.H., Lee, S.L., Chang, P.L., Juang, H.H., 2012. L-Mimosine blocks cell proliferation via upregulation of B-cell translocation gene 2 and N-myc downstream regulated gene 1 in prostate carcinoma cells. Am. J. Physiol. Cell. Physiol. 302, C676–C685.

Chung, T.W., Moon, S.K., Chang, Y.C., Ko, J.H., Lee, Y.C., Cho, G., et al., 2004. Novel and therapeutic effect of caffeic acid and caffeic acid phenyl ester on hepatocarcinoma cells: complete regression of hepatoma growth and metastasis by dual mechanism. FASEB J. 18, 1670–1681.

Claperon, A., Guedj, N., Mergey, M., Vignjevic, D., Desbois-Mouthon, C., Boissan, M., et al., 2012. Loss of EBP50 stimulates EGFR activity to induce EMT phenotypic features in biliary cancer cells. Oncogene 31, 1376–1388.

Coles, A.H., Marfella, C.G.A., Imbalzano, A.N., Steinman, H.A., Garlick, D.S., Gerstein, R.M., et al., 2008. p37(Ing1b) regulates B-cell proliferation and cooperates with p53 to suppress diffuse large B-cell lymphomagenesis. Cancer Res. 68, 8705–8714.

Collins, A.T., Berry, P.A., Hyde, C., Stower, M.J., Maitland, N.J., 2005. Prospective identification of tumorigenic prostate cancer stem cells. Cancer Res. 65, 10946–10951.

Comamala, M., Pinard, M., Theriault, C., Matte, I., Albert, A., Boivin, M., et al., 2011. Downregulation of cell surface CA125/MUC16 induces epithelial-to-mesenchymal transition and restores EGFR signalling in NIH:OVCAR3 ovarian carcinoma cells. Br. J. Cancer 104, 989–999.

Conley, S.J., Gheordunescu, E., Kakarala, P., Newman, B., Korkaya, H., Heath, A.N., et al., 2012. Antiangiogenic agents increase breast cancer stem cells via the generation of tumor hypoxia. Proc. Natl. Acad. Sci. U.S.A. 109, 2784–2789.

Contreras, C.M., Gurumurthy, S., Haynie, J.M., Shirley, L.J., Akbay, E.A., Wingo, S.N., et al., 2008. Loss of Lkb1 provokes highly invasive endometrial adenocarcinomas. Cancer Res. 68, 759–766.

Cordeiro-Silva, M. de F., Lima Oliveira, Z.F., Vasconcelos de Podesta, J.R., Gouvea, S.A., Von Zeidler, S.V., Louro, I.D., 2011. Methylation analysis of cancer-related genes in non-neoplastic cells from patients with oral squamous cell carcinoma. Mol. Biol. Rep. 38, 5435–5441.

Corradi, V., Mancini, M., Manetti, F., Petta, S., Santucci, M.A., Botta, M., 2010. Identification of the first non-peptidic small molecule inhibitor of the c-Abl/14-3-3 protein–protein interactions able to drive sensitive and Imatinib-resistant leukemia cells to apoptosis. Bioorg. Med. Chem. Lett. 20, 6133–6137.

Cotto-Rios, X.M., Jones, M.J.K., Huang, T.T., 2011. Insights into phosphorylation-dependent mechanisms regulating USP1 protein stability during the cell cycle. Cell Cycle 10, 4009–4016.

Courtwright, A., Siamakpour-Reihani, S., Arbiser, J.L., Banet, N., Hilliard, E., Fried, L., et al., 2009. Secreted Frizzle-related protein 2 stimulates angiogenesis via a calcineurin/NFAT signaling pathway. Cancer Res. 69, 4621–4628.

Cselenyi, C.S., Lee, E., 2008. Context-dependent activation or inhibition of Wnt-β-catenin signaling by Kremen. Sci. Signal. 1, pe10.

Cuevas, B.D., Abell, A.N., Johnson, G.L., 2007. Role of mitogen-activated protein kinase kinase kinases in signal integration. Oncogene 26, 3159–3171.

Culver, C., Melvin, A., Mudie, S., Rocha, S., 2011. HIF-1alpha depletion results in SP1-mediated cell cycle disruption and alters the cellular response to chemotherapeutic drugs. Cell Cycle 10, 1249–1260.

Curtis, A.E., Murphy, K.G., Chaudhri, O.B., Ramachandran, R., Young, A.M., Waxman, J., et al., 2010. Kisspeptin is released from human prostate cancer cell lines but plasma kisspeptin is not elevated in patients with prostate cancer. Oncol. Rep. 23, 1729–1734.

Curtis, C.D., Likhite, V.S., McLeod, I.X., Yates, J.R., Nardulli, A.M., 2007. Interaction of the tumor metastasis suppressor nonmetastatic protein 23 homologue h1 and estrogen receptor alpha alters estrogen-responsive gene expression. Cancer Res. 67, 10600–10607.

Dafou, D., Grun, B., Sinclair, J., Lawrenson, K., Benjamin, E.C., Hogdall, E., et al., 2010. Microcell-mediated chromosome transfer identifies EPB41L3 as a functional suppressor of epithelial ovarian cancers. Neoplasia 12, 579–589.

Dai, J., Gu, J., Lu, C., Lin, J., Stewart, D., Chang, D., et al., 2011a. Genetic variations in the regulator of G-protein signaling genes are associated with survival in late-stage non-small cell lung cancer. PLoS ONE 6, e21120.

Dai, Y., Xie, C.H., Neis, J.P., Fan, C.Y., Vural, E., Spring, P.M., 2011b. MicroRNA expression profiles of head and neck squamous cell carcinoma with docetaxel-induced multidrug resistance. Head Neck J. 33, 786–791.

Davidson, S.D., Cherry, J.P., Choudhury, M.S., Tazaki, H., Mallouh, C., Konno, S., 1999. Glyoxalase I activity in human prostate cancer. A potential marker and importance in chemotherapy. J. Urol. 161, 690–691

Davis, B.N., Hilyard, A.C., Nguyen, P.H., Lagna, G., Hata, A., 2009. Induction of microRNA-221 by platelet-derived growth factor signaling is critical for modulation of vascular smooth muscle phenotype. J. Biol. Chem. 284, 3728–3738.

Deacon, K., Onion, D., Kumari, R., Watson, S.A., Knox, A.J., 2012. Elevated SP-1 transcription factor expression and activity drives basal and hypoxia-induced vascular endothelial growth factor (VEGF) expression in non-small cell lung cancer. J. Biol. Chem. 287, 39967–39981.

Deane, R., Singh, I., Sagare, A.P., Bell, R.D., Ross, N.T., Larue, B., et al., 2012. A multimodal RAGE-specific inhibitor reduces amyloid beta-mediated brain disorder in a mouse model of Alzheimer disease. J. Clin. Invest. 122, 1377–1392.

Deans, J.P., Kalt, L., Ledbetter, J.A., Schieven, G.L., Bolen, J.B., Johnson, P., 1995. Association of 75/80-kDa phosphoproteins and the tyrosine kinases Lyn, Fyn, and Lck with the B cell molecule CD20. Evidence against involvement of the cytoplasmic regions of CD20. J. Biol. Chem. 270, 22632–22638.

De Candia, P., Akram, M., Benezra, R., Brogi, E., 2006. Id4 messenger RNA and estrogen receptor expression: inverse correlation in human normal breast epithelium and carcinoma. Human Pathol. 37, 1032–1041.

Dechering, K., Boeersma, C., Mosselman, S., 2000. Estrogen receptors alpha and beta: two receptors of a kind? Curr. Med. Chem. 7, 561–576.

Dehennaut, V., Loison, I., Pinte, S., Leprince, D., 2012. Molecular dissection of the interaction between HIC1 and SIRT1. Biochem. Biophys. Res. Commun. 421, 384–388.

De Leng, W.W.J., Keller, J.J., Luiten, S., Musler, A.R., Jansen, M., Baas, A.F., et al., 2005. STRAD in Peutz–Jeghers syndrome and sporadic cancers. J. Clin. Pathol. 58, 1091–1095.

Del Pulgar, T.G., Valdes-Mora, F., Bandres, E., Perez-Palacios, R., Espina, C., Cejas, P., et al. 2008. Cdc42 is highly expressed in colorectal adenocarcinoma and downregulates ID4 through an epigenetic mechanism. Int. J. Oncol. 33, 185-193.

Dessen, P., Le Minor, S., 2002. Updated 2012. Atlas Genet. Cytogenet. Oncol. Haematol. <http://atlasgeneticsoncology.org/Genes/GC_VASH1.html>.

Decker, S., Zirlik, K., Djebatchie, L., Hartmann, D., Ihorst, G., Schmitt-Graeff, A., et al., 2012. Trisomy 12 and elevated GLI1 and PTCH1 transcript levels are biomarkers for Hedgehog-inhibitor responsiveness in CLL. Blood 119, 997–1007.

De Hemptinne, V., Rondas, D., Vandekerckhove, J., Vancompernolle, K., 2007. Tumour necrosis factor induces phosphorylation primarily of the nitric-oxide-responsive form of glyoxalase I. Biochem. J. 407, 121–128.

De Hemptinne, V., Rondas, D., Toepoel, M., Vancompernolle, K, 2009. Phosphorylation on Thr-106 and NO-modification of glyoxalase I suppress the TNF-induced transcriptional activity of NF-kappa B. Mol. Cell. Biochem. 325, 169–178.

Deltour, S., Pinte, S., Guerardel, C., Wasylyk, B., Leprince, D., 2002. The human candidate tumor suppressor gene *HIC1* recruits CtBP through a degenerate GLDLSKK motif. Mol. Cell. Biol. 22, 4890–4901.

Denli, A.M., Tops, B.B., Plasterk, R.H., Ketting, R.F., Hannon, G.J., 2004. Processing of primary microRNAs by the microprocessor complex. Nature 432, 231–235.

Depeille, P., Young, J.J., Boguslawski, E.A., Berghuis, B.D., Kort, E.J., Resau, J.H., et al., 2007. Anthrax lethal toxin inhibits growth of and vascular endothelial growth factor release from endothelial cells expressing the human herpes virus 8 viral G protein-coupled receptor. Clin. Cancer Res. 13, 5926–5934.

De Vries, L., Zheng, B., Fischer, T., Elenko, E., Farquhar, M.G., 2000. The regulator of G protein signaling family. Ann. Rev. Pharmacol. Toxicol. 40, 235–271.

De Wilde, R.F., Ottenhof, N.A., Jansen, M., Morsink, F.H.M., de Leng, W.W.J., Offerhaus, G.J.A., et al., 2011. Analysis of LKB1 mutations and other molecular alterations in pancreatic acinar cell carcinoma. Mod. Pathol. 24, 1229–1236.

Di Agostino, S., Strano, S., Blandino, G., 2010. YAP1 (Yes-associated protein 1, 65kDa). Atlas Genet. Cytogenet. Oncol. Haematol. <http://AtlasGeneticsOncology.org/Genes/YAP1ID42855ch11q22.html>.

Dillon, R.L., Brown, S.T., Ling, C., Shioda, T., Muller, W.J., 2007. An EGR2/CITED1 transcription factor complex and the 14-3-3sigma tumor suppressor are involved in regulating ErbB2 expression in a transgenic-mouse model of human breast cancer. Mol. Cell. Biol. 27, 8648–8657.

Ding, L., Xu, Y.J., Zhang, W., Deng, Y.J., Si, M.S., Du, Y., et al., 2010. MiR-375 frequently downregulated in gastric cancer inhibits cell proliferation by targeting JAK2. Cell Res. 20, 784–793.

Ding, L., Zhang, X., Zhao, M., Qu, Z., Huang, S., Dong, M., et al., 2012. An essential role of PDCD4 in progression and malignant proliferation of gastrointestinal stromal tumors. Med. Oncol. 29, 1758–1764.

Ding, Y., Boguslawski, E.A., Berghuis, B.D., Young, J.J., Zhang, Z., Hardy, K., et al., 2008. Mitogen-activated protein kinase kinase signaling promotes growth and vascularization of fibrosarcoma. Mol. Cancer Ther. 7, 648–658.

Distler, J.H., Distaler, O., 2010. Tyrosine kinase inhibitors for the treatment of fibrotic diseases such as systemic sclerosis: towards molecular targeted therapies. Ann. Rheum. Dis. 69 (Suppl), i48–i51.

Djos, A., Martinsson, T., Kigner, P., Caren, H., 2012. The RASSF genes family members RASSF5, RASSF6 and RASSF7 show frequent DNA methylation in neuroblastoma. Mol. Cancer 11, 40.

Doench, J.G., Petersen, C.P., Sharp, P.A., 2003. siRNAs can function as miRNAs. Genes Dev. 17, 438–442.

Doherty, N.G., O'Sullivan, O.E., Healy, D.A., Murphy, M., O'Neill, A.J., Fitzpatrick, J.M., et al., 2006. TGF-β_1-induced EMT can occur independently of its proapoptotic effects and is aided by EGF receptor activation. Am. J. Physiol. Renal. Physiol. 290, F1202–F1212.

Dong, J.T., Lamb, P.W., Rinker-Schaeffer, C.W., Vukanovic, J., Ichikaa, T., Isscas, J.T., et al., 1995. KAI1, a metastasis suppressor gene for prostate cancer on human chromosome 11p11.2. Science 268, 884–886.

Dong, M., How, T., Kirkbride, K.C., Gordon, K.J., Lee, J.D., Hempel, N., et al., 2007a. The type III TGF-beta receptor suppresses breast cancer progression. J. Clin. Invest. 117, 206–217.

Dong, P., Kaneuchi, M., Watari, H., Hamada, J., Sudo, S., Ju, J., et al., 2011a. MicroRNA-194 inhibits epithelial to mesenchymal transition of endometrial cancer cells by targeting oncogene BMI-1. Mol. Cancer 10, 99.

Dong, J., Huang, S., Caikovski, M., Ji, S., McGrath, A., Custorio, M.G, et al. 2011b. ID4 regulates mammary gland development by suppressing p38MAPK activity. Development 138, 5247–5256.

Dong, S., Kang, K., Gu, T., Kardar, S., Fu, H., Lonial, S., et al., 2007b. 14-3-3 integrates pro-survival signals mediated by the AKT and MAPK pathways in ZNF198-FGFR1 transformed hematopoietic cells. Blood 110, 360–369.

Dong, S., Kang, S., Lonial, S., Khoury, H.J., Viallet, J., Chen, J., 2008. Targeting 14-3-3 sensitizes native and mutant BCR-ABL to inhibition with U0126, rapamycin and Bcl-2 inhibitor GX15-070. Leukaemia 22, 572–577.

Dorman, D., Wertz, I., Shimizu, H., Arnott, D., Frantz, G.D., Dowd, P., et al., 2004. The ubiquitin ligase COP1 is a critical negative regulator of p53. Nature 429, 86–92.

Dorrello, N.V., Peschiaroli, A., Guardavaccaro, D., Colburn, N.H., Sherman, N.E., Pagano, M., 2006. S6K1- and beta TRCP-mediated degradation of PDCD4 promotes protein translation and cell growth. Science 314, 467–471.

Douglas, A.E., Heim, J.A., Shen, F., Almada, L.L., Riobo, N.A., Fernandez-Zapico, M.E., et al., 2011. The alpha subunit of the G protein G(13) regulates activity of one or more Gli transcription factors independently of Smoothened. J. Biol. Chem. 286, 30714–30722.

Dovey, J.S., Zacharek, S.J., Kim, C.F., Lees, J.A., 2008. Bmi1 is critical for lung tumorigenesis and bronchioalveolar stem cell expansion. Proc. Natl. Acad. Sci. U.S.A. 105, 11857–11862.

Dowling, R.J., Goodwin, P.J., Stambolic, V., 2011. Understanding the benefit of metformin use in cancer treatment. BMC Medicine 9, art. no. 33.

Du, H., Yao, W., Fang, M., Wu, D., 2011. ARF triggers cell G1 arrest by a P53 independent ERK pathway. Mol. Cell. Biochem. 357, 415–422.

Du, R., Sun, W., Xia, L., Zhao, A., Yu, Y., Zhao, L., et al., 2012. Hypoxia-induced down-regulation of microRNA-34a promotes EMT by targeting the notch signaling pathway in tubular epithelial cells. PLoS ONE 7, e30771.

Dubois, T., Rommel, C., Howell, S., Steinhussen, U., Soneji, Y., Morrice, N., et al., 1997. 14-3-3 is phosphorylated by casein kinase I on residue 233. Phosphorylation at this site in vivo regulates Raf/14-3-3 interaction. J. Biol. Chem. 272, 28882–28888.

Duchaine, T.F., Wohlschlegel, J.A., Kennedy, S., Bei, Y., Conte Jr, D., Pang, K., et al., 2006. Functional proteomics reveals the biochemical niche of C. elegans DCR-1 in multiple small-RNA-mediated pathways. Cell 124, 343–354.

Duesbery, N.S., Webb, C.P., Leppla, S.H., Gordon, V.M., Klimpel, K.R., Copeland, T.D., et al., 1998. Proteolytic inactivation of MAP-kinase-kinase by anthrax lethal factor. Science 280, 734–737.

Dugas, J.C., Cuellar, T.L., Scholze, A., Ason, B., Ibrahim, A., Emery, B., et al., 2010. Dicer1 and miR-219 are required for normal oligodendrocyte differentiation and myelination. Neuron 65, 597–611.

Durkin, M.E., Ullmannova, V., Guan, M., Popescu, N.C., 2007. Deleted in liver cancer 3 (DLC-3), a novel Rho GTPase-activating protein, is downregulated in cancer and inhibits tumor cell growth. Oncogene 26, 4580–4589.

Efeyan, A., Sabatini, D.M., 2010. mTOR and cancer: many loops in one pathway. Curr. Opin. Cell. Biol. 22, 169–176.

Eggers, H., Steffens, S., Grosshennig, A., Becker, J.U., Hennenlotter, J., Stenzl, A., et al. 2012. Prognostic and diagnostic relevance of hypermethylated in cancer 1 (HIC1) CpG island methylation in renal cell carcinoma. Int. J. Oncol. 40, 1650–1658.

El-Kalla, M., Onyskiw, C., Baksh, S., 2010. Functional importance of RASSF1A microtubule localization and polymorphisms. Oncogene 29, 5729–5740.

El-Masry, O.S., Brown, B.L., Dobson, P.R.M., 2012. Effects of activation of AMPK on human breast cancer cell lines with different genetic backgrounds. Oncol. Lett. 3, 224–228.

El-Tanani, M., Platt-Higgins, A., Rudland, P.S., Campbell, F.C., 2004. Ets gene PEA3 cooperates with beta catenin-Lef-1 and c-Jun in regulation of osteopontin transcription. J. Biol. Chem. 279, 20794–20806.

Emoto, S., Ishigami, H., Yamashita, H., Yamaguchi, H., Kaisaki, S., Kitayama, J., 2012. Clinical significance of CA125 and CA72-4 in gastric cancer with peritoneal dissemination. Gastric Cancer 15, 154–161.

Endsley, M.P., Hu, Y., Deng, Y., He, X., Warejcka, D.J., Twining, S.S., et al., 2011. Maspin, the molecular bridge between the plasminogen activator system and beta 1 integrin that facilitates cell adhesion. J. Biol. Chem. 286, 24599–24607.

Engelman, J.A., Zejnullahu, K., Misutomi, T., Song, Y., Hyland, C., Park, J.O., et al., 2007. *MET* amplification leads to gefitinib resistance in lung Cancer by activating ERBB3 aignaling. Science 316, 1039–1043.

Engels, K., Giannattasio, M., Muzi-Falconi, M., Lopes, M., Ferrari, S., 2011. 14-3-3 proteins regulate exonuclease 1-dependent processing of stalled replication forks. PloS Genet. 7, e1001367.

Ergen, A., Canbay, E., Bugra, D., Zeybek, U., Yamaner, S., Bulut, T., 2012. Plasma Kisspeptin-54 levels in gastric cancer patients. Int. J. Surg. 10, 551–554.

Ernst, C., Christie, B.R., 2006. The putative neural stem cell marker, nestin, is expressed in heterogeneous cell types in the adult rat neocortex. Neuroscience 138, 183–188.

Esteller, M., 2011. Non-coding RNAs in human disease. Nat. Rev. Gen. 12, 861–874.

Esteller, M., Avizienyte, E., Corn, P.G., Lothe, R.A., Baylin, S.B., Aaltonen, L.A., et al., 2000. Epigenetic inactivation of LKB1 in primary tumors associated with the Peutz–Jeghers syndrome. Oncogene 19, 164–168.

Eves, E.M., Shapiro, P., Naik, K., Klein, U.R., Trakul, N., Rosner, M.R., 2006. Raf kinase inhibitory protein regulates Aurora B kinase and the spindle checkpoint. Mol. Cell. 23, 561–574.

Evren, S., Dermen, A., Lockwood, G., Fleshner, N., Sweet, J., 2011. mTOR-RAPTOR and 14-3-3 sigma immunohistochemical expression in high grade prostatic intraepithelial neoplasia and prostatic adenocarcinomas: a tissue microarray study. J. Clin. Pathol. 64, 683–688.

Evron, T., Philipp, M., Lu, J., Meloni, A.R., Burkhalter, M., Chen, W., et al., 2011. Growth Arrest Specific 8 (Gas8) and G protein-coupled receptor kinase 2 (GRK2) cooperate in the control of Smoothened signaling. J. Biol. Chem. 286, 27676–27686.

Fan, C., Yu, J., Liu, Y., Xu, H., Wang, E., 2012. Increased NDRG1 expression is associated with advanced T stages and poor vascularization in non-small cell lung cancer. Pathol. Oncol. Res. 18, 549–556.

Fang, J.H., Zhou, H.C., Zeng, C., Yang, J., Liu, Y., Huang, X., et al., 2011. MicroRNA-29b suppresses tumor angiogenesis, invasion, and metastasis by regulating matrix metalloproteinase 2 expression. Heptology 54, 1729–1740.

Fassina, A., Cappellesso, R., Guzzardo, V., Dalla Via, L., Piccolo, S., Ventura, L., et al., 2012. Epithelial–mesenchymal transition in malignant mesothelioma. Mod. Pathol. 25, 86–99.

Fendrich, V., Wiese, D., Waldmann, J., Lauth, M., Heverhagen, A.E., Rehm, J., et al., 2011. Hedgehog inhibition with the orally bioavailable Smo antagonist LDE225 represses tumor growth and prolongs survival in a transgenic mouse model of islet cell neoplasms. Ann. Surg. 254, 818–823.

Feng, C., Neumeister, V., Ma, W., Xu, J., Lu, L., Bordeaux, J., et al., 2012. Lin28 regulates HER2 and promotes malignancy through multiple mechanisms. Cell Cycle 11, 2486–2494.

Feng, D.D., Zhang, H., Zhang, P., Zheng, Y.S., Zhang, X.J., Han, B.W., et al., 2011a. Down-regulated miR-331-5p and miR-27a are associated with chemotherapy resistance and relapse in leukaemia. J. Cell. Mol. Med. 15, 2164–2175.

Feng, M., Huang, B., Du, Z., Xu, X., Chen, Z., 2011b. DLC-1 as a modulator of proliferation, apoptosis and migration in Burkitt's lymphoma cells. Mol. Biol. Rep. 38, 1915–1920.

Feng, S., Cong, S., Zhang, X., Bao, X., Wang, W., Li, H., et al., 2011c. MicroRNA-192 targeting retinoblastoma 1 inhibits cell proliferation and induces cell apoptosis in lung cancer cells. Nucleic Acids Res. 39, 6669–6678.

Fenton, H., Carlile, B., Montgomery, E.A., Carraway, H., Herman, J., Sahin, F., et al., 2006. LKB1 protein expression in human breast cancer. App. Immunohist. Mol. Morphol. 14, 146–153.

Fernandez-Fernandez, M.R., Veprintsev, D.B., Fersht, A.R., 2005. Proteins of the S100 family regulate the oligomerization of p53 tumor suppressor. Proc. Natl. Acad. Sci. U.S.A. 102, 4735–4740.

Ferreira, A.C., Suriano, G., Mendes, N., Gomes, B., Wen, X., Carneiro, F., et al., 2012. E-cadherin impairment increases cell survival through Notch-dependent upregulation of Bcl-2. Hum. Mol. Genet. 21, 334–343.

Ferrigno, O., Lallemand, F., Verrecchia, F., L'Hoste, S., Camonis, J., Atfi, A., et al., 2002. Yes-associated protein (YAP65) interacts with Smad7 and potentiates its inhibitory activity against TGF-beta/Smad signalling. Oncogene 21, 4879–4884.

Fillmore, C.M., Kuperwasser, C., 2008. Human breast cancer cell lines contain stem-like cells that self-renew, give rise to phenotypically diverse progeny and survive chemotherapy. Breast Cancer Res. 10, R25.

Finoux, A.L., Chartrand, P., 2008. Oncogenic and tumour suppressor microRNAs. Med. Sci. 24, 1049–1054.

Fitzgerald, M., Oshiro, M., Holtan, N., Krager, K., Cullen, J.J., Futscher, B.W., et al., 2003. Human pancreatic carcinoma cells activate maspin expression through loss of epigenetic control. Neoplasia 5, 427–436.

Fleuriel, C., Touka, M., Boulay, G., Guérardel, C., Rood, B.R., Leprince, D., 2009. HIC1 (hypermethylated in cancer 1) epigenetic silencing in tumors. Int. J. Biochem. Cell. Biol. 41, 26–33.

Foekens, J.A., Sieuwerts, A.M., Smid, M., Look, M.P., de Weerd, V., Boersma, A.W.M., et al. 2008. Four miRNAs associated with aggressiveness of lymph node-negative, estrogen receptor-positive human breast cancer. Proc. Natl. Acad. Sci. USA 105, 13021–13026.

Foley, N.H., Bray, I.M., Tivnan, A., Bryan, K., Murphy, D.M., Buckley, P.G., et al., 2010. MicroRNA-184 inhibits neuroblastoma cell survival through targeting the serine/threonine kinase AKT2. Mol. Cancer 9, 83.

Fonseca-Sanchez, M.A., Cuevas, S.R., Mendoza-Hernandez, G., Bautista-Pina, V., Ocampo, E.A., Miranda, A.H., et al., 2012. Breast cancer proteomics reveals a positive correlation between glyoxalase 1 expression and high tumor grade. Int. J. Oncol. 41, 670–680.

Fontemaggi, G., Dell'Orso, S., Trisciuoglio, D., Shay, T., Melucci, E., Fazi, F., et al. 2009. The execution of the transcriptional axis mutant p53, E2F1 and ID4 promotes tumor neo-angiogenesis. Nat. Struct. Mol. Biol. 16, 1086–1093.

Forbes, S.A., Bindal, N., Bamford, S., Cole, C., Kok, C.Y., Beare, D., et al., 2011. COSMIC: mining complete cancer genomes in the catalogue of somatic mutations in cancer. Nucleic Acids Res. 39, D945–D950.

Ford, J.C., Alkhodairy, F., Fotou, E., Sheldrick, K.S., Griffiths, D.J., Carr, A.M., 1994. 14-3-3-Protein homologues required for the DdNA damage checkpoint in fission yeast. Science 265, 533–535.

Fornari, F., Milazzo, M., Chieco, P., Negrini, M., Calin, G.A., et al., 2010. MiR-199a-3p regulates mTOR and c-Met to influence the Doxorubicin sensitivity of human hepatocarcinoma cells. Cancer Res. 70, 5184–5193.

Foronjy, R., Imai, K., Shiomi, T., Mercer, B., Sklepkiewicz, P., Thankachen, J., et al., 2010. The divergent roles of secreted Frizzled related protein-1 (SFRP1) in lung morphogenesis and emphysema. Am. J. Pathol. 177, 598–607.

Foster, K.G., Acosta-Jaquez, H.A., Romeo, Y., Ekim, B., Soliman, G.A., Carriere, A., et al., 2010. Regulation of mTOR complex 1 (mTORC1) by raptor Ser863 and multisite phosphorylation. J. Biol. Chem. 285, 80–94.

Fotovati, A., Fujii, T., Yamaguchi, M., Kage, M., Shirouzu, K., Oie, S., et al., 2006. 17 beta-estradiol induces down-regulation of Cap43/NDRG1/Drg-1, a putative differentiation-related and metastasis suppressor gene, in human breast cancer cells. Clin. Cancer Res. 12, 3010–3018.

Frank-Kamenetsky, M., Zhang, X.M., Bottega, S., Guicherit, O., Wichterle, H., Dudek, H., et al., 2002. Small-molecule modulators of Hedgehog signaling: identification and characterization of Smoothened agonists and antagonists. J. Biol. 1, 10–19.

Frankel, L.B., Christoffersen, N.R., Jacobsen, A., Lindow, M., Krogh, A., Lund, A.H., 2008. Programmed cell death 4 (PDCD4) is an important functional target of the microRNA miR-21 in breast cancer cells. J. Biol. Chem. 283, 1026–1033.

Fransvea, E., Angelotti, U., Antonaci, S., Giannelli, G., 2008. Blocking transforming growth factor-beta up-regulates E-cadherin and reduces migration and invasion of hepatocellular carcinoma cells. Hepatology 47, 1557–1566.

Friess, H., Ding, J., Kleeff, J., Fenkell, L., Rosinski, J.A., Guweidhi, A., et al., 2003. Microarray-based identification of differentially expressed growth- and metastasis-associated genes in pancreatic cancer. Cell. Mol. Life Sci. 60, 1180–1199.

Fu, L., Qin, Y.R., Xie, D., Hu, L., Kwong, D.L., Srivastava, G., et al., 2007. Characterisation of a novel tumour suppressor gene PLC delta 1 at 3p22 in oesophageal squamous cell carcinoma. Cancer Res. 67, 10720–10726.

Fu, L., Gao, Z., Zhang, X., Tsang, Y.H., Goh, H.K., Geng, H., et al., 2009. Frequent concomitant epigenetic silencing of the stress-responsive tumor suppressor gene CADM1, and its interacting partner DAL-1 in nasal NK/T-cell lymphoma. Int. J. Cancer 124, 1572–1578.

Fu, Z., Smith, P.C., Zhang, L.Z., Rubin, M.A., Dunn, R.L., Yao, Z., et al., 2003. Effects of Raf kinase inhibitor protein expression on suppression of prostate cancer metastasis. J. Natl. Cancer Inst. 95, 878–889.

Fu, Z., Kitagawa, Y., Shen, R.L., Shah, R., Mehra, R., Rhodes, D., et al., 2006. Metastasis suppressor gene Raf kinase inhibitor protein (RKIP) is a novel prognostic marker in prostate cancer. Prostate 66, 248–256.

Fujii, M., Toyoda, T., Nakanishi, H., Yatabe, Y., Sato, A., Matsudaira, Y., et al., 2012. TGF-beta synergizes with defects in the Hippo pathway to stimulate human malignant mesothelioma growth. J. Exp. Med. 209, 479–494.

Fujii, N., You, L., Xu, Z., Uematsu, K., Shan, J., He, B., et al. 2007. An antagonist of Dishevelled protein-protein interaction suppresses B-catenin–dependent tumor cell growth. Cancer Res. 67, 573–579.

Fukami, T., Fukuhara, H., Kuramochi, M., Maruyama, T., Isogai, K., Sakamoto, M., et al., 2003. Promoter methylation of the TSLC1 gene in advanced lung tumors and various cancer cell lines. Int. J. Cancer 107, 53–59.

Futscher, B.W., Oshiro, M.M., Wozniak, R.J., Holtan, N., Hanigan, C.L., Duan, H., et al., 2002. Role for DNA methylation in the control of cell type-specific maspin expression. Nat. Genet. 31, 175–179.

Gadgeel, S., Maitah, M., Ali, S., Lorusso, P., Sarkar, F., 2011. GDC-0449, a Hedgehog (HH) pathway inhibitor can attenuate epithelial mesenchymal transition (EMT) phenotype in non-small cell lung cancer (NSCLC) cells and enhance the effcts of Erlotinib and Cisplatin. J. Thoracic. Oncol. 6, S921.

Galcheva-Gargova, Z., Zhidkova, N., Geisler, S., Ozug, J., Wudyka, S., Gunay, N.S., et al., 2008. Overexpression of Merlin in B16F10 mouse melanoma cells reduces their metastatic activity: role of the cell surface heparan sulfate glycosaminoglycans. Int. J. Oncol. 32, 1237–1243.

Gan, Y., Shi, C., Inge, L., Hibner, M., Balducci, J., Huang, Y., 2010. Differential roles of ERK and Akt pathways in regulation of EGFR-mediated signaling and motility in prostate cancer cells. Oncogene 29, 4947–4958.

Gao, F., Shi, H.Y., Daughty, C., Cella, N., Zhang, M., 2004. Maspin plays an essential role in early embryonic development. Development 131, 1479–1489.

Gao, F., Wang, X., Zhu, F., Wang, Q., Zhang, X., Guo, C., et al., 2009. PDCD4 gene silencing in gliomas is associated with 5′ CpG island methylation and unfavourable prognosis. J. Cell. Mol. Med. 13, 4257–4267.

Gao, X.N., Lin, J., Li, Y.H., Gao, L., Wang, X.R., Wang, W., et al., 2011. MicroRNA-193a represses c-kit expression and functions as a methylation-silenced tumor suppressor in acute myeloid leukemia. Oncogene 30, 3416–3428.

Gauger, K.J., Chenausky, K.L., Murray, M.E., Schneider, S.S., 2011. SFRP1 reduction results in an increased sensitivity to TGF-beta signalling. BMC Cancer 11, 59.

Gaur, A.B., Holbeck, S.L., Colburn, N.H., Israel, M.A., 2011. Downregulation of Pdcd4 by mir-21 facilitates glioblastoma proliferation in vivo. Neuro-Oncol 13, 580–590.

Gautschi, O., Tepper, C.G., Purnell, P.R., Izumiya, Y., Evans, C.P., Green, T.P., et al. 2008. Regulation of Id1 expression by SRC: implications for targeting of the bone morphogenetic protein pathway in cancer. Cancer Res. 68, 2250–2258.

Geng, C., Sang, M., Yang, R., Gao, W., Zhoui, T., Wang, S., 2011. Overexpression of 14-3-3 sigma counteracts tumorigenicity by positively regulating p73 in vivo. Oncol. Lett. 2, 1177–1182.

Gregory, P.A., Bert, A.G., Paterson, E.L., Barry, S.C., Tsykin, A., Farshid, G., et al. 2008. The miR-200 family and miR-205 regulate epithelial to mesenchymal transition by targeting ZEB1 and SIP1. Nat. Cell Biol. 10, 593–601.

Georges, S.A., Biery, M.C., Kim, S.Y., Schelter, J.M., Guo, J., Chang, A.N., et al., 2008. Coordinated regulation of cell cycle transcripts by p53-inducible microRNAs, miR-192 and miR-215. Cancer Res. 68, 10105–10112.

Germeyer, A., Jauckus, J., Zorn, M., Toth, B., Capp, E., Strowitzki, T., 2011. Metformin modulates IL-8, IL-1 beta, ICAM and IGFBP-1 expression in human endometrial stromal cells. Reprod. Biomed. Online 22, 327–334.

Ghosh, G., Subramanian, I.V., Adhikari, N., Zhang, X., Joshi, H.P., Basi, D., et al., 2010. Hypoxia-induced microRNA-424 expression in human endothelial cells regulates HIF-α isoforms and promotes angiogenesis. J. Clin. Invest. 120, 4141–4154.

Giannetti, A.M., Wong, H., Dijkgraaf, G.J.P., Dueber, E.C., Ortwine, D.F., Bravo, B.J., et al., 2011. Identification, characterization, and implications of species-dependent plasma protein binding for the oral Hedgehog pathway inhibitor Vismodegib (GDC-0449). J. Med. Chem. 54, 2592–2601.

Gill, R.K., Yang, S.H., Meerzaman, D., Mechanic, L.E., Bowman, E.D., Jeon, H.S., et al. 2011. Frequent homozygous deletion of the LKB1/STK11 gene in non-small cell lung cancer. Oncogene 30, 3784–3791.

Gmeiner, W.H., Reinhold, W.C., Pommier, Y., 2010. Genome-wide mRNA and microRNA profiling of the NCI 60 Cell-line screen and comparison of FdUMP[10] with Fluorouracil, Floxuridine, and Topoisomerase 1 poisons. Mol. Cancer Ther. 9, 3105–3114.

Goc, A., Abdalla, M., Al-Azayzih, A., Somanath, P.R., 2012. Rac1 activation driven by 14-3-3 zeta dimerization promotes prostate cancer cell–matrix interactions, motility and transendothelial migration. PLoS ONE 7, e40594.

Goldberg, S.F., Miele, M.E., Hatta, N., Takata, M., Paquette-Straub, C., Freedman, L.P., et al., 2003. Melanoma metastasis suppression by Chromosome 6. Evidence for a pathway regulated by CRSP3 and TXNIP. Cancer Res. 63, 432–440.

Goldstein, I., Rotter, V., 2012. Regulation of lipid metabolism byp53 – fighting two villains with one sword. Trends Endocrinol. Metab. 23, 567–575.

Gomez-Cabello, D., Callejas, S., Benguria, A., Moreno, A., Alonso, J., Palmero, I., 2010. Regulation of the microRNA processor DGCR8 by the tumor suppressor ING1. Cancer Res. 70, 1866–1874.

Gong, C., Yao, Y., Wang, Y., Liu, B., Wu, W., Chen, J., et al., 2011. Up-regulation of miR-21 mediates resistance to Trastuzumab therapy for breast cancer. J. Biol. Chem. 286, 19127–19137.

Gong, W., Russell, M., Suzuki, K., Riabowol, K., 2006. Subcellular targeting of $p33^{ING1b}$ by phosphorylation-dependent 14-3-3 binding regulates $p21^{WAF1}$ expression. Mol. Cell. Biol. 26, 2947–2954.

Goodfellow, H., Krejcí, A., Moshkin, Y., Verrijzer, C.P., Karch, F., Bray, S.J., 2007. Gene-specific targeting of the histone chaperone asf1 to mediate silencing. Develop Cell 13, 593–600.

Goodison, S., Yuan, J., Sloan, D., Kim, R., Li, C., Popescu, N.C., et al., 2005. The RhoGAP protein DLC-1 functions as a metastasis suppressor in breast cancer cells. Cancer Res. 65, 6042–6053.

Gopal, Y.N.V., Chanchorn, E., Van Dyke, M.W., 2009. Parthenolide promotes the ubiquitination of MDM2 and activates p53 cellular functions. Mol. Cancer Ther. 8, 552–562.

Gordon, K.J., Dong, M., Chislock, E.M., Fields, T.A., Blobe, G.C., 2008. Loss of type III transforming growth factor beta receptor expression increases motility and invasiveness associated with epithelial to mesenchymal transition during pancreatic cancer progression. Carcinogenesis 29, 252–262.

Gordon, M., El-Kalla, M., Baksh, S., 2012. RASSF1 polymorphisms in cancer. Mol. Biol. Int. 2012, 365213.

Gottesman, M.M., 2002. Mechanisms of cancer drug resistance. Ann. Rev. Med. 63, 615–627.

Graham, R.A., Morrison, G.E., Chang, I., Jorga, K., Hop, C., Shin, Y., et al. 2010. Bioavailablility of the hedgehog pathway inhibitor GDC-0449 in a phase I pharmacokinetic (PK) study in healthy female subjects. J. Clin. Oncol. 28, e13009.

Gramling, M.W., Eischen, C.M., 2012. Suppression of Ras/Mapk pathway signaling inhibits Myc-induced lymphomagenesis. Cell Death Differ. 19, 1220–1227.

Granovsky, A.E., Clark, M.C., McElheny, D., Heil, G., Hong, J., Liu, X., et al., 2009. Raf kinase inhibitory protein function is regulated via a flexible pocket and novel phosphorylation-dependent mechanism. Mol. Cell. Biol. 29, 1306–1320.

Gregory, R.I., Yan, K.P., Amuthan, G., Chendrimada, T., Doratotaj, B., Cooch, N., et al., 2004. The microprocessor complex mediates the genesis of microRNAs. Nature 432, 235–240.

Grewal, S.I., Elgin, S.C., 2007. Transcription and RNA interference in the formation of heterochromatin. Nature 447, 399–406.

Grigorian, M., Tulchinsky, E., Burrone, O., Tarabykina, S., Georgiev, G., Lukanidin, E., 1994. Modulation of Mts1 expression in mouse and human normal and tumour cells. Electrophoresis 15, 3–4.

Griner, E.M., Kazanietz, M.G., 2007. Protein kinase C and other diacylglyerol effector in cancer. Nature Rev. 7, 281–294.

Grotegut, S., von Schweinitz, D., Christofori, G., Lehembre, F., 2006. Hepatocyte growth factor induces cell scattering through MAPK/Egr-1-mediated upregulation of snail. EMBO J. 25, 3534–3545.

Grotterod, I., Maelandsmo, G.M., Boye, K., 2010. Signal transduction mechanisms involved in S100A4-induced activation of the transcription factor NF-kappa B. BMC Cancer 10, 241.

Grundmann, S., Hans, F.P., Kinniry, S., Heinke, J., Helbing, T., Bluhm, F., et al., 2011. MicroRNA-100 regulates neovascularization by suppression of mammalian target of rapamycin in endothelial and vascular smooth muscle cells. Circulation 123, 999–1009.

Grünwald, V., De Graffenried, L., Russel, D., Friedrichs, W.E., Ray, R.B., Hidalgo, M., 2002. Inhibitors of mTOR reverse doxorubicin resistance conferred by PTEN status in prostate cancer cells. Cancer Res. 62, 6141–6145.

Guan, M., Zhou, X.L., Soulitzis, N., Spandidos, D.A., Popescu, N.C., 2006. Aberrant methylation and deacetylation of deleted in liver cancer-1 gene in prostate cancer: potential clinical applications. Clin. Cancer Res. 12, 1412–1419.

Gudjonsson, T., Villadsen, R., Nielsen, H.L., Ronnov-Jessen, L., Bissel, M.J., Petersen, O.W., 2002. Isolation, immortalization, and characterization of a human breast epithelial cell line with stem cell properties. Genes Dev. 16, 693–706.

Guldberg, P., Thor Straten, P., Ahrenkiel, V., Seremet, T., Kirkin, A.F., Zeuthen, J., 1999. Somatic mutation of the Peutz–Jeghers syndrome gene, LKB1/STK11, in malignant melanoma. Oncogene 18, 1777–1780.

Guo, W., Dong, Z.M., Lin, X.W., Zhang, M.H., Kuang, G., Zhu, T., 2012. Decreased expression and aberrant methylation of Raf kinase inhibitory protein gene in esophageal squamous cell carcinoma. Cancer Invest. 30, 703–711.

Gupta, P.B., Onder, T.T., Jiang, G., Tao, K., Kuperwasser, C., Weinberg, R.A., et al., 2009. Identification of selective inhibitors of cancer stem cells by high-throughput screening. Cell 138, 645–659.

Haase-Kohn, C., Wolf, S., Lenk, J., Pietzsch, J., 2011. Copper-mediated cross-linking of S100A4, but not of S100A2, results in proinflammatory effects in melanoma cells. Biochem. Biophys. Res. Commun. 413, 494–498.

Habas, R., Dawid, I.B., 2005. Dishevelled and Wnt signaling: is the nucleus the final frontier? J. Biol. 4, 2.

Haenisch, S., Laechelt, S., Bruckmueller, H., Werk, A., Noack, A., Bruhn, O., et al., 2011. Down-regulation of ATP-binding cassette C2 protein expression in HepG2 cells after rifampicin treatment is mediated by microRNA-379. Mol. Pharmacol. 80, 314–320.

Hagan, S., Al-Mulla, F., Mallon, E., Oien, K., Ferrier, R., Gusterson, B., et al., 2005. Reduction of raf-1 kinase inhibitor protein expression correlates with breast cancer metastasis. Clin. Cancer Res. 11, 7392–7397.

Hakan, C., 2004. Hypoxia upregulates the expression of the NDRG1 gene leading to its overexpression in various human cancers. BMC Genet. 5, 27.

Halder, A., Jain, M., Chaudhary, I., Kabra, M., 2010. Prevalence of 22q11.2 microdeletion in 146 patients with cardiac malformation in a referral hospital of North India. BMC Med. Genet. 11, 101.

Halder, G., Johnson, R.L., 2011. Hippo signaling: growth control and beyond. Development 138, 9–22.

Hamano, R., Miyata, H., Yamasaki, M., Sugimura, K., Tanaka, K., Kurokawa, Y., et al., 2012. High expression of Lin28 is associated with tumour aggressiveness and poor prognosis of patients in oesophagus cancer. Br. J. Cancer 106, 1415–1423.

Hammond, S.M., Caudy, A.A., Hannon, G.J., 2001. Post transcriptional gene silencing by double stranded RNA. Nature Rev. Gen. 2, 110–119.

Han, B.G., Xie, H., Chen, Q., Zhang, J.T., 2006. Sensitizing hormone-refractory prostate cancer cells to drug treatment by targeting 14-3-3 sigma. Mol. Cancer Ther. 5, 903–912.

Han, J., Pedersen, J.S., Kwon, S.C., Belair, C.D., Kim, Y.K., Yeom, K.H., et al., 2009a. Post-transcriptional cross regulation between Drosha and DGCR8. Cell 136, 75–84.

Han, L., Fatrai, S., Schuringa, J.J., 2009b. BMI1 (BMI1 polycomb ring finger oncogene). Atlas Genet. Cytogenet. Oncol. Haematol. <http://AtlasGeneticsOncology.org/Genes/BMI1ID807ch10p12.html>.

Han, S.H., Jeon, J.H., Ju, M.R., Jung, U., Kim, K.Y., Yoo, H.S., et al., 2003. VDUP1 upregulated by TGF-beta1 and 1,25-dihydorxyvitamin D3 inhibits tumor cell growth by blocking cell-cycle progression. Oncogene 22, 4035–4046.

Han, Z., Dimas, K., Tian, X., Wang, Y., Hemmi, H., Yamada, K., et al., 2009c. 14-3-3 sigma-dependent resistance to Cisplatin. Anticancer Res. 29, 2009–2014.

Hannon, G.J., 2002. RNA interference. Nature 418, 244–251.

Hansen, L.L., Jensen, L.L., Dimitrakakis, C., Michalas, S., Gilbert, F., Barber, H.R.K., et al., 2002. Allelic imbalance in selected chromosomal regions in ovarian cancer. Cancer Genet. Cytogenet. 139, 1–8.

Hao, C.F., Wei, S., Tong, Z.S., Li, S.F., Shi, Y.H., Wang, X.R., et al., 2012. The effects of RKIP gene expression on the biological characteristics of human triple-negative breast cancer cells in vitro. Tumor Biol. 33, 1159–1167.

Hao, J., Zhang, S., Zhou, Y., Hu, X., Shao, C., 2011. MicroRNA 483-3p suppresses the expression of DPC4/Smad4 in pancreatic cancer. FEBS Lett. 585, 207–213.

Hardie, D.G., Ross, F.A., Hawley, S.A., 2012. AMPK: a nutrient and energy sensor that maintains energy homeostasis. Nature Rev. Mol. Cell Biol. 13, 251–262.

Harms, J.F., Welch, D.R., Miele, M.E., 2003. KISS1 metastasis suppression and emergent pathways. Clin. Exp. Metastasis 20, 11–18.

Harris, B.Z., Lim, W.A., 2001. Mechanism and role of PDZ domains in signalling complex assembly. J. Cell Sci. 114, 3219–3231.

Harris, T.A., Yamakuchi, M., Ferlito, M., Mendell, J.T., Lowenstein, C.J., 2008. MicroRNA-126 regulates endothelial expression of vascular cell adhesion molecule 1. Proc. Natl. Acad. Sci. U.S.A. 105, 1516–1521.

Hawley, S.A., Boudeau, J., Reid, J.L., Mustard, K.J., Udd, L., Makela, T.P., et al., 2003. Complexes between the LKB1 tumor suppressor, STRAD alpha/beta and MO25 alpha/beta are upstream kinases in the AMP-activated protein kinase cascade. J. Biol. 2, 28.

Hayashi, Y., Osanai, M., Lee, G.H., 2011. Fascin-1 expression correlates with repression of E-cadherin expression in hepatocellular carcinoma cells and augments their invasiveness in combination with matrix metalloproteinases. Cancer Sci. 102, 1228–1235.

He, B., Lee, A.Y., Dadfarmay, S., You, L., Xu, Z., Reguart, N., et al., 2005. Secreted frizzled-related protein 4 is silenced by hypermethylation and induces apoptosis in beta-catenin-deficient human mesothelioma cells. Cancer Res. 65, 743–748.

Healy, J.I., Goodnow, C.C., 1998. Positive *versus* negative signaling by lymphocyte antigen receptors. Annu. Rev. Immunol. 16, 645–670.

Healy, K.D., Hodgson, L., Kim, T.Y., Shutes, A., Maddileti, S., Juliano, R.L., et al., 2008. DLC-1 suppresses non-small cell lung cancer growth and invasion by RhoGAP-dependent and independent mechanisms. Mol. Carcinog. 47, 326–337.

Hehner, S.P., Hofmann, T.G., Droge, W., Schmitz, M.L., 1999. The antiinflammatory sesquiterpene lactone parthenolide inhibits NF-kappa B by targeting the I kappa B kinase complex. J. Immunol. 163, 5617–5623.

Heishi, T., Hosaka, T., Suzuki, Y., Miyashita, H., Oike, Y., Takahashi, T., et al., 2010. Endogenous angiogenesis inhibitor vasohibin1 exhibits broad-spectrum antilymphangiogenic activity and suppresses lymph node metastasis. Am. J. Pathol. 176, 1950–1958.

Helland, A., Anglesio, M.S., George, J., Cowin, P.A., Jonstone, C.N., House, C.M., et al. 2011. Deregulation of MycN, Lin28B, and Let7 in a molecular subtype of aggressive high grade serous ovarian cancers. Plos One 6, e18064.

Hemmati, P.G., Guener, D., Gillissen, B., Wendt, J., von Haefen, C., Chinnadurai, G., et al. 2006. Bak functionally complements for loss of Bax during p14(ARF)-induced mitochondrial apoptosis in human cancer cells. Oncogene 25, 6582–6594.

Hemmati, P.G., Normand, G., Gillissen, B., Wendt, J., Doerken, B., Daniel, P.T., 2008. Cooperative effect of p21(Cip1/WAF-1) and 14-3-3 sigma on cell cycle arrest and apoptosis induction by p14(ARF. Oncogene 27, 6707–6719.

Hendrix, M.J.C., 2000. De-mystifying the mechanism(s) of maspin. Nature Med. 6, 374–376.

Hergovich, A., 2012. Mammalian Hippo signalling: a kinase network regulated by protein-protein interactions. Biochem. Soc. Trans. 40, 124–128.

Hermeking, H., 2003. The 14-3-3 cancer connection. Nature Rev. Cancer 3, 931–943.

Herzog, A., Kindermann, B., Doring, F., Daniel, H., Wenzel, U., 2004. Pleiotropic molecular effects of the pro-apoptotic dietary constituent flavone in human colon cancer cells identified by protein and mRNA expression profiling. Proteomics 4, 2455–2464.

Hesling, C., D'Incan, M., Mansard, S., Franck, F., Corbin-Duval, A., Chevenet, C., et al., 2004. In vivo and in situ modulation of the expression of genes involved in metastasis and angiogenesis in a patient treated with topical imiquimod for melanoma skin metastases. Br. J. Dermatol. 150, 761–767.

Heyn, H., Engelmann, M., Schreek, S., Ahrens, P., Lehmann, U., Kreipe, H., et al. 2011. MicroRNA miR-335 is crucial for the BRCA1 regulatory cascade in breast cancer development. Int. J. Cancer 129, 2797–2806.

Hickok, J.R., Sahni, S., Mikhed, Y., Bonini, M.G., Thomas, D.D., 2011. Nitric oxide suppresses tumor cell migration through N-Myc downstream-regulated gene-1 (NDRG1) expression. Role of chelatable iron. J. Biol. Chem. 286, 41413–41424.

Hill, V.K., Dunwell, T., Catchpoole, D., Krex, D., Brini, A.T., Griffiths, M., et al., 2011. Frequent epigenetic inactivation of KIBRA, an upstream member of the Salvador/Warts/Hippo (SWH) tumor suppressor network, is associated with specific genetic event in B-cell acute lymphocytic leukemia. Epigenetics 6, 326–332.

Hirai, K., Koizumi, K., Haraguchi, S., Hirata, T., Mikami, I., Fukushima, M., et al., 2005. Prognostic significance of the tumor suppressor gene maspin in non-small cell lung cancer. Ann. Thor. Surg. 79, 248–253.

Hirasawa, M., Noda, K., Noda, S., Suzuki, M., Ozawa, Y., Shinoda, K., et al., 2011. Transcriptional factors associated with epithelial–mesenchymal transition in choroidal neovascularization. Mol. Vis. 17, 1222–1230.

Hirsch, H.A., Iliopoulos, D., Tsichlis, P.N., Struhl, K., 2009. Metformin selectively targets cancer stem cells and acts together with chemotherapy to block tumor growth and prolong remission. Cancer Res. 69, 7507–7511.

Hisaoka, M., Tanaka, A., Hashimoto, H., 2002. Molecular alterations of h-warts/LATS1 tumor suppressor in human soft tissue sarcoma. Lab. Invest. 82, 1427–1435.

Hisaoka, M., Matsuyama, A., Nagao, Y., Luan, L., Kuroda, T., Akiyama, H., et al., 2011. Identification of altered microRNA expression patterns in synovial sarcoma. Genes Chromosomes Cancer 50, 137–145.

Ho, J., Cocolakis, E., Dumas, V.M., Posner, B.I., Laporte, P.A., Lebrun, J.J., 2005. The G protein-coupled receptor kinase-2 is a TGFbeta-inducible antagonist of TGFbeta signal transduction. EMBO J. 24, 3247–3258.

Hodgkinson, V.C., Agarwala, V., ElFadl, D., Fox, J.N., McManus, P.L., Mahapatra, T.K., et al., 2012. Pilot and feasibility study: comparative proteomic analysis by 2-DE MALDI TOF/TOF MS reveals 14-3-3 proteins as putative biomarkers of response to neoadjuvant chemotherapy in ER-positive breast cancer. J. Proteomics 75, 2745–2752.

Hoek, M., Myers, M.P., Stillman, B., 2011. An analysis of CAF-1-interacting proteins reveals dynamic and direct interactions with the KU complex and 14-3-3 proteins. J. Biol. Chem. 286, 10876–10887.

Hoekstra, M., van der Sluis, R.J., Li, Z., Oosterveer, M.H., Groen, A.K., Van Berkel, T.J.C., 2012. FXR agonist GW4064 increases plasma glucocorticoid levels in C57BL/6 mice. Mol. Cell. Endocrinol. 362, 69–75.

Holeiter, G., Heering, J., Erlmann, P., Schmid, S., Jaehne, R., Olayioye, M.A., 2008. Deleted in liver cancer 1 controls cell migration through a Dial-dependent signaling pathway. Cancer Res. 68, 8743–8751.

Holleman, A., Chung, I., Olsen, R.R., Kwak, B., Mizokami, A., Saijo, N., et al., 2011. miR-135a contributes to paclitaxel resistance in tumor cells both in vitro and in vivo. Oncogene 30, 4386–4398.

Holm, R., Ali, T., Svendsrud, D.H., Nesland, J.M., Kristensen, G.B., Lyng, H., 2009. Expression of 14-3-3 sigma in cervical squamous cell carcinomas: relationship with clinical outcome. Oncol. Rep. 22, 11–15.

Homma, Y., Emori, Y., 1995. A dual functional signal mediator showing RhoGAP and phospholipase C-δ stimulating activities. EMBO J. 14, 286–291.

Hong, H.Y., Jeon, W.K., Bae, E.J., Kim, S.T., Lee, H.J., Kim, S.J., et al., 2010. 14-3-3 sigma and 14-3-3 zeta plays an opposite role in cell growth inhibition mediated by transforming growth factor-beta 1. Mol. Cells 29, 305–309.

Hooks, S.B., Callihan, P., Altman, M.K., Hurst, J.H., Ali, M.W., Murph, M.M., 2010. Regulators of G-protein signaling RGS10 and RGS17 regulate chemoresistance in ovarian cancer cells. Mol. Cancer 9, 289.

Horiuchi, A., Iinuma, H., Akahane, T., Shimada, R., Watanabe, T., 2012. Prognostic significance of PDCD4 expression and association with microRNA-21 in each Dukes' stage of colorectal cancer patients. Oncol. Rep. 27, 1384–1392.

Hsieh, H.L., Schafer, B.W., Sasaki, N., Heizmann, C.W., 2003. Expression analysis of S100 proteins and RAGE in human tumors using tissue microarrays. Biochem. Biophys. Res. Commun. 307, 375–381.

Hu, X.T., Zhang, F.B., Fan, Y.C., Shu, X.S., Wong, A.H., Zhou, W., et al., 2009. Phospholipse C delta 1 is a novel 3.22.3 tumour suppressor involved in cytoskeleton organisation with its epigenetic silencing correlated with high stage gastric cancer. Oncogene 28, 2466–2475.

Hua, J., Chen, D., Fu, H., Zhang, R., Shen, W., Liu, S., et al., 2010. Short hairpin RNA-mediated inhibition of S100A4 promotes apoptosis and suppresses proliferation of BGC823 gastric cancer cells in vitro and in vivo. Cancer Lett. 292, 41–47.

Hua, K.Q., Yao, L.Q., Cao, Q., Huang, Y., Zhao, Y.Q., Feng, Y.J., 2006. Influence of estrogen and progestin on nm23-H1 expression in epithelial ovarian cancer cell lines via activation of phosphorylation signaling. Zhonghua fu chan ke za zhi 41, 756–761. (viewed only as abstract)

Huang, D., Ding, Y., Luo, W.M., Bender, S., Qian, C.N., Kort, E., et al., 2008. Inhibition of MAPK kinase signaling pathways suppressed renal cell carcinoma growth and angiogenesis in vivo. Cancer Res. 68, 81–88.

Huang, T.H., Wu, F., Loeb, G.B., Hsu, R., Heidersbach, A., Brincat, A., et al., 2009. Up-regulation of miR-21 by HER2/neu signaling promotes cell invasion. J. Biol. Chem. 284, 18515–18524.

Huang, W.G., Cheng, A.L., Chen, Z.C., Peng, F., Zhang, P.F., Li, M.Y., et al., 2010. Targeted proteomic analysis of 14-3-3 sigma in nasopharyngeal carcinoma. Int. J. Biochem. Cell Biol. 42, 137–147.

Huettel, B., Kanno, T., Daxinger, L., Bucher, E., van der Winden, J., Matzke, A.J., et al., 2007. RNA-directed DNA methylation mediated by DRD1 and Pol IVb: a versatile pathway for transcriptional gene silencing in plants. Biochim. Biophys. Acta 1769, 358–374.

Hui, S.T.Y., Andres, A.M., Miller, A.K., Spann, N.J., Potter, D.W., Post, N.M., et al., 2008. Txnip balances metabolic and growth signaling via PTEN disulfide reduction. Proc. Natl Acad. Sci. U.S.A. 105, 3921–3926.

Hummel, R., Watson, D.I., Smith, C., Kist, J., Michael, M.Z., Haier, J., et al., 2011. Mir-148a improves response to chemotherapy in sensitive and resistant oesophageal adenocarcinoma and squamous cell carcinoma cells. J. Gastroint. Surg. 15, 429–438.

Hurst, D.R., Edmonds, M.D., Scott, G.K., Benz, C.C., Vaidya, K.S., Welch, D.R., 2009a. Breast cancer metastasis suppressor 1 up-regulates miR-146, which suppresses breast cancer metastasis. Cancer Res. 69, 1279–1283.

Hurst, J.H., Mendpara, N., Hooks, S.B., 2009b. Regulator of G-protein signalling expression and function in ovarian cancer cell lines. Cell. Mol. Biol. Lett. 14, 153–174.

Iczkowski, K.A., 2011. Cell adhesion molecule CD44: its fnctional roles in prostate cancer. Am. J. Transl. Res. 3, 1–7.

Ikeda, M., Hirabayashi, S., Fujiwara, N., Mori, H., Kawata, A., Iida, J., et al., 2007. Ras-association domain family protein 6 induces apoptosis via both caspase-dependent and caspase-independent pathways. Exp. Cell Res. 313, 1484–1495.

Ikeda, M., Kawata, A., Nishikawa, M., Tateishi, Y., Yamaguchi, M., Nakagawa, K., et al., 2009. Hippo pathway-dependent and -independent roles of RASSF6. Sci. Signal. 2, ra59.

Ikeguchi, M., Yamaguchi, K., Kaibara, N., 2004. Clinical significance of the loss of *KiSS-1* and orphan G-Protein–Coupled receptor (*hOT7T175*) gene expression in esophageal squamous cell carcinoma. Clin. Cancer Res. 10, 1379–1383.

Imajo, M., Miyatake, K., Iimura, A., Miyamoto, A., Nishida, E., 2012. A molecular mechanism that links Hippo signalling to the inhibition of Wnt/beta-catenin signalling. EMBO J. 31, 1109–1122.

Imam, J.S., Buddavarapu, K., Lee-Chang, J.S., Ganapathy, S., Camosy, C., Chen, Y., et al., 2010. MicroRNA-185 suppresses tumor growth and progression by targeting the Six1 oncogene in human cancers. Oncogene 29, 4971–4979.

Imanaka, Y., Tsuchiya, S., Sato, F., Shimada, Y., Shimizu, K., Tsujimoto, G., 2011. MicroRNA-141 confers resistance to cisplatin-induced apoptosis by targeting YAP1 in human esophageal squamous cell carcinoma. J. Hum. Genet. 56, 270–276.

Inaguma, S., Kasai, K., Ikeda, H., 2011. GLI1 facilitates the migration and invasion of pancreatic cancer cells through MUC5AC-mediated attenuation of E-cadherin. Oncogene 30, 714–723.

Ingham, R.J., Colwill, K., Howard, C., Dsettwiler, S., Lim, C.S., Yu, J., et al., 2005. WW domains provide a platform for the assembly of multiprotein networks. Mol. Cell Biol. 25, 7092–7106.

Ingles-Esteve, J., Morales, M., Dalmases, A., Garcia-Carbonell, R., Jene-Sanz, A., Lopez-Bigas, N., et al., 2012. Inhibition of specific NF-kappa B activity contributes to the tumor suppressor function of 14-3-3 sigma in breast cancer. Plos ONE 7, e38347.

Iorio, M.V., Ferracin, M., Liu, C.G., Veronese, A., Spizzo, R., Sabbioni, S., 2005. MicroRNA gene expression deregulation in human breast cancer. Cancer Res. 65, 7065–7070.

Ishizaki, K., Fujimoto, J., Kumimoto, H., Nishimoto, Y., Shimada, Y., Shinoda, M., et al., 2002. Frequent polymorphic changes but rare tumor specific mutations of the LATS2 gene on 13q11-12 in esophageal squamous cell carcinoma. Int. J. Oncol. 21, 1053–1057.

Ismail, T.M., Zhang, S., Fernig, D.G., Gross, S., Martin-Fernandez, M.L., See, V., et al., 2010. Self-association of calcium-binding protein S100A4 and metastasis. J. Biol. Chem. 285, 914–922.

Itani, S., Kunisada, T., Morimoto, Y., Yoshida, A., Sasaki, T., Ito, S., et al., 2012. MicroRNA-21 correlates with tumorigenesis in malignant peripheral nerve sheath tumor (MPNST) via programmed cell death protein 4 (PDCD4). J. Cancer Res. Clin. Oncol. 138, 1501–1509.

Jacoby, J.J., Erez, B., Korshunova, M.V., Williams, R.R., Furutani, K., Takahashi, O., et al., 2010. Treatment with HIF-1 alpha antagonist PX-478 inhibits progression and spread of orthotopic human small cell lung cancer and lung adenocarcinoma in mice. J. Thoracic. Oncol. 5, 940–949.

Jagani, Z., Wiederschain, D., Loo, A., He, D., Mosher, R., Fordjour, P., et al., 2010. The Polycomb group protein Bmi-1 is essential for the growth of multiple myeloma cells. Cancer Res. 70, 5528–5538.

Janoueix-Lerosey, I., Lequin, D., Brugieres, L., Ribeiro, A., de Pontual, L., Combaret, V., et al., 2008. Somatic and germline activating mutations of the ALK kinase receptor in neuroblastoma. Nature 455, 967–970.

Jara-Lazaro, A.R., Akhilesh, M., Thike, A.A., Lui, P.C.W., Tse, G.M.K., Tan, P.H., 2010. Down-regulation of ING4 is associated with initiation and progression of lung cancer. Histopathology 57, 220–232.

Jazirehi, A.R., Huerta-Yepez, S., Cheng, G., Bonavida, B., 2005. Rituximab (chimeric anti-CD20 monoclonal antibody) inhibits the constitutive nuclear factor-kappa B signaling pathway in non-Hodgkin's lymphoma B-cell lines: role in sensitization to chemotherapeutic drug-induced apoptosis. Cancer Res. 65, 264–276.

Jensen, J.D., Knoop, A., Laenkholm, A.V., Grauslund, M., Jensen, M.B., Santoni-Rugiu, E., et al., 2012. PIK3CA mutations, PTEN, and pHER2 expression and impact on outcome in HER2-positive early-stage breast cancer patients treated with adjuvant chemotherapy and trastuzumab. Ann. Oncol. 23, 2034–2042.

Jeon, H.M., Jin, X., Lee, J.S., Oh, S.Y., Sohn, Y.W., Park, H.J., et al. 2008. Inhibitor of differentiation 4 drives brain tumor-initiating cell genesis through cyclin E and notch signaling. Genes Develop 22, 2028–2033.

Jeon, H.M., Sohn, Y.W., Oh, S.Y., Kim, S.H., Beck, S., Kim, S., 2011. ID4 imparts chemoresistance and cancer stemness to glioma cells by derepressing miR-9*-mediated suppression of SOX2. Cancer Res. 71, 3410–3421.

Jhanwar-Uniyal, M., 2003. BRCA1 in cancer, cell cycle and genomic stability. Front. Biosci. 8, S1107–S1117.

Ji, H., Ramsey, M.R., Hayes, D.N., Fan, C., McNamara, K., Kozlowski, P., et al., 2007. LKB1 modulates lung cancer differentiation and metastasis. Nature 448, 807–810.

Jia, C.Y., Li, H.H., Zhu, X.C., Dong, Y.W., Fu, D., Zhao, Q.L., et al., 2011. MiR-223 suppresses cell proliferation by targeting IGF-1R. PLoS ONE 6, e27008.

Jiang, Z., Wang, Z., Xu, Y., Wang, B., Huang, W., Cai, S., 2010. Analysis of RGS2 expression and prognostic significance in stage II and III colorectal cancer. Biosci. Rep. 30, 383–390.

Jin, Y., Lu, Z., Ding, K., Li, J., Du, X., Chen, C., et al., 2010. Antineoplastic mechanisms of Niclosamide in acute myelogenous leukemia stem cells: inactivation of the NF-kappa B pathway and generation of reactive oxygen species. Cancer Res. 70, 2516–2527.

Jones, J.L., Royall, J.E., Walker, R.A., 1996. E-cadherin relates to EGFR expression and lymph node metastasis in primary breast carcinoma. Br. J. Cancer 74, 1237–1241.

Jones, L., Hamilton, A.J., Voinnet, O., Thomas, C.L., Maule, A.J., Baulcombe, D.C., 1999. RNA–DNA interactions and DNA methylation in posttranscriptional gene silencing. Plant Cell 11, 2291–2301.

Joost, S., Almada, L.L., Rohnalter, V., Holz, P.S., Vrabel, A.M., Fernandez-Barrena, M.G., et al., 2012. GLI1 inhibition promotes epithelial-to-mesenchymal transition in pancreatic cancer cells. Cancer Res. 72, 88–99.

Jouppila-Matto, A., Narkio-Makela, M., Soini, Y., Pukkila, M., Sironen, R., Tuhkanen, H., et al., 2011. Twist and snail expression in pharyngeal squamous cell carcinoma stroma is related to cancer progression. BMC Cancer 11 (350)

Jung, H., Seong, H.A., Ha, H., 2007. NM23-H1 tumor suppressor and its interacting partner STRAP activate p53 function. J. Biol. Chem. 282, 35293–35307.

Jung, S., Park, R.H., Kim, S., Jeon, Y.J., Ham, D.S., Jung, M.Y., et al. 2010. Id proteins facilitate self-renewal and proliferation of neural stem cells. Stem. Cell Develop 19, 831–841.

Kang, H., Suh, J.Y., Jung, Y.S., Jung, J.W., Kim, M.K., Chung, J.H., 2011a. Peptide switch is essential for Sirt1. Mol. Cell. 44, 203–213.

Kang, H.S., Baba, T., Mandai, M., Matsumura, N., Hamanishi, J., Kharma, B., et al., 2011b. GPR54 is a target for suppression of metastasis in endometrial cancer. Mol. Cancer Ther. 10, 580–590.

Kang, M., Choi, S., Jeong, S.J., Lee, S.A., Kwak, T.K., Kim, H., et al., 2012a. Cross-talk between TGF beta 1 and EGFR signalling pathways induces TM4SF5 expression and epithelial–mesenchymal transition. Biochem. J. 443, 691–700.

Kang, M., Jeong, S.J., Park, S.Y., Lee, H.J., Kim, H.J., Park, K.H., et al., 2012b. Antagonistic regulation of transmembrane 4L6 family member 5 attenuates fibrotic phenotypes in CCL94)-treated mice. FEBS J. 279, 625–635.

Kano, M., Seki, N., Kikkawa, N., Fujimura, L., Hoshino, I., Akutsu, Y., et al., 2010. miR-145, miR-133a and miR-133b: tumor-suppressive miRNAs target FSCN1 in esophageal squamous cell carcinoma. Int. J. Cancer 127, 2804–2814.

Kaproth-Joslin, K.A., Li, X., Reks, S.E., Kelley, G.G., 2008. Phospholipase C delta 1 regulates cell proliferation and cell-cycle progression from G(1)-to S-phase by control of cyclin E-CDK2 activity. Biochem. J. 415, 439–448.

Karadedou, C.T., Gomes, A.R., Chen, J., Petkovic, M., Ho, K.K., Zwolinska, A.K., et al., 2012. FOXO3a represses VEGF expression through FOXM1-dependent and -independent mechanisms in breast cancer. Oncogene 31, 1845–1858.

Karapanagiotou, E.M., Dilana, K.D., Gkiozos, I., Gratsias, I., Tsimpoukis, S., Polyzos, A., et al., 2011. Metastin is not involved in metastatic potential of non-small cell lung cancer. Med. Oncol. 28, 559–564.

Kapinas, K., Delany, A.M., 2011. MicroRNA biogenesis and regulation of bone remodelling. Arthritis Res. Ther. 13, 220.

Katagiri, F., Nagai, K., Kida, A., Tomita, K., Oishi, S., Takeyama, M., et al., 2009. Clinical significance of plasma metastin level in pancreatic cancer patients. Oncol. Rep. 21, 815–819.

Katiyar-Agarwal, S., Morgan, R., Dahlbeck, D., Borsani, O., Villegas, A., Zhu, J.K., et al., 2006. A pathogen inducible endogenous siRNA in plant immunity. Proc. Natl. Acad. Sci. U.S.A. 103, 18002–18007.

Katoh, Y., Katoh, M., 2006. Comparative integromics on FAT1, FAT2, FAT3 and FAT4. Int. J. Mol. Med. 18, 523–528.

Katoh, Y., Katoh, M., 2008. Hedgehog signaling, epithelial-to-mesenchymal transition and miRNA (Review). Int. J. Mol. Med. 22, 271–275.

Katoh, Y., Katoh, M., 2009. Integrative genomic analyses on GLI1: positive regulation of GLI1 by Hedgehog-GLI, TGF beta-Smads, and RTK-PI3K-AKT signals, and negative regulation of GLI1 by Notch-CSL-HES/HEY, and GPCR-Gs-PKA signals. Int. J. Oncol. 35, 187–192.

Kawada, M., Usami, I., Someno, T., Watanabe, T., Abe, H., Inoue, H., et al., 2010. NBRI17671, a new antitumor compound, produced by *Acremonium* sp CR17671. J. Antibiot. 63, 237–243.

Kawai, K., Yamaga, M., Iwamae, Y., Kiyota, M., Kamata, H., Hirata, H., et al., 2004. A PLC delta(1)-binding protein, p122RhoGAP, is localized in focal adhesions. Biochem. Soc. Trans. 32, 1107–1109.

Kawatani, M., Okumura, H., Honda, K., Kanoh, N., Muroi, M., Dohmae, N., et al., 2008. The identification of an osteoclastogenesis inhibitor through the inhibition of glyoxalase I. Proc. Natl. Acad. Sci. U.S.A. 105, 11691–11696.

Kelleher, F.C., 2011. Hedgehog signaling and therapeutics in pancreatic cancer. Carcinogenesis 32, 445–451.

Keller, E.T., 2004. Metastasis suppressor genes: a role for raf kinase inhibitor protein (RKIP). Anti-cancer Drugs 15, 663–669.

Ketola, K., Hilvo, M., Hyotylainen, T., Vuoristo, A., Ruskeepaa, A.L., Oresic, M., et al., 2012. Salinomycin inhibits prostate cancer growth and migration via induction of oxidative stress. Br. J. Cancer 106, 99–106.

Khanim, F.L., Merrick, B.A.M.E., Giles, H.V., Jankute, M., Jackson, J.B., Giles, L.J., et al., 2011. Redeployment-based drug screening identifies the anti-helminthic niclosamide as anti-myeloma therapy that also reduces free light chain production. Blood Cancer J. 1, e39.

Kikuchi, S., Yamada, D., Fukami, T., Masuda, M., Sakurai-Yageta, M., Williams, Y.N., et al., 2005. Promoter methylation of DAL-1/4.1B predicts poor prognosis in non-small cell lung cancer. Clin. Cancer Res. 11, 2954–2961.

Kilinc, D., Ozdemir, O., Ozdemir, S., Korgali, E., Koksal, B., Uslu, A., et al., 2012. Alterations in promoter methylation status of tumor suppressor HIC1, SFRP2, and DAPK1 genes in prostate carcinomas. DNA Cell. Biol. 31, 826–832.

Kim, H.J., Hong, J.M., Yoon, K.A., Kim, N., Cho, D.W., Choi, J.Y., Lee, I.K., Kim, S.Y. 2012c. Early growth response 2 negatively modulates osteoclast differentiation through upregulation of Id helix-loop-helix proteins. Bone 51, 643–650.

Kim, H.L., Vander Griend, D.J., Yang, X., Benson, D.A., Dubauskas, Z., Yoshida, B.A., et al., 2001. Mitogen-activated protein kinase kinase 4 metastasis suppressor gene expression is inversely related to histological pattern in advancing human prostatic cancers. Cancer Res. 61 2833–1837

Kim, H.S., Kim, G.Y., Lim, S.J., Kim, Y.W., 2010. Loss of Raf-1 kinase inhibitory protein in pancreatic ductal adenocarcinoma. Pathology 42, 655–660.

Kim, H.S., Park, S.J., Lee, K.Y., Park, Y.K., Kim, Y.W., 2012a. Reduced Raf-1 kinase inhibitor protein expression predicts less favorable outcomes in patients with hepatic colorectal metastasis. Oncol. Rep. 28, 161–171.

Kim, K.H., Back, J.H., Zhu, Y., Arbesman, J., Athar, M., Kopelovich, L., et al., 2011. Resveratrol targets transforming growth factor-beta 2 signaling to block UV-induced tumor progression. J. Invest. Dermatol. 131, 195–202.

Kim, S.A., Choi, H.C., 2012. Metformin inhibits inflammatory response via AMPK–PTEN pathway in vascular smooth muscle cells. Biochem. Biophys. Res. Commun. 425, 866–872.

Kim, T.Y., Lee, J.W., Kim, H.P., Jong, H.S., Kim, T.Y., Jung, M., et al., 2007. DLC-1, a GTPase-activating protein for Rho, is associated with cell proliferation, morphology, and migration in human hepatocellular carcinoma. Biochem. Biophys. Res. Commun. 355, 72–77.

Kim, V.N., 2005. MicroRNA biogenesis: coordinated cropping and dicing. Nat. Rev. Mol. Cell. Biol. 6, 376–385.

Kim, Y.J., Hwang, J.S., Hong, Y.B., Bae, I., Seong, Y.S., 2012b. Transforming growth factor beta receptor I inhibitor sensitizes drug-resistant pancreatic cancer cells to Gemcitabine. Res 32, 799–806.

Kimura, H., Ng, J.M.Y., Curran, T., 2008. Transient inhibition of the hedgehog pathway in young mice causes permanent defects in bone structure. Cancer Cell. 13, 249–260.

Kimura, K., Endo, Y., Yonemura, Y., Heizmann, C.W., Schafer, B.W., Watanabe, Y., et al., 2000. Clinical significance of S100A4 and E-cadherin-related adhesion molecules in non-small cell lung cancer. Int. J. Oncol. 16, 1125–1131.

King, T.E., Pardo, A., Selman, M., 2011. Idiopathic pulmonary fibrosis. Lancet 378, 1949–1961.

Kirby, H.R., Maguire, J.J., Colledge, W.H., Davenport, A.P., 2010. International union of basic and clinical pharmacology. LXXVII Kisseptin receptor nomenclature, distribution and function. Pharmacol. Rev. 62, 565–578.

Kissil, J.L., Johnson, K.C., Eckman, M.S., Jacks, T., 2002. Merlin phosphorylation by p21-activated kinase 2 and effects of phosphorylation on merlin localization. J. Biol. Chem. 277, 10394–10399.

Klawitter, J., Shokati, T., Moll, V., Christians, U., Klawitter, J., 2010. Effects of lovastatin on breast cancer cells: a proteo-metabonomic study. Breast Cancer Res. 12, R16.

Klironomos, G., Bravou, V., Papachristou, D.J., Gatzounis, G., Varakis, J., Parassi, E., et al., 2010. Loss of inhibitor of growth (ING-4) is implicated in the pathogenesis and progression of human astrocytomas. Brain Pathol. 20, 490–497.

Klubo-Gwiezdzinska, J., Jensen, K., Costello, J., Patel, A., Hoperia, V., Bauer, A., et al., 2012. Metformin inhibits growth and decreases resistance to anoikis in medullary thyroid cancer cells. Endocrine Rel. Cancer 19, 447–456.

Ko, B.S., Lai, I.R., Chang, T.C., Liu, T.A., Chen, S.C., Wang, J., et al., 2011. Involvement of 14-3-3 gamma overexpression in extrahepatic metastasis of hepatocellular carcinoma. Human Pathol. 42, 129–135.

Ko, F.C.F., Chan, L.K., Tung, E.K.K., Lowe, S.W., Ng, I.O.L., Yam, J.W.P., 2010. Akt phosphorylation of deleted in liver cancer 1 abrogates its suppression of liver cancer tumorigenesis and metastasis. Gastroenterology 139, 1397–1407.

Kohler, C., Villar, C.B.R., 2008. Programming of gene expression by polycomb group proteins. Trends Cell Biol., 236–243.

Kohya, N., Kitajima, Y., Jiao, W., Miyazaki, K., 2003. Effects of E-cadherin transfection on gene expression of a gallbladder carcinoma cell line: Repression of MTS1/S100A4 gene expression. Int. J. Cancer 104, 44–53.

Kojima, Y., Miyoshi, H., Clevers, H.C., Oshima, M., Aoki, M., Taketo, M.M., 2007. Suppression of tubulin polymerization by the LKB1-microtubule-associated protein/microtubule affinity-regulating kinase signalling. J. Biol. Chem. 282, 23532–23540.

Kolch, W., 2005. Co-ordinating ERK/MAPK signalling through scaffolds and inhibitors. Nature Rev. Mol. Cell. Biol. 6, 827–837.

Kondo, K., Obitsu, S., Ohta, S., Matsunami, K., Otsuka, H., Teshima, R., 2010. Poly(ADP-ribose) polymerase (PARP)-1-independent apoptosis-inducing factor (AIF) release and cell death are induced by eleostearic acid and blocked by alpha-tocopherol and MEK inhibition. J. Biol. Chem. 285, 13079–13091.

Kong, D.J., Wang, Z.W., Sarkar, S.H., Li, Y.W., Banerjee, S., Saliganan, A., et al., 2008. Platelet-derived growth factor-D overexpression contributes to epithelial–mesenchymal transition of PC3 prostate cancer cells. Stem Cells 26, 1425–1435.

Kong, D.J., Li, Y.W., Wang, Z.W., Banerjee, S., Ahmad, A., Kim, H.R.C., et al., 2009. MiR-200 regulates PDGF-D-mediated epithelial–mesenchymal transition, adhesion, and invasion of prostate cancer cells. Stem Cells 27, 1712–1721.

Kong, K.L., Kwong, D.L.W., Chan, T.H.M., Law, S.Y.K., Chen, L., Li, Y., et al., 2012. MicroRNA-375 inhibits tumour growth and metastasis in oesophageal squamous cell carcinoma through repressing insulin-like growth factor 1 receptor. Gut 61, 33–42.

Kongkham, P.N., Northcott, P.A., Croul, S.E., Smith, C.A., Taylor, M.D., Rutka, J.T., 2010. The SFRP family of WNT inhibitors function as novel tumor suppressor genes epigenetically silenced in medulloblastoma. Oncogene 29, 3017–3024.

Konsavage Jr, W.M., Kyler, S.L., Rennoll, S.A., Jin, G., Yochum, G.S., 2012. Wnt/beta-catenin signaling regulates Yes-associated protein (YAP) gene expression in colorectal carcinoma cells. J. Biol. Chem. 287, 11730–11739.

Korpal, M., Lee, E.S., Hu, G., Kang, Y., 2008. The miR-200 family inhibits epithelial–mesenchymal transition and cancer cell migration by direct targeting of E-cadherin transcriptional repressors ZEB1 and ZEB2. J. Biol. Chem. 283, 14910–14914.

Kosaka, T., Yamaki, Mogi, A., Kuwano, H., 2011. Mechanisms of resistance to EGFR TKIs and development of a new generation of drugs in non-small-cell cung cancer. J. Biomed. Biotech., 165214.

Kotani, M., Detheux, M., Vandenbogaerde, A., Communi, D., Vanderwinden, J., Le Poul, E., et al., 2001. The metastasis suppressor gene KiSS-1 encodes kisspeptins, the natural ligands of the orphan G protein coupled receptor GPR54. J. Biol. Chem. 276, 34631–34636.

Kouzarides, T., 2007. Chromatin modifications and their function. Cell 128, 693–705.

Kovacevic, Z., Chikhani, S., Lovejoy, D.B., Richardson, D.R., 2011a. Novel Thiosemicarbazone iron chelators induce up-regulation and phosphorylation of the metastasis suppressor N-myc down-stream regulated gene 1: a new strategy for the treatment of pancreatic cancer. Mol. Pharmacol. 80, 598–609.

Kovacevic, Z., Sivagurunathan, S., Mangs, H., Chikhani, S., Zhang, D., Richardson, D.R., 2011b. The metastasis suppressor, N-myc downstream regulated gene 1 (NDRG1), upregulates p21 via p53-independent mechanisms. Carcinogenesis 32, 732–740.

Kovi, R.C., Paliwal, S., Pande, S., Grossman, S.R., 2010. An ARF/CtBP2 complex regulates BH3-only gene expression and p53-independent apoptosis. Cell Death Diff. 17, 513–521.

Kranenburg, O., Gebbink, M.F.B.G., Voest, E.F., 2004. Stimulation of angiogenesis by Ras proteins. Biochim. Biophys. Acta. 1654, 23–37.

Kremerskothen, J., Plaas, C., Buther, K., Finger, I., Veltel, S., Matanis, T., et al. 2003. Characterization of KIBRA, a novel WW domain-containing protein. Biochem. Biophys. Res. Commun. 300, 862–867.

Krug, S., Huth, J., Goeke, F., Buchholz, M., Gress, T.M., Goeke, R., et al., 2012. Knock-down of Pdcd4 stimulates angiogenesis via up-regulation of angiopoietin-2. Biochim. Biophys. Acta. Mol. Cell. Res. 1823, 789–799.

Kuehbacher, A., Urbich, C., Dimmeler, S., 2008. Targeting microRNA expression to regulate angiogenesis. Trends Pharmacol. Sci. 29, 12–15.

Kuljaca, S., Liu, T., Dwarte, T., Kavallaris, M., Haber, M., Norris, M.D., et al., 2009. The cyclin-dependent kinase inhibitor, p21WAF1, promotes angiogenesis by repressing gene transcription of thioredoxin-binding protein 2 in cancer cells. Carcinogenesis 30, 1865–1871.

Kulshreshtha, R., Ferracin, M., Wojcik, S.E., Garzon, R., Alder, H., Agosto-Perez, F.J., et al., 2007. A microRNA signature of hypoxia. Mol. Cell. Biol. 27, 1859–1867.

Kumamoto, K., Fujita, K., Kurotani, R., Saito, M., Unoki, M., Hagiwara, N., et al., 2009. ING2 is upregulated in colon cancer and increases invasion by enhanced MMP13 expression. Int. J. Cancer 125, 1306–1315.

Kumar, D., Shadrach, J.L., Wagers, A.J., Lassar, A.B. 2009. Id3 is a direct transcriptional target of Pax7 in quiescent satellite cells. Mol. Biol. Cell 20, 3170–3177.

Kumar, R., Wang, R.A., Mazumdar, A., Talukder, A.H., Mandal, M., Yang, Z., et al., 2002. A naturally occurring MTA1 variant sequesters oestrogen receptor-alpha in the cytoplasm. Nature 418, 654–657.

Kuramitsu, Y., Baron, B., Yoshino, S., Zhang, X., Tanaka, T., Yashiro, M., et al., 2010. Proteomic differential display analysis shows up-regulation of 14-3-3 sigma protein in human scirrhous-type gastric carcinoma cells. Anticancer Res. 30, 4459–4465.

Kurisetty, V.V., Johnston, P.G., Johnston, N., Erwin, P., Crowe, P., Fernig, D.G., et al., 2008. RAN GTPase is an effector of the invasive/metastatic phenotype induced by osteopontin. Oncogene 27, 7139–7149.

Kuzontkoski. P.M., Mulligan-Kehoe, M.J., Harris, B.T., Israel, M.A. (2010). Inhibitor of DNA binding-4 promotes angiogenesis and growth of glioblastoma multiforme by elevating matrix GLA levels. Oncogene 29, 3793–3802.

Kwok, B.H., Koh, B., Ndubuisi, M.I., Elofsson, M., Crews, C.M., 2001. The anti-inflammatory natural product parthenolide from the medicinal herb Feverfew directly binds to and inhibits IkappaB kinase. Chem. Biol. 8, 759–766.

Lai, D., Chen, Y., Wang, F., Jiang, L., Wei, C., 2012. LKB1 controls the pluripotent state of human embryonic stem cells. Cell Program 14, 164–170.

Lai, Z.C., Wei, X.M., Shimizu, T., Ramos, E., Rohrbaugh, M., Nikolaidis, N., et al., 2005. Control of cell proliferation and apoptosis by Mob as tumor suppressor, Mats. Cell 120, 675–685.

Lakshmi, M.S., Sherbet, G.V., 1997. Expression of transmembrane glycoprotein CD44 and metastasis associated gene 18A2/mts1 gene in B16 murine melanoma. Anticancer Res. 17, 3451–3456.

Lakshmikuttyamma, A., Takahashi, N., Pastural, E.,Torlakovic, E., Amin, H.M., Garcia-Manero, G., et al., (2009). RIZ1 is potential CML tumor suppressor that is down-regulated during disease progression. J Hematol Oncol. 2, art. no. 28.

Lal, A., Thomas, M.P., Altschuler, G., Navarro, F., O'Day, E., Li, X.L., et al., 2011. Capture of microRNA-bound mRNAs identifies the tumor suppressor miR-34a as a regulator of growth factor signaling. Plos Genet. 7, e1002363.

Lallemand, D., Curto, M., Saotome, I., Giovannini, M., McClatchey, A.I., 2003. NF2 deficiency promotes tumorigenesis and metastasis by destabilizing adherens junctions. Genes Develop. 17, 1090–1100.

Lambert, K.E., Huang, H., Mythreye, K., Blobe, G.C., 2011. The type III transforming growth factor-beta receptor inhibits proliferation, migration, and adhesion in human myeloma cells. Mol. Biol. Cell 22, 1463–1472.

Laprise, P., Langlois, M.J., Boucher, M.J., Jobin, C., Rivard, N., 2004. Down-regulation of MEK/ERK signaling by E-cadherin-dependent PI3K/Akt pathway in differentiating intestinal epithelial cells. J. Cell. Physiol. 199, 32–59.

Lavergne, E., Hendaoui, I., Coulouarn, C., Ribault, C., Leseur, J., Eliat, P.A., et al., 2011. Blocking Wnt signaling by SFRP-like molecules inhibits in vivo cell proliferation and tumor growth in cells carrying active beta-catenin. Oncogene 30, 423–433.

Lazennec, G., Bresson, D., Lucas, A., chauveau, C., Vignon, F., 2001. ER beta inhibits proliferation and invasion of breast cancer cells. Endocrinology 142, 4120–4130.

Le, N.T.V., Richardson, D.R., 2004. Iron chelators with high antiproliferative activity up-regulate the expression of a growth inhibitory and metastasis suppressor gene: a link between iron metabolism and proliferation. Blood 104, 2967–2975.

Lee, C.W., Wong, L.L.Y., Tse, Y.E.T., Liu, H.F., Leong, V.Y.L., Lee, J.M.F., et al., 2012. AMPK promotes p53 acetylation via phosphorylation and inactivation of SIRT1 in liver cancer cells. Cancer Res. 72, 4394–4404.

Lee, H.A., Park, I., Byun, H.J., Jeoung, D., Kim, Y.M., Lee, H., 2011. Metastasis suppressor KAI1/CD82 attenuates the matrix adhesion of human prostate cancer cells by suppressing fibronectin expression and beta(1) integrin ctivation. Cell. Physiol. Biochem. 27, 575–586.

Lee, J.H., Welch, D.R., 1997a. Identification of highly expressed genes in metastasis-suppressed chromosome 6/human malignant melanoma hybrid cells using subtractive hybridization and differential display. Int. J. Cancer 71, 1035–1044.

Lee, J.H., Welch, D.R., 1997b. Suppression of metastasis in human breast carcinoma MDA-MB-435 Cells after transfection with the metastasis suppressor Gene, *KiSS-1*. Cancer Res. 57, 2384–2387.

Lee, K.H., Goan, Y.G., Hsiao, M., Lee, C.H., Jian, S.H., Lin, J.T., et al., 2009. MicroRNA-373 (miR-373) post-transcriptionally regulates large tumor suppressor, homolog 2 (LATS2) and stimulates proliferation in human esophageal cancer. Exp. Cell. Res. 315, 2529–2538.

Lee, K.N., Kang, H.-S., Jeon, J.-H., Kim, E.-M., Yoon, S.-R., Song, H., et al., 2005. VDUP1 is required for the development of natural killer cells. Immunity 22, 195–208.

Lee, R.C., Hammell, C.M., Ambros, V., 2006. Interacting endogenous and exogenous RNAi pathways in caenorhabditis elegans. RNA 12, 589–597.

Lee, S., Jung, J.W., Park, S.B., Roh, K., Lee, S.Y., Kim, J.H., et al., 2011. Histone deacetylase regulates high mobility group A2-targeting microRNAs in human cord blood-derived multipotent stem cell ageing. Cell Mol. Life Sci. 68, 325–336.

Lee, S.J., Jang, H., Park, C., 2012. Maspin increases Ku70 acetylation and Bax-mediated cell death in cancer cells. Int. J. Mol. Med. 29, 225–230.

Lee, Y., Ahn, C., Han, J.J., Choi, H., Kim, J., Yim, J., et al., 2003. The nuclear RNase III Drosha initiates microRNA processing. Nature 425, 415–419.

Lefkowitz, R.J., Rajagopal, K., Whalen, E.J., 2006. New roles for beta-arrestins in cell signaling: not just seven-transmembrane receptors. Mol. Cell 24, 643–652.

Lehmann, M., Noack, D., Wood, M., Perego, M., Knaus, U.G., 2009. Lung epithelial injury by B-Anthracis lethal toxin is caused by MKK-dependent loss of cytoskeletal integrity. PloS ONE 4, e4755.

Lei, Q.Y., Zhang, H., Zhao, B., Zha, Z.Y., Bai, F., Pei, X.H., et al., 2008. TAZ promotes cell proliferation and epithelial–mesenchymal transition and is inhibited by the hippo pathway. Mol. Cell Biol. 28, 2426–2436.

Leivonen, S.K., Makela, R., Ostling, P., Kohonen, P., Haapa-Paananen, S., Kleivi, K., et al., 2009. Protein lysate microarray analysis to identify microRNAs regulating estrogen receptor signaling in breast cancer cell lines. Oncogene 28 (44), 3926–3936.

Lerner, A., Grafi-Cohen, M., Napso, T., Azzam, N., Fares, F., 2012. The indolic diet-derivative, 3,3'-diindolylmethane, induced apoptosis in human colon cancer cells through upregulation of NDRG1. J. Biomed Biotech., 256178.

Leung, D.W., Tompkins, C., Brewer, J., Ball, A., Coon, M., Morris, V., et al., 2004. Phospholipase C delta-4 overexpression upregulates Erb1/2 expression, Erk signalling pathway, and proliferation in MCF-7 cells. Mol. Cancer 3, 15.

Li, J., Li, G., 2010. Cell cycle regulator ING4 is a suppressor of melanoma angiogenesis that is regulated by the metastasis suppressor BRMS1. Cancer Res. 70, 10445–10453.

Li, B., Carey, M., Workman, J.L., 2007a. The role of chromatin during transcription. Cell 128, 707–719.

Li, B., Tsao, S.W., Li, Y.Y., Wang, X., Ling, M.T., Wong, Y.C., et al., 2009a. Id-1 promotes tumorigenicity and metastasis of human esophageal cancer cells through activation of PI3K/AKT signaling pathway. Int. J. Cancer 125, 2576–2585.

Li, B., Jia, Z., Wang, T., Wang, W., Zhang, C., Chen, P., et al., 2012a. Interaction of Wnt/beta-catenin and notch signaling in the early stage of cardiac differentiation of P19CL6 cells. J. Cell Biochem. 113, 629–639.

Li, C., Heidt, D.G., Dalerba, P., Burant, C.F., Zhang, L., Adsay, V., et al., 2007a. Identification of pancreatic cancer stem cells. Cancer Res. 67, 1030–1037.

Li, C., Liu, VWS, Chan, D.W., Yao, K.M., Ngan, H.Y.S., 2012b. LY294002 and metformin cooperatively enhance the inhibition of growth and the induction of apoptosis of ovarian cancer cells. Int. J. Gynaecol. Cancer 22, 15–22.

Li, D., Zhou, K., Wang, S., Shi, Z., Yang, Z., 2010a. Recombinant adenovirus encoding vasohibin prevents tumor angiogenesis and inhibits tumor growth. Cancer Sci. 101, 448–452.

Li, D.J., Deng, G., Mao, Z.Q., Yao, H., Li, C., Peng, F., et al., 2009b. Identificating 14-3-3 sigma as a lymph node metastasis-related protein in human lung squamous carcinoma. Cancer Lett. 279, 65–73.

Li, D.Q., Pakala, S.B., Reddy, S.D.N., Ohshiro, K., Peng, S.H., Lian, Y., et al., 2010b. Revelation of p53-independent function of MTA1 in DNA damage response via modulation of p21waf1–PCNA pathway. J. Biol. Chem. 285, 10044–10052.

Li, F., Papworth, M., Minczuk, M., Rohde, C., Zhang, Y., Rogozin, S., et al., 2007c. Chimeric DNA methyltransferases target DNA methylation to specific DNA sequences and repress expression of target genes. Nucl. Acids Res. 35, 100–112.

Li, H., Collado, M., Villasante, A., Strati, K., Ortega, S., Cañamero, M., et al., 2009c. The Ink4/Arf locus is a barrier for iPS cell reprogramming. Nature 460, 1136–1139.

Li, H., Wolfe, A., Septer, S., Edwards, G., Zhong, X., Abdulkarim, A.B., et al., 2012c. Deregulation of Hippo kinase signalling in human hepatic malignancies. Liver Int. 32, 38–47.

Li, H.W.R., Leung, S.W., Chan, C.S.S., Yu, M.M.Y., Wong, Y.F., 2007d. Expression of maspin in endometrioid adenocarcinoma of endometrium. Oncol. Rep. 17, 393–398.

Li, H.Z., Wang, Y., Gao, Y., Shao, J., Zhao, X.L., Deng, W.M., et al., 2008. Effects of Raf kinase inhibitor protein expression on metastasis and progression of human epithelial ovarian cancer. Mol. Cancer Res. 6, 917–928.

Li, H.Z., Gao, Y., Zhao, X.L., Liu, Y.X., Sun, B.C., Yang, J., et al., 2009d. Effects of Raf kinase inhibitor protein expression on metastasis and progression of human breast cancer. Mol. Cancer Res. 7, 832–840.

Li, J., Zhou, J., Chen, G., Wang, H., Wang, S., Xing, H., et al., 2006. Inhibition of ovarian cancer metastasis by adeno-associated virus-mediated gene transfer of nm23H1 in an orthotopic implantation model. Cancer Gene. Ther. 13, 266–272.

Li, J., Zhang, Y., Kuruba, R., Gao, X., Gandhi, C.R., Xie, W., et al., 2011a. Roles of microRNA-29a in the antifibrotic effect of farnesoid X receptor in hepatic stellate cells. Mol. Pharmacol. 80, 191–200.

Li, L., Davie, J.R., 2010. The role of Sp1 and Sp3 in normal and cancer cell biology. Ann. Anat. 192, 275–283.

Li, M.A., He, L., 2012. MicroRNAs as novel regulators of stem cell pluripotency and somatic cell reprogramming. Bioessays 34, 670–680.

Li, Q., Chen, H., 2011. Transcriptional silencing of N-Myc downstream-regulated gene 1 (NDRG1) in metastatic colon cancer cell line SW620. Clin. Exp. Metastasis 28, 127–135.

Li, M., Jin, Y., Sun, W.J., Yu, Y., Bai, J., Tong, D.D., et al., 2009e. Reduced expression and novel splice variants of ING4 in human gastric adenocarcinoma. J. Pathol. 219, 87–95.

Li, S., Chen, P.L., Subramanian, T., Chinnadurai, G., Tomlinson, G., Osborne, C.K., et al., 1999. Binding of CtIP to the BRCT repeats of BRCA1 involved in the transcription regulation of p21 is disrupted upon DNA damage. J. Biol. Chem. 274, 11334–11338.

Li, T., Li, H., Zhang, Y.X., Liu, J.Y., 2011b. Identification and analysis of seven H2O2-responsive miRNAs and 32 new miRNAs in the seedlings of rice (*Oryza sativa* L. ssp indica). Nucl. Acids Res. 39, 2821–2833.

Li, X., Zhang, Y., Zhang, H., Liu, X., Gong, T., Li, M., et al., 2011c. MiRNA-223 promotes gastric cancer invasion and metastasis by targeting tumor suppressor EPB41L3. Mol. Cancer Res. 9, 824–833.

Li, X.H., Kikuchi, K., Zheng, Y., Noguchi, A., Takahashi, H., Nishida, T., et al., 2011d. Downregulation and translocation of nuclear ING4 is correlated with tumorigenesis and progression of head and neck squamous cell carcinoma. Oral oncol. 47, 217–223.

Li, Y., VandenBoom II, T.G., Wang, Z., Kong, D., Ali, S., Philip, P.A., et al., 2010c. miR-146a suppresses invasion of pancreatic cancer cells. Cancer Res. 70, 1486–1495.

Li, Y., Zhu, X., Gu, J., Dong, D., Yao, J., Lin, C., et al., 2010d. Anti-miR-21 oligonucleotide sensitizes leukemic K562 cells to arsenic trioxide by inducing apoptosis. Cancer Sci. 101, 948–954.

Li, Y., Maitah, M.Y., Ahmad, A., Kong, D., Bao, B., Sarkar, F.H., 2012d. Targeting the Hedgehog signaling pathway for cancer therapy. Expert Opin. Ther. Targets 16, 49–66.

Li, Y.M., Zhu, X.J., Gu, J.Y., Hu, H.Y., Dong, D.W., Yao, J.L., et al., 2010e. Anti-miR-21 oligonucleotide enhances chemosensitivity of leukemic HL60 cells to arabinosylcytosine by inducing apoptosis. Hematology 15, 215–221.

Li, Z., Xiang, J., Zhang, W., Fan, S., Wu, M., Li, X., et al., 2009f. Nanoparticle delivery of anti-metastatic NM23-H1 gene improves chemotherapy in a mouse tumor model. Cancer Gene. Ther. 16, 423–429.

Li, Z., Hu, S., Wang, J., Cai, J., Xiao, L., Yu, L., et al., 2010g. MiR-27a modulates MDR1/P-glycoprotein expression by targeting HIPK2 in human ovarian cancer cells. Gynaecol. Oncol. 119, 125–130.

Li, Z., Zhang, H., Chen, Y., Fan, L., Fang, J., 2012e. Forkhead transcription factor FOXO3a protein activates nuclear factor kappa B through B-cell lymphoma/leukemia 10 (BCL10) protein and promotes tumor cell survival in serum deprivation. J. Biol. Chem. 287, 17737–17745.

Li, Z.Y., Xie, Y.F., Sheng, W.H., Miao, J.C., Xiang, J., Yang, J.C., 2010h. Tumor-suppressive effect of adenovirus-mediated inhibitor of growth 4 gene transfer in breast carcinoma cells in vitro and in vivo. Cancer Biother. Radiopharm. 25, 427–437.

Liang, R., Khanna, A., Muthusamy, S., Li, N., Sarojini, H., Kopchick, J.J., et al., 2011a. Post-transcriptional regulation of IGF1R by key microRNAs in long-lived mutant mice. Ageing Cell 10, 1080–1088.

Liang, S., He, L., Zhao, X., Miao, Y., Gu, Y., Guo, C., et al., 2011b. MicroRNA Let-7f inhibits tumor invasion and metastasis by targeting MYH9 in human gastric cancer. PloS ONE 6, e18409.

Liang, X., Wang, P., Gao, Q., Xiang, T., Tao, X., 2010a. Endogenous LKB1 knockdown accelerates G(1)/S transition through p53 and p16 pathways. Cancer Biol. Ther. 9, 156–160.

Liang, Z., Wu, H., Xia, J., Li, Y., Zhang, Y., Huang, K., et al., 2010b. Involvement of miR-326 in chemotherapy resistance of breast cancer through modulating expression of multidrug resistance-associated protein 1. Biochem. Pharmacol. 79, 817–824.

Liao, X., Siu, M.K.Y., Chan, K.Y.K., Wong, E.S.Y., Ngan, H.Y.S., Chan, Q.K.Y., et al., 2008. Hypermethylation of RAS effector related genes and DNA methyltransferase 1 expression in endometrial carcinogenesis. Int. J. Cancer 123, 296–302.

Liao, X., Siu, M.K.Y., Au, C.W.H., Wong, E.S.U., Chan, H.Y., Ip, P.P.C., et al., 2009. Aberrant activation of hedgehog signaling pathway in ovarian cancers: effect on prognosis, cell invasion and differentiation. Carcinogenesis 30, 131–140.

Liao, Y.C., Shih, Y.P., Lo, S.H., 2008. Mutations in the focal adhesion targeting region of deleted in liver cancer-1 attenuate their expression and function. Cancer Res. 68, 7718–7722.

Lim, J., Jang, G., Kang, S., Lee, G., Nga, D.T.T., Phuong, D.T.L., et al., 2011. Cell-permeable NM23 blocks the maintenance and progression of established pulmonary metastasis. Cancer Res. 71, 7216–7225.

Lin, F., Li, Y.Y., Cao, J., Fan, S.F., Wen, J.F., Zhu, G.H., et al., 2011a. Overexpression of osteopontin in hepatocellular carcinoma and its relationships with metastasis, invasion of tumor cells. Mol. Biol. Rep. 38, 5205–5210.

Lin, J.N., Lin, V.C.H., Rau, K.M., Shieh, P.C., Kuo, D.H., Shieh, J.C., et al., 2010. Resveratrol modulates tumor cell proliferation and protein translation via SIRT1-dependent AMPK activation. J. Agric. Food Chem. 58, 1584–1592.

Lin, K.H., Wang, W.J., Wu, Y.H., Cheng, S.Y., 2002. Activation of antimetastatic nm23-H1 gene expression by estrogen and its alpha-receptor. Endocrinology 143, 467–475.

Lin, Y.S., Tsai, P.H., Kandaswami, C.C., Cheng, C.H., Ke, F.C., Lee, P.P., et al., 2011b. Effects of dietary flavonoids, luteolin, and quercetin on the reversal of epithelial–mesenchymal transition in A431 epidermal cancer cells. Cancer Sci. 102, 1829–1839.

Lin, Y.T., Ding, J.Y., Li, M.Y., Yeh, T.S., Wang, T.W., Yu, J.Y., 2012. YAP regulates neuronal differentiation through Sonic hedgehog signaling pathway. Exp. Cell Res. 318, 1877–1888.

Ling, C., Su, V.M.T., Zuo, D., Muller, W.J., 2012. Loss of the 14-3-3σ tumor suppressor is a critical event in erbB2-mediated tumor progression. Cancer Discov. 2, 68–81.

Ling, M.T., Lau, T.C., Zhou, C., Chua, C.W., Kwok, W.K., Wang, Q., et al., 2005. Overexpression of Id-1 in prostate cancer cells promotes angiogenesis through the activation of vascular endothelial growth factor (VEGF). Carcinogenesis 26, 1668–1676.

Ling, M.T., Kwok, W.K., Fung, M.K., Wang, X.H., Wong, Y.C. 2006. Proteasome mediated degradation of Id-1 is associated with TNF alpha-induced apoptosis in prostate cancer cells. Carcinogenesis 27, 205–215.

Linher-Melville, K., Zantinge, S., Singh, G., 2012. Liver kinase B1 expression (LKB1) is repressed by estrogen receptor alpha (ER alpha) in MCF-7 human breast cancer cells. Biochem. Biophys. Res. Commun. 417, 1063–1068.

Liu, P., Wilson, M.J., 2012. miR-520c and miR-373 upregulate MMP9 expression by targeting mTOR and SIRT1, and activate the Ras/Raf/MEK/Erk signaling pathway and NF-kappa B factor in human fibrosarcoma cells. J. Cell Physiol. 227, 867–876.

Liu, B., Fan, Z., Edgerton, S.M., Deng, X.S., Alimova, I.N., Lind, S.E., et al., 2009. Metformin induces unique biological and molecular responses in triple negative breast cancer cells. Cell Cycle 8, 2031–2040.

Liu, C., Kelnar, K., Liu, B., Chen, X., Calhoun-Davis, T., Li, H., et al., 2011a. The microRNA miR-34a inhibits prostate cancer stem cells and metastasis by directly repressing CD44. Nature Med. 17, 211–215.

Liu, H.Y., Scholz, C., Zang, C., Schefe, J.H., Habbel, P., Regierer, A.C., et al., 2012a. Metformin and the mTOR inhibitor everolimus (RAD001) sensitize breast cancer cells to the cytotoxic effect of chemotherapeutic drugs in vitro. Anticancer Res. 32, 1627–1637.

Liu, L., Siu, F.M., Che, C.M., Xu, A., Wang, Y., 2012b. Akt blocks the tumor suppressor activity of LKB1 by promoting phosphorylation-dependent nuclear retention through 14-3-3 proteins. Am. J. Transl. Res. 4, 175–186.

Liu, L.H., Li, H., Li, J.P., Zhong, H., Zhang, H.C., Chen, J., et al., 2011b. miR-125b suppresses the proliferation and migration of osteosarcoma cells through down-regulation of STAT3. Biochem. Biophys. Res. Commun. 416, 31–38.

Liu, L.P., Ho, R.L.K., Chen, G.G., Lai, P.B.S., 2012c. Sorafenib inhibits hypoxia-inducible factor-1 alpha synthesis. implications for antiangiogenic activity in hepatocellular carcinoma. Clin. Cancer Res. 18, 5662–5671.

Liu, T.A., Jan, Y.J., Ko, B.S., Chen, S.C., Liang, S.M., Hung, Y.L., et al., 2011c. Increased expression of 14-3-3 beta promotes tumor progression and predicts extrahepatic metastasis and worse survival in hepatocellular carcinoma. Am. J. Pathol. 179, 2698–2708.

Liu, W., Iiizumi-Gairani, M., Okuda, H., Kobayashi, A., Watabe, M., et al., 2011d. KAI1 gene is engaged in NDRG1 gene-mediated metastasis suppression through the ATF3-NF kappa B complex in human prostate cancer. J. Biol. Chem. 286, 18949–18959.

Liu, W., Xing, F., Iiizumi-Gairani, M., Okuda, H., Watabe, M., Pai, S.K., et al., 2012d. N-myc downstream regulated gene 1 modulates Wnt-β-catenin signalling and pleiotropically suppresses metastasis. EMBO Mol. Med. 4, 93–108.

Liu, W.J., Monahan, K.B., Pfefferle, A.D., Shimamura, T., Sorrentino, J., Chan, K.T., et al., 2012e. LKB1/STK11 inactivation leads to expansion of a prometastatic tumor subpopulation in melanoma. Cancer Cell 21, 751–764.

Liu, X., Lee, K., Herbison, A.E., 2008. Kisspeptin excites gonadotropin-releasing hormone neurons through a phospholipase C/calcium-dependent pathway regulating multiple ion channels. Endocrinology 149, 4605–4614.

Liu, X., Wang, C., Chen, Z., Jin, Y., Wang, Y., Kolokythas, A., et al., 2011e. MicroRNA-138 suppresses epithelial–mesenchymal transition in squamous cell carcinoma cell lines. Biochem. J. 440, 23–31.

Liu, Y., Liu, H., Han, B., Zhang, J.T., 2006. Identification of 14-3-3σ as a contributor to drug resistance in human breast cancer cells using functional proteomic analysis. Cancer Res. 66, 3248–3254.

Liu, Y., Tian, R.F., Li, Y.M., Liu, W.P., Cao, L., Yang, X.L., et al., 2010. The expression of seven 14-3-3 isoforms in human meningioma. Brain Res. 1336, 98–102.

Liu, Y., Geng, Y., Li, K., Wang, F., Zhou, H., Wang, W., et al., 2012f. Comparative proteomic analysis of the function and network mechanisms of MASPIN in human lung cells. Exp. Ther. Med. 3, 470–474.

Liu, Y., Mayo, M.W., Nagji, A.S., Smith, P.W., Ramsey, C.S., Li, D., et al., 2012g. Phosphorylation of RelA/p65 promotes DNMT-1 recruitment to chromatin and represses transcription of the tumor metastasis suppressor gene BRMS1. Oncogene 31, 1143–1154.

Liu, Y.I., Bai, W.T., Luo, W., Zhang, D.X., Yan, Y., Xu, Z.K., et al., 2011f. Downregulation of NDRG1 promotes invasion of human gastric cancer AGS cells through MMP-2. Tumor. Biol. 32, 99–105.

Liu, Y.L., Yu, L.Q., Wang, Y.W., Zhang, Y.L., Wang, Y.C., Zhang, G.M., 2012h. Expression of tumor suppressor gene ING4 in ovarian carcinoma is correlated with microvessel density. J. Cancer Res. Clin. Oncol. 138, 647–655.

Lo, S.H., 2004. Tensin. Int. J. Biochem. Cell Biol. 36, 31–34.

Lo, P.H.Y., Xie, D., Chan, K.C., Xu, F.P., Kuzmin, I., Lerman, M.I., et al., 2007. Reduced expression of RASSF1A in esophageal and nasopharyngeal carcinomas significantly correlates with tumor stage. Cancer Lett. 257, 199–205.

Logsdon, C.D., Simeone, D.M., Binkley, C., Arumugam, T., Greenson, J.K., Giordano, T.J., et al., 2003. Molecular profiling of pancreatic adenocarcinoma and chronic pancreatitis identifies multiple genes differentially regulated in pancreatic cancer. Cancer Res. 63, 2649–2657.

Londesborough, A., Vaahtomeri, K., Tiainen, M., Katajisto, P., Ekman, N., Vallenius, T., et al., 2008. LKB1 in endothelial cells is required for angiogenesis and TGF beta-mediated vascular smooth muscle cell recruitment. Development 135, 2331–2338.

Lopez-Girona, A., Furnari, B., Mondesert, O., Russell, P., 1999. Nuclear localization of Cdc25 is regulated by DNA damage and a 14-3-3 protein. Nature 397, 172–175.

Lopez-Lago, M.A., Thodima, V.J., Guttapalli, A., Chan, T., Heguy, A., Molina, A.M., et al., 2010. Genomic deregulation during metastasis of renal cell carcinoma implements a myofibroblast-like program of gene expression. Cancer Res. 70, 9682–9692.

Lorenz, K., Lohse, M.J., Quitterer, U., 2003. Protein kinase C switches the Raf kinase inhibitor from Raf-1 to GRK-2. Nature 426, 574–579.

LoRusso, P.M., Rudin, C.M., Reddy, J.C., Tibes, R., Weiss, G.J., Borad, M.J., et al., 2011, Phase I trial of hedgehog pathway inhibitor GDC-0449 in patients with refractory, locally-advanced or metastatic solid tumors. Clin. Cancer Res. 17, 2502–2511.

Lou, Y., Cui, Z., Wang, F., Yang, X., Qian, J., 2011. miR-21 down-regulation promotes apoptosis and inhibits invasion and migration abilities of OVCAR3 cells. Clin. Invest. Med. 34, E281–E289.

Lowery, A.J., Miller, N., Dwyer, R.M., Kerin, M.J., 2010. Dysregulated miR-183 inhibits migration in breast cancer cells. BMC Cancer 10, 502.

Lu, C., Han, H.D., Mangala, L.S., Ali-Fehmi, R., Newton, C.S., Ozbun, L., et al., 2010. Regulation of tumor angiogenesis by EZH2. Cancer Cell 18, 185–197.

Lu, D., Choi, M.Y., Yu, J., Castro, J.E., Kipps, T.J., Carson, D.A., 2011a. Salinomycin inhibits Wnt signaling and selectively induces apoptosis in chronic lymphocytic leukemia cells. Proc. Natl. Acad. Sci. U.S.A. 108, 13253–13257.

Lu, F., Dai, D.L., Martinka, M., Ho, V., Li, G., 2006. Nuclear ING2 expression is reduced in human cutaneous melanomas. Br. J. Cancer 95, 80–86.

Lu, M.L., Chen, F., Wang, Q.W., Wang, K.S., Pan, Q.H., Zhang, X., 2012. Downregulation of inhibitor of growth 3 is correlated with tumorigenesis and progression of hepatocellular carcinoma. Oncol. Lett. 4, 47–52.

Lu, W., Lin, C., Roberts, M.J., Waud, W.R., Piazza, G.A., Li, Y., 2011b. Niclosamide suppresses cancer cell growth by inducing Wnt co-receptor LRP6 degradation and inhibiting the Wnt/beta-catenin pathway. PloS ONE 6, e29290.

Luetzner, N., Arce, J.De.-C., Roesl, F., 2012. Gene expression of the tumour suppressor LKB1 is mediated by Sp1, NF-Y and FOXO transcription factors. PloS ONE 7, e32590.

Luo, J., Feng, J., Lu, J., Wang, Y., Tang, X., Xie, F., et al., 2010. Aberrant methylation profile of 14-3-3 sigma and its reduced transcription/expression levels in Chinese sporadic female breast carcinogenesis. Med. Oncol. 27, 791–797.

Lyden, D., Young, A.Z., Zagzag, D., Yan, W., Gerald, W., O'Reilly, R. et al. (1999). Id1 and Id3 are required for neurogenesis, angiogenesis and vascularization of tumour xenografts. Nature 401, 670–677.

Ma, W., Lee, S.E., Guo, J., Qu, W., Hudson, B.L., Schmidt, A.M., et al., 2007. RAGE Ligand Upregulation of VEGF Secretion in ARPE-19 Cells. Invest. Ophthalmol. Vis. Sci. 48, 1355–1361.

Maass, N., Hojo, T., Ueding, M., Luttges, J., Kloppel, G., Jonat, W., et al., 2001. Expression of the tumor suppressor gene maspin in human pancreatic cancers. Clin. Cancer Res. 7, 812–817.

Maass, N., Nagasaki, K., Ziebart, M., Mundhenke, C., Jonat, W., 2002. Expression and regulation of tumor suppressor gene maspin in breast cancer. Clin. Breast Cancer 3, 281–287.

Machado, R., Rafael, F., Trevisan, P., Koshiyama, D.B., Pilla, C.B., Gazzola, Z., et al., 2010. 22q11.2 deletion syndrome and complex congenital heart defects. Rev. Assoc. Med. Brasil 57, 62–65.

Madison, D.L., Lundblad, J.R., 2010. C-terminal binding protein and poly(ADP)ribose polymerase 1 contribute to repression of the p21(waf1/cip1) promoter. Oncogene 29, 6027–6039.

Maekawa, T., Sano, Y., Shinagawa, T., Rahman, Z., Sakuma, T., Nomura, S., et al., 2008. ATF-2 controls transcription of Maspin and GADD45 alpha genes independently from p53 to suppress mammary tumors. Oncogene 27, 1045–1054.

Mahoney, M.G., Simpson, A., Jost, M., Noé, M., Kari, C., Pepe, D., et al., 2002. Metastasis-associated protein (MTA)1 enhances migration, invasion, and anchorage-independent survival of immortalized human keratinocytes. Oncogene 21, 2161–2170.

Maitah, M.Y., Ali, S., Ahmad, A., Gadgeel, S., Sarkar, F.H., 2011. Upregulation of Sonic hedgehog contributes to TGF-beta 1-induced epithelial to mesenchymal transition in NSCLC cells. PloS ONE 6, e16068.

Malashkevich, V.N., Dulyaninova, N.G., Ramagopal, U.A., Liriano, M.A., Varney, K.M., Knight, D., et al., 2010. Phenothiazines inhibit S100A4 function by inducing protein oligomerization. Proc. Natl. Acad. Sci. U.S.A. 107, 8605–8610.

Malyukova, A.V., Loginov, W.I., Hodyrev, D.S., Kadyrova, E.L., Pronina, I.V., Ivanova, T.A., et al., 2004. Methylation of the putative tumor suppressor gene RASSF1A in primary cervical tumors. Mol. Biol. 38, 857–864.

Mancini, M., Corradi, V., Petta, S., Barbieri, E., Manetti, F., Botta, M., et al., 2011. A new nonpeptidic inhibitor of 14-3-3 induces apoptotic cell death in chronic myeloid leukemia sensitive or resistant to Imatinib. J. Pharmacol. Exp. Ther. 336, 596–604.

Mao, X., Seidlitz, E., Truant, R., Hitt, M., Ghosh, H.P., 2004. Re-expression of TSLC1 in a non-small-cell lung cancer cell line induces apoptosis and inhibits tumor growth. Oncogene 23, 5632–5642.

Marcato, P., Dean, C.A., Pan, D., Araslanova, R., Megan Gillis, M., Joshi, M., et al., 2011. Aldehyde dehydrogenase activity of breast cancer stem cells is primarily due to isoform ALDH1A3 and its expression is predictive of metastasis. Stem Cells 29, 32–45.

Marenholz, I., Volz, A., Ziegler, A., Davies, A., Ragoussis, I., Korge, B.P., et al., 1996. Genetic analysis of the epidermal differentiation complex (EDC) on human chromosome 1q21: chromosomal orientation, new markers, and a 6-Mb YAC contig. Genomics 37, 295–302.

Margulis, V., Maity, T., Zhang, X.Y., Cooper, S.J., Copland, J.A., Wood, C.G., 2008. Type III transforming growth factor-beta (TGF-beta) receptor mediates apoptosis in renal cell carcinoma independent of the canonical TGF-beta signaling pathway. Clin. Cancer Res. 14, 5722–5730.

Marioni, G., Blandamura, S., Giacomelli, L., Calgaro, N., Segato, P., Leo, G., et al., 2005. Nuclear expression of maspin is associated with a lower recurrence rate and a longer disease-free interval after surgery for squamous cell carcinoma of the larynx. Histopathology 46, 576–582.

Marreiros, A., Czolij, R., Yardley, G., Crossley, M., Jackson, P., 2003. Identification of regulatory regions within the KAI1 promoter: a role for binding of AP1, AP2 and p53. Gene 302, 155–164.

Marsit, C.J., McClean, M.D., Furniss, C.S., Kelsey, K.T., 2006. Epigenetic inactivation of the SFRP genes is associated with drinking, smoking and HPV in head and neck squamous cell carcinoma. Int. J. Cancer 119, 1761–1766.

Martchenko, M., Jeong, S.Y., Cohen, S.N., 2010. Heterodimeric integrin complexes containing beta 1-integrin promote internalization and lethality of anthrax toxin. Proc. Natl. Acad. Sci. U.S.A. 107, 15583–15588.

Martín, C., Uribe, K.B., Gómez-Bilbao, G., Ostolaza, H., 2011. Adenylate cyclase toxin promotes internalisation of integrins and raft components and decreases macrophage adhesion capacity. PLoS ONE 6, e17383.

Martinez-Glez, V., Franco-Hernandez, C., Gonzalez-Gomez, P., Isla, A., De Campos, J.M., Vaquero, J., et al., 2007. DAPK1 promoter hypermethylation in brain metastases and peripheral blood. Noeplasma 54, 123–126.

Martinho, O., Granja, S., Jaraquemada, T., Caeiro, C., Miranda-Goncalves, V., Honavar, M., et al., 2012. Downregulation of RKIP is associated with poor outcome and malignant progression in gliomas. PloS ONE 7, e30769.

Martuszewska, D., Ljungberg, B., Johansson, M., Landberg, G., Oslakovic, C., et al., 2009. Tensin3 is a negative regulator of cell migration and all four tensin family members are downregulated in human kidney cancer. PLoS ONE 4, e4350.

Maru, D.M., Singh, R.R., Hannah, C., Albarracin, C.T., Li, Y.X., Abraham, R., et al., 2009. MicroRNA-196a is a potential marker of progression during Barrett's metaplasia–dysplasia–invasive adenocarcinoma sequence in esophagus. Am. J. Pathol. 174, 1940–1948.

Maruyama, Y., Ono, M., Kawahara, A., Yokoyama, T., Basaki, Y., Kage, M., et al., 2006. Tumor growth suppression in pancreatic cancer by a putative metastasis suppressor gene Cap43/NDRG1/Drg-1 through modulation of angiogenesis. Cancer Res. 66, 6233–6242.

Masaki, S., Masutani, H., Yoshihara, E., Yodoi, J., 2012. Deficiency of thioredoxin binding protein-2 (TBP-2) enhances TGF-beta signaling and promotes epithelial to mesenchymal transition. PloS ONE 7, e39900.

Mashimo, T., Watabe, M., Hirota, S., Hosobe, S., Miura, K., Tegtmeyer, P.J., et al., 1998. The expression of the KAI1 gene, a tumor metastasis suppressor, is directly activated by p53. Proc. Natl. Acad. Sci. U.S.A. 95, 11307–11311.

Mashimo, T., Bandyopadhyay, S., Goodarzi, G., Watabe, M., Pai, S.K., Gross, S.C., et al., 2000. Activation of the tumor metastasis suppressor gene, KAI1, by etoposide is mediated by p53 and c-Jun genes. Biochem. Biophys. Res. Commun. 274, 370–376.

Masri, S., Liu, Z., Phung, S., Wang, E., Yuan, Y.C., Chen, S., 2010. The role of microRNA-128a in regulating TGFbeta signaling in letrozole-resistant breast cancer cells. Br. Cancer Res. Treat. 124, 89–99.

Matakatsu, H., Blair, S.S., 2006. Separating the adhesive and signaling functions of the Fat and Dachsous protocadherins. Development 133, 2315–2324.

Matallanas, D., Romano, D., Yee, K., Meissl, K., Kucerova, L., Piazzolla, D., et al., 2007. RASSRA elicits apoptosis through an MST2 pathway directing proapoptotic transcription by the P73 tumor suppressor protein. Mol. Cell 27, 962–975.

Matallanas, D, Birtwistle, M., Ramano, D., Zebisch, A., Rauch, J., Von Kriegsheim, A., et al., 2011. Raf family kinases. Old dogs have learned new tricks. Genes Cancer 2, 232–260.

Matsui, H., Takatsu, Y., Tanaka, A., Asami, T., Nishizawa, N., Kiba, A., et al., 2010. Potent and efficient testosterone suppression by chronic administration of novel metastin analogues, TAK-448 and TAK-683, in male rats. EJC Suppl. 8, 66.

Matsuura, M., Suzuki, T., Saito, T., 2010. Osteopontin is a new target molecule for ovarian clear cell carcinoma therapy. Cancer Sci. 101, 1828–1833.

Matusan-Ilijas, K., Damante, G., Fabbro, D., Dordevic, G., Hadzisejdic, I., Grahovac, M., et al., 2011. Osteopontin expression correlates with nuclear factor-kappa B activation and apoptosis downregulation in clear cell renal cell carcinoma. Pathol. Res. Pract. 207, 104–110.

Matzke, A.J., Neuhuber, F., Park, Y.D., Ambros, P.F., Matzke, M.A., 1994. Homology-dependent gene silencing in transgenic plants: epistatic silencing loci contain multiple copies of methylated transgenes. Mol. Gen. Genet. 244, 219–229.

Matzke, M., Kanno, T., Daxinger, L., Huettel, B., Matzke, A.J., 2009. RNA mediated chromatin based silencing in plants. Curr. Opin. Cell Biol. 21, 367–376.

Maw. M.K., Fujimoto, J., Tamaya, T. 2009. Overexpression of inhibitor of DNA-binding (ID)-1 protein related to angiogenesis in tumor advancement of ovarian cancers. BMC Cancer 9, Art. no. 430.

Maw, M.K., Fujimoto, J., Tamaya, T. 2010. Role of inhibitor of DNA binding-1 protein is related to angiogenesis in the tumor advancement of uterine endometrial cancers. Exp. Ther. Med. 1, 351-356.

Maxwell, S.A., Cherry, E.M., Bayless, K.J., 2011. Akt, 14-3-3 zeta zeta, and vimentin mediate a drug-resistant invasive phenotype in diffuse large B-cell lymphoma. Leuk. Lymphoma 52, 849–864.

McEleny, K., Coffey, R., Morrissey, C., Fitzpatrick, J.M., Watson, R.W.G., 2004. Caffeic acid phenethyl ester-induced PC-3 cell apoptosis is caspase-dependent and mediated through the loss of inhibitors of apoptosis proteins. Br. J. Urol. Int. 94, 402–406.

McInnes, K.J., Brown, K.A., Hunger, N.I., Simpson, E.R., 2012. Regulation of LKB1 expression by sex hormones in adipocytes. Int. J. Obes. 36, 982–985.

Mehenni, H., Lin-Marq, N., Buchet-Poyau, K., Reymond, A., Collart, M.A., Picard, D., et al., 2005. LKB1 interacts with and phosphorylates PTEN: a functional link between two proteins involved in cancer predisposing syndromes. Human Mol. Genet. 14, 2209–2219.

Meister, G., Tuschl, T., 2004. Mechanisms of gene silencing by double-stranded RNA. Nature 431, 343–349.

Meloni, A.R., Fralish, G.B., Kelly, P., Salahpour, A., Chen, J.K., Wechsler-Reya, R.J., et al., 2006. Smoothened signal transduction is promoted by G protein-coupled receptor kinase 2. Mol. Cell Biol. 26, 7550–7560.

Mencia, N., Selga, E., Noe, V., Ciudad, C.J., 2011. Underexpression of miR-224 in methotrexate resistant human colon cancer cells. Biochem. Pharmcol. 82, 1572–1582.

Meng, F., Henson, R., Wehbe-Janek, H., Ghoshal, K., Jacob, S.T., Patel, T., 2007. MicroRNA-21 regulates expression of the PTEN tumor suppressor gene in human hepatocellular cancer. Gastroenterology 133, 647–658.

Meng, Y., Wang, Q.G., Wang, J.X., Zhu, S.T., Jiao, Y., Li, P., et al., 2011. Epigenetic inactivation of the SFRP1 gene in esophageal squamous cell carcinoma. Dig. Dis. Sci. 56, 3195–3203.

Meteoglu, I., Meydan, N., Erkus, M. 2009. Id-1: Regulator of EGFR and VEGF and potential target for colorectal cancer therapy. J Exp. Clin. Cancer Res. 27, Art. no. 69.

Metge, B.J., Frost, A.R., King, J.A., Dyess, D.L., Welch, D.R., Samant, R.S., et al., 2008. Epigenetic silencing contributes to the loss of BRMS1 expression in breast cancer. Clin. Exp. Metastasis 25, 753–763.

Mhawech-Fauceglia, P., Herrmann, F.R., Andrews, C., South, S., Beck, A., Lele, S., et al., 2009. 14-3-3 sigma expression and prognostic value in patients with epithelial ovarian carcinoma: a high throughput tissue microarray analysis. Eur. J. Surg. Oncol. 35, 763–767.

Mian, I., Pierre-Louis, W.S., Dole, N., Gilberti, R.M., Dodge-Kafka, K., Tirnauer, J.S., 2012. LKB1 destabilizes microtubules in myoblasts and contributes to myoblast differentiation. PloS ONE 7, e31583.

Mihic-Probst, D., Kuster, A., Kilgus, S., Bode-Lesniewska, B., Ingold-Heppner, B., Leung, C., et al., 2007. Consistent expression of the stem cell renewal factor BMI-1 in primary and metastatic melanoma. Int. J. Cancer 121, 1764–1770.

Mikami, S., Katsube, K.I., Oya, M., Ishida, M., Kosaka, T., Mizuno, R., et al., 2011. Expression of Snail and Slug in renal cell carcinoma: E-cadherin repressor Snail is associated with cancer invasion and prognosis. Lab. Invest. 91, 1443–1458.

Minn, A.H., Hafele, C., Shalev, A., 2005. Thioredoxin-interacting protein is stimulated by glucose through a carbohydrate response element and induces beta-cell apoptosis. Endocrinology 146, 2397–2405.

Minor, J., Wang, X., Zhang, F., Song, J., Jimeno, A., et al., 2012. Methylation of microRNA-9 is a specific and sensitive biomarker for oral and oropharyngeal squamous cell carcinomas. Oral Oncol. 48, 73–78.

Misawa, A., Katayama, R., Koike, S., Tomida, A., Watanabe, T., Fujita, N., 2010. AP-1-dependent miR-21 expression contributes to chemoresistance in cancer stem cell-like SP cells. Oncol. Res. 19, 23–33.

Mischke, D., Korge, B.P., Marenholz, I., Volz, A., Ziegler, A., 1996. Genes encoding structural proteins of epidermal cornification and S100 calcium-binding proteins form a gene complex ("epidermal differentiation complex") on human chromosome 1q21. J. Invest. Dermatol. 106, 989–992.

Mishra, M., Karagiannis, J., Sevugan, M., Singh, P., Balasubramanian, M.K., 2005. The 14-3-3 protein Rad24p modulates function of the Cdc14p family phosphatase Clp1p/Flp1p in fission yeast. Curr. Biol. 15, 1376–1383.

Mitchell, D.C., Stafford, L.J., Li, D., Bar-Eli, M., Liu, M., 2007. Transcriptional regulation of KiSS-1 gene expression in metastatic melanoma by specificity protein-1 and its coactivator DRIP-130. Oncogene 26, 1739–1747.

Mizuno, Y., Tokuzawa, Y., Ninomiya Y., Yagi K, Yatsuka-Kanesaki Y., Suda, T., et al. (2009). miR-210 promotes osteoblastic differentiation through inhibition of AcvR1b. FEBS Lett., 583(13), 2263–2268.

Moggs, J.G., Grandi, P., Quivy, J.P., Jónsson, Z.O., Hübscher, U., Becker, P.B., et al., 2000. A CAF-1-PCNA-mediated chromatin assembly pathway triggered by sensing DNA damage. Mol. Cell Biol. 20, 1206–2018.

Mohsin, S.K., Zhang, M., Clark, G.M., Allred, D.C., 2003. Maspin expression in invasive breast cancer: association with other prognostic factors. J. Pathol. 199, 432–435.

Moreira, J.M.A., Ohlsson, G., Rank, F.E., Celis, J.E., 2005. Down-regulation of the tumor suppressor protein 14-3-3 sigma is a sporadic event in cancer of the breast. Mol. Cell Proteomics 4, 555–569.

Moreno, A., Palacios, A., Orgaz, J.L., Jimenez, B., Blanco, F.J., Palmero, I., 2010. Functional impact of cancer-associated mutations in the tumor suppressor protein ING4. Carcinogenesis 31, 1932–1938.

Morgenstern, B., Atchley, W.R. 1999. Evolution of bHLH transcription factors: Modular evolution by domain shuffling? Mol. Biol. Evolution 16, 1654–1663.

Moriyama-Kita, M., Endo, Y., Yonemura, Y., Heizmann, C.W., Miyamori, H., Sato, H., et al., 2005. S100A4 regulates E-cadherin expression in oral squamous cell carcinoma. Cancer Lett. 230, 211–218.

Morrow, K.A., Das, S., Metge, B.J., Ye, K., Mulekar, M.S., Tucker, J.A., et al., 2011. Loss of tumor suppressor Merlin in advanced breast cancer is due to post-translational regulation. J. Biol. Chem 286, 40376–40385.

Morton, J.P., Jamieson, N.B., Karim, S.A., Athineos, D., Ridgway, R.A., Nixon, C., et al., 2010. LKB1 haploinsufficiency cooperates with Kras to promote pancreatic cancer through suppression of p21-dependent growth arrest. Gastroenterology 139, 586–597.

Moskwa, P., Buffa, F.M., Pan, Y., Panchakshari, R., Gottipati, P., Muschel, R.J., et al., 2011. miR-182-mediated downregulation of BRCA1 impacts DNA repair and sensitivity to PARP inhibitors. Mol. Cell 41, 210–220.

Mueller, M.T., Hermann, P.C., Witthauer, J., Rubio-Viqueira, B., Leicht, S.F., Huber, S., et al., 2009. Combined targeted treatment to eliminate tumorigenic cancer stem cells in human pancreatic cancer. Gastroenterology 137, 1102–1113.

Muinos-Gimeno, M., Espinosa-Parrilla, Y., Guidi, M., Kagerbauer, B., Sipilae, T., Maron, E., et al., 2011. Human microRNAs miR-22, miR-138-2, miR-148a, and miR-488 are associated with panic disorder and regulate several anxiety candidate genes and related pathways. Biol. Psychol. 69, 526–533.

Muley, A., Majumder, S., Kolluru, G.K., Parkinson, S., Viola, H., Hool, L., et al., 2010. Secreted Frizzled-related protein 4 an angiogenesis inhibitor. Am. J. Pathol. 176, 1505–1516.

Murillo, M.M., del Castillo, G., Sanchez, A., Fernandez, M., Fabregat, I., 2005. Involvement of EGF receptor and c-Src in the survival signals induced by TGF-beta 1 in hepatocytes. Oncogene, 244580–244587.

Murray, M.J., Halsall, D.J., Hook, C.E., Williams, D.M., Nicholson, J.C., Coleman, N., 2011. Identification of microRNAs from the miR-371 similar to 373 and miR-302 clusters as potential serum biomarkers of malignant germ cell tumors. Am. J. Clin. Pathol. 135, 119–125.

Nadal, A., Ropero, A.B., Laribi, O., Maillet, M., Fuentes, E., Soria, B., 2000. Nongenomic actions of estrogens and xenoestrogens by binding at a plasma membrane receptor unrelated to estrogen receptor α and estrogen receptor β. Proc. Natl. Acad. Sci. U.S.A. 97, 11603–11608.

Nagai, K., Doi, R., Katagiri, F., Ito, T., Kida, A., Koizumi, M., et al., 2009. Prognostic value of metastin expression in human pancreatic cancer. J. Exp. Clin. Cancer Res. 28, 9.

Nagai, M.A., Gerhard, R., Fregnani, J.H.T.G., Nonogaki, S., Rierger, R.B., Netto, M.M., et al., 2011. Prognostic value of NDRG1 and SPARC protein expression in breast cancer patients. Breast Cancer Res. Treat. 126, 1–14.

Nagaraja, A.K., Creighton, C.J., Yu, Z., Zhu, H., Gunaratne, P.H., Reid, J.G., et al., 2010. A link between mir-100 and FRAP1/mTOR in clear cell ovarian cancer. Mol. Endocrinol. 24, 447–463.

Nagji, A.S., Liu, Y., Stelow, E.B., Stukenborg, G.J., Jones, D.R., 2010. BRMS1 transcriptional repression correlates with CpG island methylation and advanced pathological stage in non-small cell lung cancer. J. Pathol. 221, 229–236.

Nagy, N., Brenner, C., Markadieu, N., Chaboteaux, C., Camby, I., Schafer, B.W., et al., 2001. S100A2, a putative tumor suppressor gene, regulates in vitro squamous cell carcinoma migration. Lab. Invest. 81, 599–612.

Naidoo, K., Jones, R., Dmitrovic, B., Wijesuriya, N., Kocher, H., Hart, I.R., et al., 2012. Proteome of formalin-fixed paraffin-embedded pancreatic ductal adenocarcinoma and lymph node metastases. J. Pathol. 226, 756–763.

Nakamura, Y., Oshima, K., Naoi, Y., Nakayama, T., Kim, S.J., Shimazu, K., et al., 2012. 14-3-3 sigma expression is associated with poor pathological complete response to neoadjuvant chemotherapy in human breast cancers. Breast Cancer Res. Treat. 134, 229–236.

Nakazawa, Y., Kamijo, T., Koike, K., Noda, T., 2003. ARF tumour suppressor induces mitochondria dependent apoptosis by modudlation of mitochondrial Bcl-2 family proteins. J. Biol. Chem. 278, 27888–27895.

Nalls, D., Tang, S.N., Rodova, M., Srivastava, R.K., Shankar, S., 2011. Targeting epigenetic regulation of miR-34a for treatment of pancreatic Cancer by inhibition of pancreatic cancer stem cells. Plos ONE 6, e24099.

Nambotin, S.B., Lefrancois, L., Sainsily, X., Berthillon, P., Kim, M., Wands, J.R., et al., 2011. Pharmacological inhibition of Frizzled-7 displays anti-tumor properties in hepatocellular carcinoma. J. Hepatol. 54, 288–299.

Nath-Sain, S., Marignani, P.A., 2009. LKB1 catalytic activity contributes to estrogen receptor alpha signalling. Mol. Biol. Cell. 20, 2785–2795.

Neal, C.L., Yu, D., 2010. 14-3-3 zeta as a prognostic marker and therapeutic target for cancer. Expert Opin. Ther. Targets 14, 1343–1354.

Neal, C.L., Xu, J., Li, P., Mori, S., Yang, J., Neal, N.N., et al., 2012a. Overexpression of 14-3-3 zeta in cancer cells activates PI3K via binding the p85 regulatory subunit. Oncogene 31, 897–906.

Neal, C.L., Henderson, V., Smith, B.N., McKeithen, D., Graham, T., Vo, B.T., et al., 2012b. Snail transcription factor negatively regulates maspin tumor suppressor in human prostate cancer cells. BMC Cancer 12, 336.

Neupane, D., Korc, M., 2008. 14-3-3 sigma modulates pancreatic cancer cell survival and invasiveness. Clin. Cancer Res. 14, 7614–7623.

Ngamkitidechakul, C., Warejcka, D.J., Burke, J.M., O'Brien, W.J., Twining, S.S., 2003. Sufficiency of the reactive site loop of maspin for induction of cell–matrix adhesion and inhibition of cell invasion – conversion of ovalbumin to a maspin-like molecule. J. Biol. Chem. 278, 31796–31806.

Nickoloff, B.J., Lingen, M.W., Chang, B.D., Shen, M., Swift, M., Curry, J., et al., 2004. Tumor suppressor maspin is up-regulated during keratinocyte senescence, exerting a paracrine antiangiogenic activity. Cancer Res. 64, 2956–2961.

Niehrs, C., 2006. Function and biological roles of the Dickkopf family of Wnt modulators. Oncogene 25, 7469–7481.

Nishida, N., Mimori, K., Fabbri, M., Yokobori, T., Sudo, T., Tanaka, F., et al., 2011. MicroRNA-125a-5p is an independent prognostic factor in gastric cancer and inhibits the proliferation of human gastric cancer cells in combination with Trastuzumab. Clin. Cancer Res. 17, 2725–2733.

Nishikori, M. 2005. Classical and alternative NF-kB activation pathways and their roles in lymphoid malignancies. J. Clin. Exp. Hematopathol. 45, 15–24.

Nishinaka, Y., Masutani, H., Oka, S., Matsuo, Y., Yamaguchi, Y., Nishio, K., et al., 2004. Importin α1 (Rch1) mediates nuclear translocation of thioredoxin-binding protein-2/vitamin D3-up-regulated protein 1. J. Biol. Chem. 279, 37559–37565.

Nishio, S., Ushijima, K., Tsuda, N., Takemoto, S., Kawano, K., Yamaguchi, T., et al., 2008. Cap43/NDRG1/Drg-1 is a molecular target for angiogenesis and a prognostic indicator in cervical adenocarcinoma. Cancer Lett. 264, 36–43.

Nishiyama, A., Matsui, M., Iwata, S., Hirota, K., Masutani, H., Nakamura, H., et al., 1999. Identification of thioredoxin-binding protein-2/vitamin D3 up-regulated protein 1 as a negative regulator of thioredoxin function and expression. J. Biol. Chem. 274, 21645–21650.

Niu, J., Shi, Y., Tan, G., Yang, C.H., Fan, M., Pfeffer, L.M., et al., 2012. DNA damage induces NF-kappa B-dependent microRNA-21 up-regulation and promotes breast cancer cell invasion. J. Biol. Chem. 287, 21783–21795.

Noetzel, E., Veeck, J., Niederacher, D., Galm O., Horn, F., Hartmann, A., Knuechel, R., Dahl, E. 2008. Promoter methylation-associated loss of ID4 expression is a marker of tumour recurrence in human breast cancer. BMC Cancer 8, Art. no. 154.

Nomoto, K., Tomita, N., Miyake, M., Xhu, D.B., LoGerfo, P.R., Weinstein, I.B., 1995. Expression of phospholipases gamma1, beta 1 and delta 1in primary human colon carcinoma and colon carcinoma cell lines. Mol. Carcinog. 12, 146–152.

Norton, J.D., Deed, R.W., Craggs, G., Sablitzky, F. 1998. Id helix-loop-helix proteins in cell growth and diffeerentiation. Trends Cell Biol. 8, 58–65.

Notarbartolo, M., Giannitrapani, L., Vivona, N., Poma, P., Labbozzetta, M., Florena, A.M., et al., 2011. Frequent alteration of the Yin Yang 1/Raf-1 kinase inhibitory protein ratio in hepatocellular carcinoma. OMICS 15, 267–272.

Novina, C.D., Sharp, P.A., 2004. The RNAi revolution. Nature 430, 161–184.

Oates, A.J., Barraclough, R., Rudland, P.S., 1996. The identification of osteopontin as a metastasis-related gene product in a rodent mammary tumour model. Oncogene 13, 97–104.

Obrador-Hevia, A., Serra-Sitjar, M., Rodriguez, J., Villalonga, P., Fernandez de Mattos, S., 2012. The tumour suppressor FOXO3 is a key regulator of mantle cell lymphoma proliferation and survival. Br. J. Haematol. 156, 334–345.

O'Brien, C.A., Pollet, A., Gallinger, S., Dick, J.E., 2007. A human colon caner cell capable of initiating tumour growth in immunodeficient mice. Nature 445, 106–110.

Ochiai, K., Watanabe, M., Ueki, H., Huang, P., Fujii, Y., Nasu, Y., et al., 2011. Tumor suppressor REIC/Dkk-3 interacts with the dynein light chain, Tctex-1. Biochem. Biophys. Res. Commun. 412, 391–395.

Ogasawara, S., Maesawa, C., Yamamoto, M., Akiyama, Y., Wada, K., Fujisawa, K., et al., 2004. Disruption of cell-type-specific methylation at the Maspin gene promoter is frequently involved in undifferentiated thyroid cancers. Oncogene 23, 1117–1124.

Ogden, S.K., Fei, D.L., Schilling, N.S., Ahmed, Y.F., Hwa, J., Robbins, D.J., 2008. G protein G alpha(i) functions immediately downstream of smoothened in Hedgehog signalling. Nature 456, 967–970.

Ogihara, T., Isobe, T., Ichimura, T., Taoka, M., Funaki, M., Sakoda, H., et al., 1997. 14-3-3 protein binds to insulin receptor substrate-1, one of the binding sites of which is in the phophotyrosine binding domain. J. Biol. Chem. 272, 25267–25274.

Ogmundsdottir, H.M., Gudlaugsdottir, S., Bjornsson, J., Jonasdottir, S., 1996. Altered expression of CA-125 in breast carcinomas. APMIS 104, 47–53.

Ogunjimi A. A., Zeqiraj E., Ceccarelli D. F., Sicheri F., Wrana J. L., David L., 2012. Structural basis for specificity of TGFbeta family receptor small molecule inhibitors. Cell Signal. 24, 476–483.

Oh, J.S., Kim, J.J., Byun, J.Y., Kim, I.A., 2010. Lin28–Let7 modulates radiosensitivity of human cancer cells with activation of K-ras. Int. J. Radiat. Oncol. Biol. Phys. 76, 5–8.

Ohashi, S., Natsuizaka, M., Wong, G.S., Michaylira, C.Z., Grugan, K.D., Stairs, D.B., et al., 2010. Epidermal growth factor receptor and mutant p53 expand an esophageal cellular subpopulation capable of epithelial-to-mesenchymal transition through ZEB transcription factors. Cancer Res. 70, 4174–4184.

Ohike, N., Maass, N., Mundhenke, C., Biallek, M., Zhang, M., Jonat, W., et al., 2003. Clinicopathological significance and molecular regulation of maspin expression in ductal adenocarcinoma of the pancreas. Cancer Lett. 199, 193–200.

Ohno, T., Shirakami, Y., Shimizu, M., Kubota, M., Sakai, H., Yasuda, Y., et al., 2012. Synergistic growth inhibition of human hepatocellular carcinoma cells by acyclic retinoid and GW4064, a farnesoid X receptor ligand. Cancer Lett. 323, 215–222.

Ohta, K., Ito, A., Kuriyama, S., Lupo, G., Kosaka, M., Ohnuma, S.I., et al., 2011. Tsukushi functions as a Wnt signaling inhibitor by competing with Wnt2b for binding to transmembrane protein Frizzled4. Proc. Natl. Acad. Sci. U.S.A. 108, 14962–14967.

Ohtaki, T., Shintani, Y., Honda, S., Matsumoto, H., Hori, A., Kanehashi, K., et al., 2001. Metastasis suppressor gene KiSS-1 encodes peptide ligand of a G-protein-coupled receptor. Nature 411, 613–617.

Okada, T., Lopez-Lago, M., Giancotti, F.G., 2005. Merlin/NF-2 mediates contact inhibition of growth by suppressing recruitment of Rac to the plasma membrane. J. Cell Biol. 171, 361–367.

Okuda, K., Sasaki, H., Hikosaka, Y., Kawano, O., Moriyama, S., Yano, M., et al., 2011. LKB1 gene alterations in surgically resectable adenocarcinoma of the lung. Surg. Today 41, 107–110.

Okumura, H., Kita, Y., Yokomakura, N., Uchikado, Y., Setoyama, T., Sakurai, H., et al., 2010. Nuclear expression of 14-3-3 sigma is related to prognosis in patients with esophageal squamous cell carcinoma. Anticancer Res. 30, 5175–5179.

Olbrich, T., Ziegler, E., Tuerk, G., Schubert, A., Emons, G., Gruendker, C., 2010. Kisspeptin-10 inhibits bone-directed migration of GPR54-positive breast cancer cells: evidence for a dose-window effect. Gynecol. Oncol. 119, 571–578.

Onori, P., DeMorrow, S., Gaudio, E., Franchitto, A., Mancinelli, R., Venter, J., et al., 2009. Caffeic acid phenethyl ester decreases cholangiocarcinoma growth by inhibition of NF-kappa B and induction of apoptosis. Int. J. Cancer 125, 565–576.

Osada, T., Chen, M., Yang, X.Y., Spasojevic, I., Vandeusen, J.B., Hsu, D., et al., 2011. Antihelminth compound Niclosamide downregulates Wnt signaling and elicits antitumor responses in tumors with activating APC mutations. Cancer Res. 71, 4172–4182.

Osoegawa, A., Kometani, T., Nosaki, K., Ondo, K., Hamatake, M., Hirai, F., et al., 2011. LKB1 mutations frequently detected in mucinous bronchioloalveolar carcinoma. Jpn. J. Clin. Oncol. 41, 1132–1137.

Ou, W., Ye, S., Yang, W., Wang, Y., Ma, Q., Yu, C., et al., 2012. Enhanced antitumor effect of cisplatin in human NSCLC cells by tumor suppressor LKB1. Cancer Gene. Ther. 19, 489–498.

Ouatas, T., Halverson, D., Steeg, P.S., 2003. Dexamethasone and medroxyprogesterone acetate elevate Nm23-H1 metastasis suppressor gene expression in metastatic human breast carcinoma cells: new uses for old compounds. Clin. Cancer Res. 9, 3763–3772.

Pahl, H.L., 1999. Activators and target genes of Rel/ NF-κB transcription factor. Oncogene 18, 6853–6866.

Palamarchuk, A., Efanov, A., Maximov, V., Aqeilan, R.I., Croce, C.M., Pekarsky, Y., 2005. Akt phosphorylates and regulates Pdcd4 tumor suppressor protein. Cancer Res. 65, 11282–11286.

Palmieri, D., Halverson, D.O., Ouatas, T., Horak, C.E., Salerno, M., Johnson, J., et al., 2005. Medroxyprogesterone acetate elevation of Nm23-H1 metastasis suppressor expression in hormone receptor-negative breast cancer. J. Natl. Cancer Inst. 97, 632–642.

Palmieri, D., Horak, C.E., Lee, J.H., Halverson, D.O., Steeg, P.S., 2006. Translational approaches using metastasis suppressor genes. J. Bioenergy Biomembr. 38, 151–161.

Pan, S., Wu, X., Jiang, J., Gao, W., Wan, Y., Cheng, D., et al., 2010. Discovery of NVP-LDE225, a potent and selective smoothened antagonist. ACS Med. Chem. Lett. 1, 130–134.

Pandey, D.P., Picard, D., 2009. miR 22 inhibits estrogen signaling by directly targeting the estrogen receptor alpha mRNA. Mol. Cell. Biol. 29 (13), 3783–3790.

Pandit, K.V., Corcoran, D., Yousef, H., Yarlagadda, M., Tzouvelekis, A., Gibson, K.F., et al., 2010. Inhibition and role of let-7d in idiopathic pulmonary fibrosis. Am. J. Respir. Crit. Care Med. 182, 220–229.

Pang, S.T., Hsieh, W.C., Chuang, C.K., Chao, C.H., Weng, W.H., Juang, H.H., 2009. Thioredoxin-interacting protein: an oxidative stress-related gene is upregulated by glucose in human prostate carcinoma cells. J. Mol. Endocrinol. 42, 205–214.

Pannifer, A.D., Wong, T.Y., Schwarzenbacher, R., Renatus, M., Petosa, C., Bienkowska, J., et al., 2001. Crystal structure of the anthrax lethal factor. Nature 414, 229–233.

Pantuck, A.J., Zeng, G., Belldegrun, A.S., Figlin, R.A., 2003. Pathobiology, prognosis, and targeted therapy for renal cell carcinoma. Exploiting the hypoxia-induced pathway. Clin. Cancer Res. 9, 4641–4652.

Park, J., Kang, S.I., Lee, S.Y., Zhang, X.F., Kim, M.S., Beers, L.F., et al., 2010. Tumor suppressor Ras association domain family 5 (RASSF5/NORE1) mediates death receptor ligand-induced apoptosis. J. Biol. Chem. 285, 35029–35038.

Park, J.J., Jin, Y.B., Lee, Y.J., Lee, J.S., Lee, Y.S., Ko, Y.G., et al., 2012. KAI1 suppresses HIF-1 alpha and VEGF expression by blocking CDCP1-enhanced Src activation in prostate cancer. BMC Cancer 12, 81.

Park, S.M., Gaur, A.B., Lengyel, E., Peter, M.E., 2008. The miR-200 family determines the epithelial phenotype of cancer cells by targeting the E-cadherin repressors ZEB1 and ZEB2. Genes Dev. 22, 894–907.

Parker, C., Sherbet, G.V., 1992. Modulators of intracellular $Ca2^{+}$ and the calmodulin inhibitor W-7 alter the expression of the metastasis associated genes mts1 and nm23 in metastatic variants of the B16 murine melanoma. Melanoma Res. 2, 337–343.

Partanen, J.I., Tervonen, T.A., Myllynen, M., Lind, E., Imai, M., Katajisto, P., et al., 2012. Tumor suppressor function of liver kinase B1 (Lkb1) is linked to regulation of epithelial integrity. Proc. Natl. Acad. Sci. U.S.A. 109, E388–E397.

Passiatore, G., Gentilella, A., Rom, S., Pacifici, M., Bergonzini, V., Peruzzi, F. 2011. Induction of Id-1 by FGF-2 involves activity of EGR-1 and sensitises neuroblastoma cells to cell death. J. Cell. Physiol. 226, 1763–1770.

Paulissen, G., Rocks, N., Gueders, M.M., Crahay, C., Quesada-Calvo, F., Bekaert, S., et al., 2009. Role of ADAM and ADAMTS metalloproteinases in airway diseases. Respir. Res 10, 127.

Pearson, M.A., Reczek, D., Bretscher, A., Karplus, P.A., 2000. Structure of the ERM protein moesin reveals the FERM domain fold masked by an extended actin binding tail domain. Cell 101, 259–270.

Pencil, S.D., Toh, Y., Nicolson, G.L., 1993. Candidate metastasis-associated genes of the rat 13762NF mammary adenocarcinoma. Breast Cancer Res. Treat. 25, 165–174.

Peng, X., Wang, Y., Kolli, S., Deng, J., Li, L., Wang, Z., Raj, J.U., Gou, D. 2012. Physical and functional interaction between the ID1 and p65 for activation of NF-kappa B. Am. J. Physiol. Cell. Physiol. 303, C267–C277.

Petersen, M., Thorikay, M., Deckers, M., van Dinther, M., Grygielko, E.T., Gellibert, F., et al., 2008. Oral administration of GW788388, an inhibitor of TGF-beta type I and II receptor kinases, decreases renal fibrosis. Kidney Int. 73, 705–715.

Petti, C., Vegetti, C., Molla, A., Bersani, I., Cleris, L., Mustaard, K.J., et al., 2012. AMPK activators inhibit the proliferation of human melanomas bearing the activated MAPK pathay. Melanoma Res. 22, 341–350.

Philip, N., Bassett, A., 2011. Cognitive, behavioural and psychiatric phenotype in 22q11.2 deletion syndrome. Behav. Genet. 41, 403–412.

Piao, Z.H., Yoon, S.R., Kim, M.S., Jeon, J.H., Lee, S.H., Kim, T.D., et al., 2009. VDUP1 potentiates Ras-mediated angiogenesis via ROS production in endothelial cells. Cell. Mol. Biol. 55 (suppl.), OL1096–OL1103.

Pichiorri, F., Suh, S.S., Rocci, A., De Luca, L., Taccioli, C., Santhanam, R., et al., 2010. Downregulation of p53-inducible microRNAs 192, 194, and 215 impairs the p53/MDM2 autoregulatory loop in multiple myeloma development. Cancer Cell 18, 367–381.

Pierson, C.R., McGowen, R., Grignon, D., Sakr, W., Dey, J., Sheng, S.J., 2002. Maspin is up-regulated in premalignant prostate epithelia. Prostate 53, 255–262.

Pietersen, A.M., Evers, B., Prasad, A.A., Tanger, E., Cornelissen-Steijger, P., Jonkers, J., et al., 2008. Bmi1 regulates stem cells and proliferation and differentiation of committed cells in mammary epithelium. Curr. Biol. 18, 1094–1099.

Pineda, R., Aguilar, E., Pinilla, L., Tena-Sempere, M., 2010. Physiological roles of the kisspeptin/GPR54 system in the neuroendocrine control of reproduction. Prog. Brain Res. 181, 55–77.

Pinte, S., Stankovic-Valentin, N., Deltour, S., Rood, B.R., Guérardel, C., Leprince, D., 2004. The tumor suppressor gene HIC1 (hypermethylated in cancer 1) is a sequence-specific transcriptional repressor. Definition of its consensus binding sequence and analysis of its DNA binding and repressive properties. J. Biol. Chem. 279, 38313–38324.

Place, R.F., Li, L.C., Pookot, D., Noonan, E.J., Dahiya, R., 2008. MicroRNA-373 induces expression of genes with complementary promoter sequences. Proc. Natl. Acad. Sci. U.S.A. 105, 1608–1613.

Pohlers, D., Brenmoehl, J., Löffler, I., Cocolakis, E., Dumas, V.M., Posner, B.I., et al., 2009. TGF-beta and fibrosis in different organs – molecular pathway imprints. Biochim. Biophys. Acta 1792, 746–756.

Polanski, R., Maguire, M., Nield, P.C., Jenkins, R.E., Park, B.K., Krawczynska, K., et al., 2011. MDM2 interacts with NME2 (non-metastatic cells 2, protein) and suppresses the ability of NME2 to negatively regulate cell motility. Carcinogenesis 32, 1133–1142.

Polizio, A.H., Chinchilla, P., Chen, X., Manning, D.R., Riobo, N.A., 2011. Sonic Hedgehog activates the GTPases Rac1 and RhoA in a Gli-independent manner through coupling of Smoothened to G(i) proteins. Sci. Signal. 4, pt7.

Pollak, M.N., 2012. Investigating metformin for cancer prevention and treatment: the end of the beginning. Cancer Discov. 2, 778–790.

Popova, T., Espina, V., Bailey, C., Liotta, L., Petricoin, E., Popov, S., 2009. Anthrax infection inhibits the AKT signaling involved in the E-cadherin-mediated adhesion of lung epithelial cells. FEMS Immunol. Med. Microbiol. 56, 129–142.

Popow-Wozniak, A., Mazur, A.J., Mannherz, H.G., Malicka-Blaszkiewicz, M., Nowak, D., 2012. Cofilin overexpression affects actin cytoskeleton organization and migration of human colon adenocarcinoma cells. Histochem. Cell. Biol. 138, 725–736.

Porter, B.E., Lee, C., Liu, C., Risbud, R., 2011. Role of TGF-beta signaling in the regulation of miR21 in glia and neurons. 41st Annual Meeting of the Society-for-Neuroscience, 41 (abstract).

Poulikakos, P.I., Xiao, G.H., Gallagher, R., Jablonski, S., Jhanwar, S.C., Testa, J.R., 2006. Re-expression of the tumor suppressor NF2/merlin inhibits invasiveness in mesothelioma cells and negatively regulates FAK. Oncogene 25, 5960–5968.

Powell, D.W., Rane, M.J., Chen, Q., Singh, S., McLeish, K.R., 2002. Identification of 14-3-3 zeta as a protein kinase B/akt substrate. J. Biol. Chem. 277, 21639–21642.

Powell, D.W., Rane, M.J., Joughin, B.A., Kalmukova, R., Hong, J.G., Tidor, B., et al., 2003. Proteomic indentifical of 14-3-3ζ as a mitogen-activated protein kinase-activated protein kinase 2 substrate. Mol. Cell. Biol. 15, 5376–5387.

Powers, M.A., Fay Marta, M., Factor, R.E., Welm, A.L., Ullman, K.S., 2011. Protein arginine methyltransferase 5 accelerates tumor growth by arginine methylation of the tumor suppressor programmed cell death 4. Cancer Res. 71, 5579–5587.

Prasad, C.P., Rath, G., Mathur, S., Bhatnagar, D., Parshad, R., Ralhan, R., 2009. Expression analysis of E-cadherin, Slug and GSK3 beta in invasive ductal carcinoma of breast. BMC Cancer 9, 325.

Price, M.A., Kalderon, D., 2002. Proteolysis of the Hedgehog signaling effector cubitus interruptus requires phosphorylation by glycogen synthase kinase 3 and casein kinase 1. Cell 108, 823–835.

Pullikuth, A.K., Catlin, A.D., 2010. Extracellular signal-regulated kinase promotes Rho-dependent focal adhesion formation by suppressing p190A RhoGAP. Mol. Cell. Biol. 30, 3233–3248.

Qi, C., Zhu, Y.T., Hu, L., Zhu, Y.J., 2009. Identification of Fat4 as a candidate tumor suppressor gene in breast cancers. Int. J. Cancer 124, 793–798.

Qian, H., Yu, J., Li, Y., Wang, H., Song, C., Zhang, X., et al., 2007a. RNA interference of metastasis associated gene 1 inhibits metastasis of B16F10 melanoma cells in a C57BL/6 mouse model. Biol. Cell. 99, 573–581.

Qian, X., Li, G., Asmussen, H.K., Asnaghi, L., Vass, W.C., Braverman, R., et al., 2007b. Oncogenic inhibition by a deleted in liver cancer gene requires cooperation between tensin binding and Rho-specific GTPase-activating protein activities. Proc. Natl. Acad. Sci. U.S.A. 104, 9012–9017.

Qian, X.L., Karpova, T., Sheppard, A.M., McNally, J., Lowy, D.R., 2004. E-cadherin-mediated adhesion inhibits ligand-dependent activation of diverse receptor tyrosine kinases. EMBO J. 23, 1739–1748.

Qin, L., Zhang, M., 2010. Maspin regulates endothelial cell adhesion and migration through an integrin signalling pathway. J. Biol. Chem. 285, 32360–32369.

Rabbani, S.A., Valentino, M.L., Arakelian, A., Ali, S., Boschelli, F., 2010. SKI-606 (Bosutinib) blocks prostate cancer invasion, growth, and metastasis in vitro and in vivo through regulation of genes involved in cancer growth and skeletal metastasis. Mol. Cancer Ther. 9, 1147–1157.

Radhakrishnan, V.M., Martinez, J.D., 2010. 14-3-3 gamma ind.uces oncogenic transformation by stimulating MAP kinase and PI3K signalling. PLoS ONE 5, e11433.

Radhakrishnan, V.M., Jensen, T.J., Cui, H., Futscher, B.W., Martinez, J.D., 2011a. Hypomethylation of the 14-3-3 sigma promoter leads to increased expression in non-small cell lung cancer. Genes Chromosomes Cancer 50, 830–836.

Radhakrishnan, V.M., Putnam, C.W., Qi, W., Martinez, J.D., 2011b. P53 suppresses expression of the 14-3-3gamma oncogene. BMC Cancer 11, 378.

Rajagopalan, S., Sade, R.S., Townsley, F.M., Fersht, A.R., 2010. Mechanistic differences in the transcriptional activation of p53 by 14-3-3 isoforms. Nucl. Acids Res. 38, 893–906.

Rak, J., Mitsuhashi, Y., Sheehan, C., Tamir, A., Viloria-Petit, A., Filmus, J., et al., 2000. Oncogenes and tumor angiogenesis: differential modes of vascular endothelial growth factor up-regulation in ras-transformed epithelial cells and fibroblasts. Cancer Res. 60, 490–498.

Ramakrishnan, R., Donahue, H., Garcia, D., Tan, J., Shimizu, N., Rice, A.P., et al. 2011. Epstein-Barr virus BART9 miRNA modulates LMP1 levels and affects growth rate of nasal NK T cell lymphomas. PloS One 6, e27271.

Ramaswamy, B., Lu, Y., Teng, K.Y., Nuovo, G., Li, X., Shapiro, C.L., et al., 2012. Hedgehog signalling is a novel therapeutic target in tamoxifen-resistant breast cancer aberrantly activated by PI3K/Akt pathway. Cancer Res. 72, 5048–5059.

Ranganathan, S., Walsh, E.S., Godwin, A.K., Tew, K.D., 1993. Cloning and characterization of human colon glyoxalase-I. J. Biol. Chem. 268, 5661–5667.

Rath, O., Park, S., Tang, H.H., Banfield, M.J., Brady, R.L., Lee, Y.C., et al., 2008. The RKIP (Raf-1 Kinase Inhibitor Protein) conserved pocket binds to the phosphorylated N-region of Raf-1 and inhibits the Raf-1-mediated activated phosphorylation of MEK. Cell Signal. 20, 935–941.

Rath, P., Miller, D.C., Litofsky, N.S., Anthony, D.C., Feng, Q., Franklin, C., et al., 2011. Isolation and characterization of a population of stem-like progenitor cells from an atypical meningioma. Exp. Mol. Pathol. 90, 179–188.

Rattan, R., Graham, R.P., Maguire, J.L., Giri, S., Shridhar, V., 2011. Metformin suppresses ovarian cancer growth and metastasis with enhancement of cisplatin cytotoxicity in vivo. Neoplasia 13, 483–491.

Rattner, A., Hsieh, J.C., Smallwood, P.M., Gilbert, D.J., Copeland, N.G., Jenkins, N.A., et al., 1997. A family of secreted proteins contains homology to the cysteine-rich ligand-binding domain of frizzled receptors. Proc. Natl. Acad. Sci. U.S.A. 94, 2859–2863.

Raval, A., Tanner, S.M., Byrd, J.C., Angerman, E.V., Perko, J.D., Chen, S.S., et al., 2007. Downregulation of death-associated protreinkinase (DAPK1) in chronic lymphocytic leukemai. Cell 129, 879–890.

Ravelli, A., Massobrio, M., Tesarik, J., 1998. Nongenomic actions of steroid hormones in reproductive tissues. Endocrine. Rev. 19, 3–17.

Ravindranath, A., Yuen, H.F., Chan, K.K., Grills, C., Fennell, D.A., Lappin, T.R., et al., 2011. Wnt-beta-catenin-Tcf-4 signalling-modulated invasiveness is dependent on osteopontin expression in breast cancer. Br. J. Cancer 105, 542–551.

Rawson, J.B., Manno, M., Mrkonjic, M., Daftary, D., Dicks, E., Buchanan, D.D., et al., 2011. Promoter methylation of Wnt antagonists DKK1 and SFRP1 is associated with opposing tumor subtypes in two large populations of colorectal cancer patients. Carcinogenesis 32, 741–747.

Rayala, S.K., den Hollander, P., Manavathi, B., Talukder, A.H., Song, C., Peng, S., et al., 2006. Essential role of KIBRA in co-activator function of dynein light chain 1 in mammalian cells. J. Biol. Chem. 281, 19092–19099.

Rayner, K., Chen, Y.X, Hibbert, B., White, D., Miller, H., Postel, E.H., et al., 2007. NM23-H2, an estrogen receptor beta-associated protein, shows diminished expression with progression of atherosclerosis. Am. J. Physiol. Regul. Integr. Com. Physiol. 292, R743–750.

Rayner, K., Chen, Y.X., Hibbert, B., White, D., Miller, H., Postel, E.H., et al., 2008. Discovery of NM23-H2 as an estrogen receptor beta-associated protein: role in estrogen-induced gene transcription and cell migration. J. Steroid. Biochem. Mol. Biol. 108, 72–81.

Recino, A., Sherwood, V., Flaxman, A., Cooper, W.N., Latif, F., Ward, A., et al., 2010. Human RASSF7 regulates the microtubule cytoskeleton and is required for spindle formation, Aurora B activation and chromosomal congression during mitosis. Biochem. J. 430, 207–213.

Reeves, M.E., Baldwin, S.W., Baldwin, M.L., Chen, S., Moretz, J.M., Aragon, R.J., et al., 2010. Ras-association domain family 1C protein promotes breast cancer cell migration and attenuates apoptosis. BMC Cancer 10, 562.

Reifenberger, J., Knobbe, C.B., Sterzinger, A.A., Blaschke, B., Schulte, K.W., Ruzicka, T., et al., 2004. Frequent alterations of Ras signaling pathway genes in sporadic malignant melanomas. Int. J. Cancer 109, 377–384.

Reis, S., Pontes Jr., J., Antunes, A.A., Dall'Oglio, M.F., Dip, N., Passerotti, C.C., 2012. Mir-21 may act as an oncomir by targeting RECK, a matrix metalloproteinase regulator, in prostate cancer. BMC Urology. 12, 14.

Reiter, E., Lefkowitz, R.J., 2006. GRKs and b-arrestins: roles in receptor silencing, trafficking and signalling. Trends. Endocrinol. Metab. 17, 159–165.

Ren, H.Z., Pan, G.Q., Wang, J.S., Wen, J.F., Wang, K.S., Luo, G.Q., et al., 2010a. Reduced stratifin expression can serve as an independent prognostic factor for poor survival in patients with esophageal squamous cell carcinoma. Dig. Dis. Sci. 55, 2552–2560.

Ren, Y., Zhou, X., Mei, M., Yuan, X.B., Han, L., Wang, G.X., et al., 2010b. MicroRNA-21 inhibitor sensitizes human glioblastoma cells U251 (PTEN-mutant) and LN229 (PTEN-wild type) to taxol. BMC Cancer 10, 27.

Ren, Y., Cheung, H.W., von Maltzhan, G., Agrawal, A., Cowley, G.S., Weir, B.A., et al., 2012. Targeted tumor-penetrating siRNA nanocomplexes for credentialing the ovarian cancer oncogene ID4. Sci. Transl. Med. 4, 147ra112.

Renault, M.A., Roncalli, J., Tongers, J., Misener, S., Thorne, T., Jujo, K., et al., 2009. The Hedgehog transcription factor Gli3 modulates angiogenesis. Circ. Res. 105, 818–826.

Richardson, D.R., 2005. Molecular mechanisms of iron uptake by cells and the use of iron chelators for the treatment of cancer. Curr. Med. Chem. 12, 2711–2729.

Richardson, D.R., Sharpe, P.C., Lovejoy, D.B., Senaratne, D., Kalinowski, D.S., Islam, M., et al., 2006. Dipyridyl thiosemicarbazone chelators with potent and selective antitumor activity form iron complexes with redox activity. J. Med. Chem. 49, 6510–6521.

Richter, A.M., Pfeifer, G.P., Dammann, R.H., 2009. The RASSF proteins in cancer; from epigenetic silencing to functional characterization. Biochim. Biophys. Act 1796, 114–128.

Richter, A.M., Schagdarsurengin, U., Rastetter, M., Steinmann, K., Dammann, R.H., 2010. Protein kinase A-mediated phosphorylation of the RASSF1A tumour suppressor at Serine 203 and regulation of RASSF1A function. Eur. J. Cancer 46, 2986–2995.

Ridgway, P., Almouzni, G., 2000. CAF-1 and the inheritance of chromatin states: at the crossroads of DNA replication and repair. J. Cell. Sci. 113, 2647–2658.

Rmali, K.A., Puntis, M.C., Jiang, W.G., 2005. Prognostic values of tumor endothelial markers in patients with colorectal cancer. World J. Gastroenterol. 11, 1283–1286.

Roa, J., Aguilar, E., Dieguez, C., Pinilla, L., Tena-Sempere, M., 2008a. New frontiers in kisspeptin/GPR54 physiology as fundamental gatekeepers of reproductive function. Frontiers Neuroendocrinol. 29, 48–69.

Roa, J., Vigo, E., García-Galiano, D., Castellano, J.M., Navarro, V.M., Pineda, R., et al., 2008b. Desensitization of gonadotropin responses to kisspeptin in the female rat: analyses of LH and FSH secretion at different developmental and metabolic states. Am. J. Physiol. Endocrinol. Metab. 294, 1088–1096.

Roa, W., Brunet, B., Guo, L., Amanie, J., Fairchild, A., Gabos, Z., et al., 2010. Identification of a new microRNA expression profile as a potential cancer screening tool. Clin. Invest. Med. 33, E124–E132.

Robbins, D.J., Nybakken, K.E., Kobayashi, R., Sisson, J.C., Bishop, J.M., Therond, P.P., 1997. Hedgehog elicits signal transduction by means of a large complex containing the kinesin-related protein costal2. Cell 90, 225–234.

Rocaro, AM, Leleu, X., Sacco, A., Jia, X., Melhem, M., Moreau, A.S., et al., 2008. Dual targeting of the proteasome regulates survival and homing in Waldenstrom macroglobulinemia. Blood 111, 4752–4763.

Rocha, G.Z., Dias, M.M., Ropelle, E.R., Osorio-Costa, F., Rossato, F.A., Vercesi, A.E., et al., 2011. Metformin amplifies chemotherapy-induced AMPK activation and antitumoral growth. Clin. Cancer Res. 17, 3993–4005.

Rodon, J., Tawbi, H.A., Thomas, A.L., Amakye, D.D., Granvil, C., Shou, Y., et al., 2011. Phase I study of the smoothened (SMO) antagonist LDE225 in patients with advanced solid tumours. Preliminary safety, pharmacokinetics (PK), pharmacodynamics (PD) and antitumour activity in medulloblastoma. Neuro-Oncology 13, 87–88.

Roessler, L., Ezhou, L., Budhu, A., Chen, Y., Zhao, X., Ji, J., et al., 2012. Integrative genomic identification of genes on 8p associated with hepatocellular carcinoma progression and patient survival. Gastroenterology 142, 957–966.

Roldan, G., Delgado, L., Muse, I.M., 2006. Tumoral expression of BRCA1, estrogen receptor alpha and ID4 protein in patients with sporadic breast cancer. Cancer Biol. Ther. 5, 505–510.

Roldo, C., Missiaglia, E., Hagan, J.P., Falconi, M., Capelli, P., Bersani, S., et al., 2006. MicroRNA expression abnormalities in pancreatic endocrine and acinar tumors are associated with distinctive pathologic features and clinical behavior. J. Clin. Oncol. 24, 4677–4684.

Romanuik, T.L., Ueda, T., Le, N., Haile, S., Yong, T.M.K., Thomson, T., et al., 2009. Novel biomarkers for prostate cancer including noncoding transcripts. Am. J. Pathol. 175, 2264–2276.

Rominger, C.M., Bee, W.L.T., Copeland, R.A., Davenport, E.A., Gilmartin, A., Gontarek, R., et al., 2009. Evidence for allosteric interactions of antagonist binding to the smoothened receptor. J. Pharmacol. Exp. Ther. 329, 995–1005.

Ropero, A.B., Soria, B., Nadal, A., 2002. A nonclassical estrogen membrane receptor triggers rapid differential actions in the endocrine pancreas. Mol. Endocrinol. 16, 497–505.

Rosner, M.R., 2007. MAP kinase meets mitosis: a role for raf kinase inhibitory protein in spindle checkpoint regulation. Cell. Div. 2, 1.

Rothschild, S.I., Kappeler, A., Ratschiller, D., Betticher, D.C., Tschan, M.P., Gugger, M., et al. 2011. The stem cell gene "inhibitor of differentiation 1" (ID1) is frequently expressed in non-small cell lung cancer. Lung Cancer 71, 306–311.

Rouault, J.P., Falette, N., Guehenneux, F., Guillot, C., Rimokh, R., Wang, Q., et al., 1996. Identification of BTG2, an antiproliferative p53-dependent component of the DNA damage cellular response pathway. Nat. Genet. 14, 482–486.

Roux, K.J., Amici, S.A., Fletcher, B.S., Notterpek, L., 2005. Modulation of epithelial morphology, monolayer permeability and cell migration by growth arrest specific3/peripheral myelin protein 22. Mol. Cell. Biol. 16, 1142–1151.

Rowan, A., Bataille, V., MacKie, R., Healy, D., Bicknell, D., Bodmer, W., et al., 1999. Somatic mutations in the Peutz–Jeghers (LKB1/STKII) gene in sporadic malignant melanomas. J. Invest. Dermatol. 112, 509–511.

Roy, B.C., Kohno, T., Iwakawa, R., Moriguchi, T., Kiyono, T., Morishita, K., et al., 2010. Involvement of LKB1 in epithelial–mesenchymal transition (EMT) of human lung cancer cells. Lung Cancer 70, 136–145.

Ru, G.Q., Wang, H.J., Xu, W.J., Zhao, Z.S., 2011. Upregulation of Twist in gastric carcinoma associated with tumor invasion and poor prognosis. Pathol. Oncol. Res. 17, 341–347.

Ruan, J., Zhang, L., Yan, L., Liu, Y., Yue, Z., Chen, L., et al., 2012. Inhibition of hypoxia-induced epithelial mesenchymal transition by luteolin in non-small cell lung cancer cells. Mol. Med. Rep. 6, 232–238.

Rudin, C.M., Hann, C.L., Laterra, J., Yauch, R.L., Callahan, C.A., Fu, L., et al., 2009. Brief report: treatment of medulloblastoma with hedgehog pathway inhibitor GDC-0449. New Engl. J. Med. 361, 1173–1178.

Rudland, P.S., Platt-Higgins, A., El-Tanani, M., Rudland, S.D.S., Barraclough, R., Winstanley, J.H.R., et al., 2002. Prognostic significance of the metastasis-associated protein osteopontin in human breast cancer. Cancer Res. 62, 3417–3427.

Rudland, S.D.S., Martin, L., Roshanlall, C., Winstanley, J., Leinster, S., Platt-Higgins, A., et al., 2006. Association of S100A4 and osteopontin with specific prognostic factors and survival of patients with minimally invasive. breast cancer. Clin. Cancer Res. 12, 1192–1200.

Rulli, A., Carli, L., Romani, R., Baroni, T., Giovannini, E., Rosi, G., et al., 2001. Expression of glyoxalase I and II in normal and breast cancer tissues. Breast Cancer Res. Treat. 66, 67–72.

Sack, U., Walther, W., Scudiero, D., Selby, M., Aumann, J., Lemos, C., et al., 2011a. S100A4-induced cell motility and metastasis is restricted by the Wnt/beta-catenin pathway inhibitor calcimycin in colon cancer cells. Mol. Biol. Cell 22, 3344–3354.

Sack, U., Walther, W., Scudiero, D., Selby, M., Kobelt, D., Lemm, M., et al., 2011b. Novel effect of antihelminthic Niclosamide on S100A4-mediated metastatic progression in colon cancer. J. Natl. Cancer Inst. 103, 1018–1036.

Saha, A.K., Xu, X.J., Balon, T.W., Brandon, A., Kraegen, E.W., Ruderman, N.B., 2011. Insulin resistance due to nutrient excess Is it a consequence of AMPK downregulation? Cell Cycle 10, 3447–3451.

Sahlgren, C., Gustafsson, M.V., Jin, S., Poellinger, L., Lendahl, U., 2008. Notch signaling mediates hypoxia-induced tumor cell migration and invasion. Proc. Natl. Acad. Sci. U.S.A. 105, 6392–6397.

Saini, S., Yamamura, S., Majid, S., Shahryari, V., Hirata, H., Tanaka, Y., et al., 2011. MicroRNA-708 induces apoptosis and suppresses tumorigenicity in renal cancer cells. Cancer Res. 71, 6208–6219.

Saito, D., Maeshimad, Y., Nasu, T., Yamasaki, H., Tanabe, K., Sugiyama, H., et al., 2011. Amelioration of renal alterations in obese type 2 diabetic mice by vasohibin-1, a negative feedback regulator of angiogenesis. Ren. Physiol. 300, F873–F886.

Salih, D.A.M., Brunet, A., 2008. FoxO transcription factors in the maintenance of cellular homeostasis during aging. Curr. Opin. Cell Biol. 20, 126–136.

Salim, S., Asghar, M., Taneja, M., Hovatta, I., Wu, Y.L., Saha, K., et al., 2011. Novel role of RGS2 in regulation of antioxidant homeostasis in neuronal cells. FEBS Lett. 585, 1375–1381.

Samant, R.S., Seraj, M.J., Saunders, M.M., Sakamaki, T.S., Shevde, L.A., Harms, J.F., et al., 2000. Analysis of mechanisms underlying BRMS1 suppression of metastasis. Clin. Exp. Metastasis 18, 683–693.

Samant, R.S., Clark, D.W., Fillmore, R.A., Cicek, M., Metge, B.J., Chandramouli, K.H., et al., 2007. Breast cancer metastasis suppressor I (BRMSI) inhibits osteopontin transcription by abrogating NF-kappa B activation. Mol. Cancer 6, 6.

Samanta, J., Kessler, J.A., 2004. Interactions between ID and OLIG proteins mediate the inhibitory effects of BMP4 on oligodendroglial differentiation. Development 131, 4131–4142.

Sanchez, A.M., Flamini, M.I., Baldacci, C., Gogliad, L., Genazzani, A.R., Simoncini, T., 2010. Estrogen receptor alpha promotes breast cancer cell motility and invasion via focal adhesion kinase and N-WASP. Mol. Endocrinol. 24, 2114–2125.

Sánchez-Tilló, E., Lázaro, A., Torrent, R., Cuatrecasas, M., Vaquero, E.C., Castells, A., et al., 2010. ZEB1 represses E-cadherin and induces an EMT by recruiting the SWI/SNF chromatin-remodeling protein BRG1. Oncogene 29, 3490–3500.

Sang, M., Li, Y., Ozaki, T., Ono, S., Ando, K., Yamamoto, H., et al., 2006. p73-dependent induction of 14-3-3 sigma increases the chemo-sensitivity of drug-resistant human breast cancers. Biochem. Biophys. Res. Commun. 347, 327–333.

Sankpal, U.T., Goodison, S., Abdelrahim, M., Basha, R., 2011. Targeting sp1 transcription factors in prostate cancer therapy. Med. Chem. 7, 518–525.

Santarius, T., Bignell, G.R., Greenman, C.D., Widaa, S., Chen, L.N., Mahoney, C.L., et al., 2010. GLO1 – a novel amplified gene in human cancer. Genes. Chrom. Cancer 49, 711–725.

Sarbassov, D.D., Guertin, D.A., Ali, S.M., Sabtini, D.M., 2005. Phosphorylation and regulation of Akt/PKB by the rictor–mTOR complex. Science 307, 1098–1101.

Sarver, A.L., Li, L., Subramanian, S., 2010. MicroRNA miR-183 functions as an oncogene by targeting the transcription factor EGR1 and promoting tumor cell migration. Cancer Res. 70, 9570–9580.

Sayed, D., He, M., Hong, C., Gao, S., Rane, S., Yang, Z., et al., 2010. MicroRNA-21 is a downstream effector of AKT that mediates its antiapoptotic effects via suppression of Fas ligand. J. Biol. Chem. 285, 20281–20290.

Sbai, O., Devi, T.S., Melone, M.A.B., Feron, F., Khrestchatisky, M., Singh, L.P., et al., 2010. RAGE–TXNIP axis is required for S100B-promoted Schwann cell migration, fibronectin expression and cytokine secretion. J. Cell. Sci. 123, 4332–4339.

Schäfer, B.W., Wicki, R., Engelkamp, D., Mattei, M.G., Heizmann, C.W., 1995. Isolation of a YAC clone covering a cluster of nine S100 genes on human chromosome 1q21: rationale for a new nomenclature of the S100 calcium-binding protein family. Genomics 25, 638–643.

Schilling, S.H., Hjelmeland, A.B., Radiloff, D.R., Liu, I.M., Wakeman, T.P., Fielhauer, J.R., et al., 2009. NDRG4 is required for cell cycle progression and survival in glioblastoma cells. J. Biol. Chem. 284, 25160–25169.

Schlingensiepen, K.H., Jaschinski, F., Lang, S.A., Moser, C., Geissler, E.K., Schlitt, H.J., et al., 2011. Transforming growth factor-beta 2 gene silencing with trabedersen (AP 12009) in pancreatic cancer. Cancer Sci. 102, 1193–1200.

Schmid, T., Bajer, M.M., Blees, J.S., Eifler, L.K., Milke, L., Ruebsamen, D., et al., 2011. Inflammation-induced loss of Pdcd4 is mediated by phosphorylation-dependent degradation. Carcinogenesis 32, 1427–1433.

Schnidar, H., Eberl, M., Klingler, S., Mangelberger, D., Kasper, M., Hauser-Kronberger, C., et al., 2009. Epidermal growth factor receptor signaling synergizes with Hedgehog/GLI in oncogenic transformation via activation of the MEK/ERK/JUN pathway. Cancer Res. 69, 1284–1292.

Schreck, K.C., Taylor, P., Marchionni, L., Gopalakrishnan, V., Bar, E.E., Gaiano, N., et al., 2010. The notch target hes1 directly modulates gli1 expression and hedgehog signaling: a potential mechanism of gherapeutic resistance. Clin. Cancer Res. 16, 6060–6070.

Schultz, J., Ibrahim, S.M., Vera, J., Kunz, M., 2009. 14-3-3 sigma gene silencing during melanoma progression and its role in cell cycle control and cellular senescence. Mol. Cancer 8, 53.

Scobie, H.M., Young, J.A., 2005. Interactions between anthrax toxin receptors and protective antigen. Curr. Opin. Microbiol. 8, 106–112.

Scott, G., Ahmad, I., Howard Km Maclean, D., Oliva, C., Warrington, S., Wilbraham, D., et al., 2013. Double-blind, randomized, placebo-controlled study of safety, tolerability, pharmacokinetics and pharmacodynamics of TAK-683, an investigational metastin analogue in healthy men. Br. J. Clin. Pharmacol. 75, 381–391.

Sebova, K., Zmetakova, I., Bella, V., Kajo, K., Stankovicova, I., Kajabova, V., et al., 2012. RASSF1A and CDH1 hypermethylation as potential epimarkers in breast cancer. Cancer Biomark. 10, 13–26.

Sekimata, M., Kabuyama, Y., Emori, Y., Homma, Y., 1999. Morphological changes and detachment of adherent cells induced by p122, a GTPase-activating protein for Rho. J. Biol. Chem. 274, 17757–17762.

Selvanathan, S.P., Thakurta, A.G., Dhaksshnamoorthy, J., Zhou, M., Vennstra, T.D., Dhar, R., 2010. Schizosaccharomyces pombe Dss1p is a DNA damage checkpoint protein that recruits Rad24p, Cdc25p, and Rae1p to DNA double-strand breaks. J. Biol. Chem. 285, 14122–14133.

Semaan, A., Qazi, A.M., Seward, S., Chamala, S., Bryant, C.S., Kumar, S., et al., 2011. MicroRNA-101 inhibits growth of epithelial ovarian cancer by relieving chromatin-mediated transcriptional repression of p21(waf1/cip1). Phamaceut. Res. 28, 3079–3090.

Serrano, M., 2000. The INK4a/ARF locus in murine tumorigenesis. Carcinogenesis 21, 865–869.

Setogawa, T., Shinozaki-Yabana, S., Masuda, T., Matsuura, K., Akiyama, T., 2006. The tumor suppressor LKB1 induces p21 expression in collaboration with LMO4, GATA-6, and Ldb1. Biochem. Biophys. Res. Commun. 343, 1186–1190.

Sahab, Z.J., Man, Y.G., Semaan, S.M., Newcomer, R.G., Byers, S.W., Sang, Q.X.A., 2010. Alteration in protein expression in estrogen receptor alpha-negative human breast cancer tissues indicates a malignant and metastatic phenotype. Clin. Exp. Metastasis 27, 493–503.

Shams, T.M., Samaka, R.M., Shams, M.E., 2006. Maspin protein expression: a special feature of papillary thyroid carcinoma. J. Egypt. Nat. Cancer Inst. 18, 274–280. [Seen only as Abstract]

Shank, J.J., Yang, K., Ghannam, J., Cabrera, L., Johnston, C.J., Reynolds, R.K., et al., 2012. Metformin targets ovarian cancer stem cells in vitro and in vivo. Gynecol. Oncol. 127, 390–397.

Sharma, G.D., He, J., Bazan, H.E., 2003. p38 and ERK1/2 coordinate cellular migration and proliferation in epithelial wound healing: evidence of cross-talk activation between MAP kinase cascades. J. Biol. Chem. 278, 21989–21997.

Sharma, G., Mirza, S., Parshad, R., Srivastava, A., Gupta, S.D., Pandya, P., et al., 2011. Clinical significance of Maspin promoter methylation and loss of its protein expression in invasive ductal breast carcinoma: correlation with VEGF-A and MTA1 expression. Tumor. Biol. 32, 23–32.

Sharma, P., Chinaranagari, S., Patel, D., Carey, J., Chaudhary, J., 2012. Epigenetic inactivation of inhibitor of differentiation 4 (Id4) correlates with prostate cancer. Cancer Med. 1, 176–186.

Sharpless, N.E., DePinho, R.A., 1999. The INK4A/ARF locus and its two gene products. Curr. Opin. Genet. Dev. 9, 22–30.

Shaw, R.J., Paez, J.G., Curto, M., Yaktine, A., Pruitt, W.M., Saotome, I., et al., 2001. The Nf2 tumor suppressor, merlin, functions in Rac-dependent signalling. Develop. Cell 1, 63–72.

Shellman, Y.G., Park, Y.L., Marr, D.G., Casper, K., Xu, Y., Fujita, M., et al., 2003. Release of vascular endothelial growth factor from a human melanoma cell line, WM35, is induced by hypoxia but not ultraviolet radiation and is potentiated by activated Ras mutation. J. Invest. Dermatol. 121, 910–917.

Sheng, S., Pemberton, P.A., Sager, R., 1994. Production, purification, and characterization of recombinant maspin proteins. J. Biol. Chem. 269, 30988–30993.

Sheng, X.J., Zhou, Y.Q., Song, Q.Y., Zhou, D.M., Liu, Q.C., 2012. Loss of breast cancer metastasis suppressor 1 promotes ovarian cancer cell metastasis by increasing chemokine receptor 4 expression. Oncol. Rep. 27, 1011–1018.

Sherbet, G.V., 2001. Calcium signalling in Cancer: The Role of Calcium Binding Proteins in Cell Proliferation, Invasion and Metastasis. CRC Press, Boca Raton, FL.

Sherbet, G.V., 2003. Genetic Recombination in Cancer. Academic Press/Elsevier, London.

Sherbet, G.V., 2006. DNA ploidy and the progression of cancer. In: Sherbet, G.V. (Ed.), The Molecular and Cellular Pathology of Cancer Progression and Prognosis. Research Signpost, Trivandrum, pp. 173–198.

Sherbet, G.V., 2009. Metastasis promoter S100A4 is a potentially valuable molecular target for cancer therapy. Cancer Lett. 280, 15–30.

Sherbet, G.V., 2011a. Growth Factors and Their Receptors in Cell Differentiation, Cancer and Cancer Therapy. Elsevier,

Sherbet, G.V., 2011b. S100A4 (S100 calcium binding protein A4). Atlas Genet. Cytogenet. Oncol. Haematol. 15, 877–886.

Sherbet, G.V., Lakshmi, M.S., 1997. The Genetics of Cancer: Genes Associated with Cancer Invasion, Metastasis and Cell Proliferation. Academic Press, London.

Sherbet, G.V., Patil, D.B., 2006. Molecular progression of ovarian cancer. In: Sherbet, G.V. (Ed.), The Molecular and Cellular Pathology of Cancer Progression and Prognosis. Research Signpost, Trivandrum, pp. 130–172.

Shiba-Ishii, A., Kano, J., Morishita, Y., Sato, Y., Minami, Y., Noguchi, M., 2011. High expression of stratifin is a universal abnormality during the course of malignant progression of early-stage lung adenocarcinoma. Int. J. Cancer 129, 2445–2453.

Shiohama, A., Sasaki, T., Noda, S., Minoshima, S., Shimizu, N., 2003. Molecular cloning and expression analysis of a novel gene DGCR8 located in the DiGeorge syndrome chromosomal region. Biochem. Biophys. Res. Commun. 304, 184–190.

Sherwood, V., Recino, A., Jeffries, A., Ward, A., Chalmers, A.D., 2010. The N-terminal RASSF family: a new group of Ras-association-domain-containing proteins, with emerging links to cancer formation. Biochem. J. 425, 303–311.

Shi, W.Y., Xiao, D., Wang, L., Dong, L.H., Yan, Z.X., Shen, Z.X., et al., 2012. Therapeutic metformin/AMPK activation blocked lymphoma cell growth via inhibition of mTOR pathway and induction of autophagy. Cell. Death. Dis. 3, e275.

Shih, Y.P., Takada, Y., Lo, S.H., 2012. Silencing of DLC1 upregulates PAI-1 expression and reduces migration in normal prostate cells. Mol. Cancer Res. 10, 34–39.

Shin, H., Kim, J.H., Lee, Y.S., Lee, Y.C., 2012. Change in gene expression profiles of secreted frizzled-related proteins (SFRPs) by sodium butyrate in gastric cancers: induction of promoter demethylation and histone modification causing inhibition of Wnt signaling. Int. J. Oncol. 40, 1533–1542.

Shin, I., Kim, S., Song, H., Kim, H.R.C., Moon, A., 2005. H-ras-specific activation of Rac-MKK3/6-p38 pathway – its critical role in invasion and migration of breast epithelial cells. J. Biol. Chem. 280, 14675–14683.

Shin, K.H., Bae, S.D., Hong, H.S., Kim, R.H., Kang, M.K., Park, N.H., 2011. miR-181a shows tumor suppressive effect against oral squamous cell carcinoma cells by downregulating K-ras. Biochem. Biophys. Res. Commun. 404, 896–902.

Shinawi, T., Hill, V., Dagklis, A., Baliakas, P., Stamatopoulos, K., Agathanggelou, A., et al., 2012. KIBRA gene methylation is associated with unfavorable biological prognostic parameters in chronic lymphocytic leukemia. Epigenetics 7, 211–215.

Shoji, S., Tang, X.Y., Umemura, S., Itoh, J., Takekoshi, S., Shima, M., et al., 2009. Metastin inhibits migration and invasion of renal cell carcinoma with overexpression of metastin receptor. Eur. Oncol. 55, 441–451.

Shoji, S., Tang, X.Y., Sato, H., Usui, Y., Uchida, T., Terachi, T., 2010. Metastin has potential as a suitable biomarker and novel effective therapy for cancer metastasis (review). Oncol. Lett. 1, 783–788.

Shorning, B.Y., Griffiths, D., Clarke, A.R., 2011. Lkb1 and Pten synergise to suppress mTOR-mediated tumorigenesis and epithelial–mesenchymal transition in the mouse bladder. PloS ONE 6, e16209.

Si, M.L., Zhu, S., Wu, H., Lu, Z., Wu, F., Mo, Y.Y., 2007. miR-21-mediated tumor growth. Oncogene 26, 2799–2803.

Sibold, S., Roh, V., Keogh, A., Studer, P., Tiffon, C., Angst, E., et al., 2007. Hypoxia increases cytoplasmic expression of NDRG1, but is insufficient for its membrane localization in human hepatocellular carcinoma. FEBS Lett. 58, 989–994.

Silber, J., Lim, D.A., Petritsch, C., Persson, A.I., Maunakea, A.K., Yu, M., et al., 2008. miR-124 and miR-137 inhibit proliferation of glioblastoma multiforme cells and induce differentiation of brain tumor stem cells. BMC Med. 6, 14.

Simooka, H., Oyama, T., Sano, T., Horiguchi, J., Nakajima, T., 2004. Immunohistochemical analysis of 14-3-3 sigma and related proteins in hyperplastic and neoplastic breast lesions, with particular reference to early carcinogenesis. Pathol. Int. 54, 595–602.

Singh, N.K., Rao, P., Asea, A., 2008. Silencing F metastasis-associated gene 1 (MTA1) stimulates Hsp70 cellular release and neurite extension in neuroblastoma cells. Heat Shock Proteins 3, 273–282.

Singh, S.K., Hawkins, C., Clarke, I.D., Squire, J.A., Bayani, J., Hide, T., et al., 2004. Identification of human brain tumour initiating cells. Nature 432, 396–401.

Skulic, A., Migden, M.R., Oro, A.E., Dirix, L., Lewis, K.D., Hainsworth, J.D., et al., 2012. Efficacy and safety of vismodegrib in advanced basal cell carcinoma. New Engl. J. Med. 366, 2171–2179.

Skvara, H., Kalthoff, F., Meingassner, J.G., Wolff-Winiski, B., Aschauer, H., Kelleher, J.F., et al., 2011. Topical treatment of basal cell carcinomas in nevoid basal cell carcinoma syndrome with a Smoothened inhibitor. J. Invest. Dermatol. 131, 1735–1744.

Slipicevic, A., Holm, R., Emilsen, E., Rosnes, A.K.R., Welch, D.R., Maelandsmo, G.M., et al., 2012. Cytoplasmic BRMS1 expression in malignant melanoma is associated with increased disease-free survival. BMC Cancer 12, 73.

Sloss, C.M., Wang, F., Liu, R., Xia, L., Houston, M., Ljungman, D., et al., 2008. Proteasome inhibition activates epidermal growth factor receptor (EGFR) and EGFR-independent mitogenic kinase signaling pathways in pancreatic cancer cells. Clin. Cancer Res. 14, 5116–5123.

Solanas, G., Porta-de-la-Riva, M., Agusti, C., Casagolda, D., Sanchez-Aguilera, F., Larriba, M.J., et al., 2008. E-cadherin controls β-catenin and NF-κB transcriptional activity in mesenchymal gene expression. J. Cell Sci. 121, 2224–2234.

Solomon, L.A., Munkarah, A.R., Schimp, V.L., Arabi, M.H., Morris, R.T., Nassar, H., et al., 2006. Maspin expression and localization impact on angiogenesis and prognosis in ovarian cancer. Gynecol. Oncol. 101, 385–389.

Son, A., Nakamura, H., Okuyama, H., Si, O.k.a., Yoshihara, E., Liu, W., et al., 2008. Dendriticcells derived from TBP-2-deficient mice are defective in inducing T cell responses. Eur. J. Immunol. 38, 1358–1367.

Song, C.W., Lee, H., Dings, R.P.M., Williams, B., Powers, J., Dos Santos, T., et al., 2012a. Metformin kills and radiosensitizes cancer cells and preferentially kills cancer stem cells. Sci. Rep. 2, 362.

Song, G., Zeng, H., Li, J., Xiao, L., He, Y., Tang, Y., et al., 2010a. MiR-199a regulates the tumor suppressor mitogen-activated protein kinase kinase kinase 11 in gastric cancer. Biol. Pharm. Bull. 33, 1822–1827.

Song, G.J., Leslie, K.L., Barrick, S., Bougoin, S., Taboas, J.M., Bisello, A., 2012b. EBP50 promotes focal adhesion turnover and vascular smooth muscle cells migration. J. Mol. Cell Cardiol. 53, 809–819.

Song, J.J., Liu, Q., Li, Y., Yang, Z.S., Yang, L., Xiang, T.X., et al., 2012c. Epigenetic inactivation of PLCD1 in chronic myeloid leukemia. Int. J. Mol. Med. 30, 179–184.

Song, Y.H., Oda, Y., Hori, M., Kuroiwa, K., Ono, M., Hosoi, F., et al., 2010b. N-myc downstream regulated gene-1/Cap43 may play an important role in malignant progression of prostate cancer, in its close association with E-cadherin. Human Pathol. 41, 214–222.

Sood, A.K., Fletcher, M.S., Gruman, L.M., Coffin, J.E., Jabbari, S., Khalkhali-Ellis, Z., et al., 2002. The paradoxical expression of maspin in ovarian carcinoma. Clin. Cancer Res. 8, 2924–2932.

Sorbera, L.A., 2004. Metelimumab – agent for scleroderma – human anti-TGF-beta 1 monoclonal antibody. Drugs Future 29, 1081–1083.

Spector, N.L., Yarden, Y., Smith, B., Lyass, L., Trusk, P., Pry, K., et al., 2007. Activation of AMP-activated protein kinase by human EGF receptor 2/EGF receptor tyrosine kinase inhibitor protects cardiac cells. Proc. Natl. Acad. Sci. U.S.A. 104, 10607–10612.

Spiekerkoetter, E., Guignabert, C., de Jesus Perez, V., Alastalo, T.P., Powers, J.M., Wang, L., et al., 2009. S100A4 and bone morphogenetic protein-2 codependently induce vascular smooth muscle cell migration via phospho-extracellular signal-regulated kinase and chloride intracellular channel 4. Circ. Res. 105, 639–647.

Sridhar, S.C., Miranti, C.K., 2006. Tetraspanin KAI1/CD82 suppresses invasion by inhibiting integrin-dependent crosstalk with c-Met receptor and Src kinases. Oncogene 25, 2367–2378.

Srivastava, S.K., Bhardwaj, A., Singh, S., Arora, S., Wang, B., Grizzle, W.E., et al., 2011. MicroRNA-150 directly targets MUC4 and suppresses growth and malignant behavior of pancreatic cancer cells. Carcinogenesis 32, 1832–1839.

Stallings, JD, Zeng, YX, Narvaez, F, Rebecchi, MJ., 2008. Phospholipase C-delta(1) expression is linked to proliferation, DNA synthesis, and cyclin E levels. J. Biol. Chem. 283, 13992–14001.

Stamenkovic, I, Yu, Q., 2010. Merlin, a "Magic" linker between the extracellular cues and intracellular signaling pathways that regulate cell motility, proliferation, and survival. Curr. Protein Pept. Sci. 11, 471–484.

Stark, A.M., Tongers, K., Maass, N., Mehdorn, H.M., Held-Feindt, J., 2005. Reduced metastasis-suppressor gene mRNA-expression in breast cancer brain metastases. J. Cancer Res. Clin. Oncol. 131, 191–198.

St Croix, B., Rago, C., Velculescu, V., Traverso, G., Romans, K.E., Montgomery, E., et al., 2000. Genes expressed in human tumor endothelium. Science 289, 1197–1202.

Stein, S., Thomas, E.K., Herzog, B., Westfall, M.D., Rocheleau, J.V., Jackson, R.S., et al., 2004. NDRG1 is necessary for p53-dependent apoptosis. J. Biol. Chem. 279, 48930–48940.

Stein, U., Arlt, F., Walther, W., Smith, J., Waldman, T., Harris, E.D., et al., 2006. The metastasis-associated gene S100A4 is a novel target of beta-catenin/T-cell factor signaling in colon cancer. Gastroenterology 131, 1486–1500.

Stein, U., Arlt, F., Smith, J., Sack, U., Herrmann, P., Walther, W., et al., 2011. Intervening in beta-catenin signaling by Sulindac inhibits S100A4-dependent colon cancer metastasis. Neoplasia 13, 131–144.

Strand, M.F., Wilson, S.R., Dembinski, J.L., Holsworth, D.D., Khvat, A., Okun, I., et al., 2011. A novel synthetic Smoothened antagonist transiently inhibits pancreatic adenocarcinoma xenografts in a mouse model. PloS ONE 6, e19904.

Straza, M.W., Paliwal, S., Kovi, R.C., Rajeshkumar, B., Trenh, P., Parker, D., et al. 2010. Therapeutic targeting of C-terminal binding protein in human cancer. Cell Cycle 9, 3740–3750.

Stengel, K., Zheng, Y., 2011. Cdc42 in oncogenic transformation, invasion, and tumorigenesis. Cell Signal. 23, 1415–1423.

Strappel, M.M., Vincent, A., Mukherjee, R., Campbell, N.R., Chen, S.H., Konstantopoulos, K., et al., 2012. Mucin 16 (cancer antigen 125) expression in human tissues and cell lines and correlation with clinical outcome in adenocarcinomas of the pancreas, esophagus, stomach, and colon. Human Pathol. 43, 1755–1763.

Strazisar, M., Mlakar, V., Rott, T., Glavac, D., 2009. Somatic alterations of the serine/threonine kinase LKB1 gene in squamous cell (SCC) and large cell (LCC) lung carcinoma. Cancer Invest. 27, 407–416.

Strzelczyk, B., Szulc, A., Rzepko, R., Kitowska, A., Skokowski, J., Szutowicz, A., et al., 2009. Identification of high-risk stage II colorectal tumors by combined analysis of the NDRG1 gene expression and the depth of tumor invasion. Ann. Surg. Oncol. 16, 1287–1294.

Su, C.H., Zhao, R., Zhang, F., Qu, C., Chen, B., Feng, Y.H., et al., 2011. 14-3-3 sigma exerts tumor-suppressor activity mediated by regulation of COP1 stability. Cancer Res. 71, 884–894.

Subramanian, C., Jarzembowski, J.A., Opipari Jr, A.W., Castle, V.P., Kwok, R.P.S., 2011. HDAC6 deacetylates Ku70 and regulates Ku70-Bax binding in neuroblastoma. Neoplasia 13, 726–734.

Sugimoto, S., Maass, N., Takimoto, Y., Sato, K., Minei, S., Zhang, M., et al., 2004. Expression and regulation gene maspin in human of tumor suppressor bladder cancer. Cancer Lett. 203, 209–215.

Suh, S.O., Chen, Y., Zaman, M.S., Hirata, H., Yamamura, S., Shahryari, V., et al., 2011. MicroRNA-145 is regulated by DNA methylation and p53 gene mutation in prostate cancer. Carcinogenesis 32, 772–778.

Sum, E.Y.M., Peng, B., Yu, X., Chen, J.J., Byrne, J., Lindeman, G.J., et al., 2002. The LIM domain protein LMO4 interacts with the cofactor CtIP and the tumor suppressor BRCA1 and inhibits BRCA1 activity. J. Biol. Chem. 277, 7849–7856.

Sun, W., Tan, X., Shi, Y., Xu, G., Mao, R., Gu, X., et al., 2010. USP11 negatively regulates TNF-α induced NK-κB activation by targeting on IκBα. Cell Signal. 22, 386–394.

Sun, W., Guo, M.M., Han, P., Lin, J.Z., Liang, F.Y., Tan, G.M., et al., 2012. Id-1 and the p65 subunit of NF-kB promote migration of nasopharyngeal carcinoma cells and are correlated with poor prognosis. Carcinogenesis 33, 810–817.

Sundaram, P., Hultine, S., Smith, L.M., Dews, M., Fox, J.L., Biyashev, D., et al., 2011. p53-responsive miR-194 inhibits thrombospondin-1 and promotes angiogenesis in colon cancers. Cancer Res. 71, 7490–7501.

Susuki, D., Kimura, S., Naganuma, S., Tsuchiyama, K., Tanaka, T., Kitamura, N., et al., 2011. Regulation of microRNA expression by hepatocyte growth factor in human head and neck squamous cell carcinoma. Cancer Sci. 102, 2164–2171.

Suzuki, K., Juelich, T., Lim, H., Ishida, T., Watanebe, T., et al., 2008. Closed chromatin architecture is induced by an RNA duplex targeting the HIV-1 promoter region. J. Biol. Chem. 283, 23353–23363.

Suzuki, Y., Oonishi, T., Kudo, T., Doi, H., 2012. LKB1, TP16, EGFR, and KRAS somatic mutations in lung adenocarcinomas from a Chiba Prefecture, Japan cohort. Drug Disc. Ther. 6, 24–30.

Svedlund, J., Edblom, S.K., Marquez, V.E., Akerstrom, G., Bjorklund, P., Westin, G., 2012. Hypermethylated in cancer 1 (HIC1), a tumor suppressor gene epigenetically deregulated in hyperparathyroid tumors by histone H3 lysine modification. J. Clin. Endocrinol. Metab. 97, E1307–E1315.

Syed, V., Mukherjee, K., Lyons-Weiler, J., Lau, K.M., Mashima, T., Tsuruo, T., et al., 2005. Identification of ATF-3, caveolin-1, DLC-1, and NM23-H2 as putative antitumorigenic, progesterone-regulated genes for ovarian cancer cells by gene profiling. Oncogene 24, 1774–1787.

Tabata, T., Tsukamoto, N., Fooladi, A.A.I., Yamanaka, S., Furukawa, T., Ishida, M., et al., 2009. RNA interference targeting against S100A4 suppresses cell growth and motility and induces apoptosis in human pancreatic cancer cells. Biochem. Biophys. Res. Commun. 390, 475–480.

Takahashi, M., Sugiura, T., Abe, M., Ishii, K., Shirasuna, K., 2007. Regulation of c-Met signaling by the tetraspanin KAI-1/CD82 affects cancer cell migration. Int. J. Cancer 121, 1919–1929.

Takahashi, Y., Iwai, M., Kawai, T., Arakawa, A., Ito, T., Sakurai-Yageta, M., et al., 2012. Aberrant expression of tumor suppressors CADM1 and 4.1B in invasive lesions of primary breast cancer. Breast Cancer 19, 242–252.

Takeda, M., Nagasaka, T., Dong-Sheng, S., Nishie, H., Oka, T., Yamada, E., et al., 2011. Expansion of CpG methylation in the SFRP2 promoter region during colorectal tumorigenesis. Acta Med. Okayama 65, 169–177.

Takeda, T., Kikuchi, E., Mikami, S., Suzuki, E., Matsumoto, K., Miyajima, A., et al., 2012. Prognostic role of KiSS-1 and possibility of therapeutic modality of metastin, the final peptide of the KiSS-1 gene, in urothelial carcinoma. Mol. Cancer Ther. 11, 853–863.

Tan, A.R., Alexe, G., Reiss, M., 2009a. Transforming growth factor-beta signaling: emerging stem cell target in metastatic breast cancer? Br. Cancer Res. Treat. 115, 453–495.

Tan, D.S., Lambros, M.B., Rayter, S., Natrajan, R., Vatcheva, R., Gao, Q., et al., 2009b. PPM1D is a potential therapeutic target in ovarian clear cell carcinomas. Clin. Cancer Res. 15, 2269–2280.

Tan, S., Li, R., Ding, K., Lobie, P.E., Zhu, T., 2011. miR-198 inhibits migration and invasion of hepatocellular carcinoma cells by targeting the HGF/c-MET pathway. FEBS Lett. 585, 2229–2234.

Tanaka, T., Arai, M., Wu, S., Kanda, T., Miyauchi, H., Imazeki, F., et al., 2011. Epigenetic silencing of microRNA-373 plays an important role in regulating cell proliferation in colon cancer. Oncol. Rep. 26, 1329–1335.

Tanaka, Y., Kato, K., Mibu, R., Uchida, S., Asanoma, K., Suzuki, K., et al., 2008. Medroxyprogesterone acetate inhibits proliferation of colon cancer cell lines by modulating cell cycle-related protein expression. Menopause-J. North Am. Menopause. Soc. 15, 442–453.

Tang, H., Park, S., Sun, S.C., Trumbly, R., Ren, G., Tsung, E., et al., 2010. RKIP inhibits NF-kappa B in cancer cells by regulating upstream signaling components of I kappa B kinase complex. FEBS Lett. 584, 662–668.

Tang, J.T., Wang, J.L., Du, W., Hong, J., Zhao, S.L., Wang, Y.C., et al., 2011a. MicroRNA 345, a methylation-sensitive microRNA is involved in cell proliferation and invasion in human colorectal cancer. Carcinogenesis 32, 1207–1215.

Tang, Q.L., Zhao, Z.Q., Li, J.C., Liang, Y., Yin, J.Q., Zou, C.Y., et al., 2011b. Salinomycin inhibits osteosarcoma by targeting its tumor stem cell. Cancer Lett. 311, 113–121.

Tang, T., Tang, J.Y., Li, D., Reich, M., Callahan, C.A., Fu, L., et al., 2011c. Targeting superficial or nodular basal cell carcinoma with topically formulated small molecule inhibitor of Smoothened. Clin. Cancer Res. 17, 3378–3387.

Taniguchi, H., Horinaka, M., Yoshida, T., Yano, K., Goda, A.E., Yasuda, S., et al., 2012. Targeting the glyoxalase pathway enhances TRAIL efficacy in cancer cells by downregulating the expression of antiapoptotic molecules. Mol. Cancer Ther. 11, 2294–2300.

Tao, J., Lu, Q., Wu, D., Li, P., Xu, B., Qing, W., et al., 2011. microRNA-21 modulates cell proliferation and sensitivity to doxorubicin in bladder cancer cells. Oncol. Rep. 25, 1721–1729.

Tarasov, V., Jung, P., Verdoodt, B., Lodygin, D., Epanchintsev, A., Menssen, A., et al., 2007. Differential regulation of microRNAs by p53 revealed by massively parallel sequencing – miR-34a is a p53 target that induces apoptosis and G(1)-arrest. Cell Cycle 6, 1586–1593.

Tatton, L., Morley, G.M., Chopra, R., Khwaja, A., 2003. The Src-selective kinase inhibitor also inhibits Kit and Bcr-Abl tyrosine kinases. J. Biol. Chem. 278, 4847–4853.

Tawbi, H.A., Rodon Ahnert, J., Dummer, R., Thomas, A.L., Granvil, C., Shou, Y., et al. 2011. Phase I study of LED225 in advanced solid tumours: updated analysis of safety, preliminary efficacy, pharmacokinetic-pharmacodynamic correlation. J. Clin. Oncol. 29, 3062.

Teng, I.W., Hou, P.C., Lee, K.D., Chu, P.Y., Yeh, K.T., Jin, V.X., et al., 2011. Targeted methylation of two tumor suppressor genes is sufficient to transform mesenchymal stem cells into cancer stem/initiating cells. Cancer Res. 71, 4653–4663.

Thaler, S., Haehnel, P.S., Schad, A., Dammann, R., Schuler, M., 2009. RASSF1A mediates p21(Cip1/Waf1)-dependent cell cycle arrest and senescence through modulation of the Raf–MEK–ERK pathway and inhibition of Akt. Cancer Res. 69, 1748–1757.

Thomas, Z.I., Gibson, W., Sexton, J.Z.K., Aird, K.M.L., Ingram, S.M., Aldrich, A., et al., 2011. Targeting GLI1 expression in human inflammatory breast cancer cells enhances apoptosis and attenuates migration. Br. J. Cancer 104, 1575–1586.

Thoren, K.L., Krantz, B.A., 2011. The unfolding story of anthrax toxin translocation. Mol. Microbiol. 80, 588–595.

Tiainen, M., Vaahtomeri, K., Ylikorkala, A., Makela, T.P., 2002. Growth arrest by the LKB1 tumor suppressor: Induction of p21(WAF1/CIP1). Human Mol. Genet. 11, 1497–1504.

Tison, B.E., Nicholas, S.K., Abramson, S.L., Hanson, I.C., Paul, M.E., Seeborg, F.O., et al., 2011. Autoimmunity in a cohort of 130 pediatric patients with partial DiGeorge syndrome. J. Allergy Clin. Immunol. 128, 1115–1117.

Tivnan, A., Tracey, L., Buckley, P.G., Alcock, L.C., Davidoff, A.M., Stallings, R.L., 2011. MicroRNA-34a is a potent tumor suppressor molecule in vivo in neuroblastoma. BMC Cancer 11, 33.

Toffoli, S., Delaive, E., Dieu, M., Feron, O., Raes, M., Michiels, C., 2009. NDRG1 and CRK-I/II are regulators of endothelial cell migration under intermittent hypoxia. Angiogenesis 12, 339–354.

Toh, Y., Nicolson, G.L., 2009. The role of the MTA family and their encoded proteins in human cancers: Molecular functions and clinical implications. Clin. Exp. Metastasis 26, 215–227.

Toh, Y., Pencil, S.D., Nicolson, G.L., 1994. A novel candidate metastasis-associated gene, mta1, differentially expressed in highly metastatic mammary adenocarcinoma cell lines. cDNA cloning, expression, and protein analyses. J. Biol. Chem. 269, 22958–22963.

Toh, Y., Pencil, S.D., Nicolson, G.L., 1995. Analysis of the complete sequence of the novel metastasis-associated candidate gene, mta1, differentially expressed in mammary adenocarcinoma and breast cancer cell lines. Gene 159, 97–104.

Tokuda, H., Kozawa, O., Uematsu, T., 2000. Basic fibroblast growth factor stimulates vascular endothelial growth factor release in osteoblasts: divergent regulation by p42/p44 mitogen-activated protein kinase and p38 mitogen-activated protein kinase. J. Bone Mineral Res. 15, 2371–2379.

Tomimaru, Y., Eguchi, H., Nagano, H., Wada, H., Tomokuni, A., Kobayashi, S., et al., 2010. MicroRNA-21 induces resistance to the anti-tumour effect of interferon-alpha/5-fluorouracil in hepatocellular carcinoma cells. Br. J. Cancer 103, 1617–1626.

Tommasi, S., Besaratinia, A., Wilczynski, S.P., Pfeifer, G.P., 2011. Loss of Rassf1a enhances p53-mediated tumor predisposition and accelerates progression to aneuploidy. Oncogene 30, 690–700.

Torres, A., Torres, K., Paszkowski, T., Radej, S., Staskiewicz, G.J., Ceccaroni, M., et al. 2011. Highly increased maspin expression corresponds with up-regulation of miR-21 in endometrial cancer: a preliminary report. Int. J. Gynecol. Cancer 21, 8–14.

Toyoshima-Morimoto, F., Taniguchi, E., Nishida, E., 2002. Plk1 promotes nuclear translocation of human Cdc25C during prophase. EMBO J. 3, 341–348.

Trachtman, H., Fervenza, F.C., Gipson, D.S., Heering, P., Jayne, D.R.W., Peters, H., et al., 2011. A phase 1, single-dose study of fresolimumab, an anti-TGF-beta antibody, in treatment-resistant primary focal segmental glomerulosclerosis. Kidney Int. 79, 1236–1243.

Trakul, N., Menard, R.E., Schade, G.R., Qian, Z., Rosner, M.R., 2005. Raf kinase inhibitory protein regulates Raf-1 but not B-Raf kinase activation. J. Biol. Chem. 280, 24931–24940.

Tremblay, M.R., Nevalainen, M., Nair, S.J., Porter, J.R., Castro, A.C., Behnke, M.L., et al., 2008. Semisynthetic cyclopamine analogues as potent and orally bioavailable hedgehog pathway antagonists. J. Med. Chem. 51, 6646–6649.

Tremblay, M.R., Lescarbeau, A., Grogan, M.J., Tan, E., Lin, G., Austad, B.C., et al., 2009. Discovery of a potent and orally active hedgehog pathway antagonist (IPI-926). J. Med. Chem. 52, 4400–4418.

Tripathi, V., Popescu, N.C., Zimonjic, D.B., 2013. DLC1 induces expression of E-cadherin in prostate cancer cells through Rho pathway and suppresses invasion. Oncogene, February 4, doi: 10.1038/onc.2013.7.

Trojan, J., Brieger, A., Raedle, J., Esteller, M., Zeuzem, S., 2000. 5'-CpG island methylation of the LKB1/STK11 promoter and allelic loss at chromosome 19p13.3 in sporadic colorectal cancer. Gut 47, 272–276.

Tryndyak, V.P., Beland, F.A., Pogribny, I.P., 2010. E-cadherin transcriptional downregulation by epigenetic and microRNA-200 family alterations is related to mesenchymal and drug-resistant phenotypes in human breast cancer cells. Int. J. Cancer 126, 2575–2583.

Tsai, W.C., Hsu, P.W.C., Lai, T.C., Chau, G.Y., Lin, C.W., Chen, C.M., et al., 2009. MicroRNA-122, a tumor suppressor microRNA that regulates intrahepatic metastasis of hepatocellular carcinoma. Hepatology 49, 1571–1582.

Tsoporis, J.N., Izhar, S., Proteau, G., Slaughter, G., Parker, T.G., 2012. S100B-RAGE dependent VEGF secretion by cardiac myocytes induces myofibroblast proliferation. J. Mol. Cell Cardiol. 52, 464–473.

Tsui, K.H., Chung, L.C., Feng, T.H., Chang, P.L., Juang, H.H., 2012. Upregulation of prostate-derived Ets factor by luteolin causes inhibition of cell proliferation and cell invasion in prostate carcinoma cells. Int. J. Cancer 130, 2812–2823.

Tsukamoto, Y., Nakada, C., Noguchi, T., Tanigawa, M., Lam Tung, N., Uchida, T., et al., 2010. MicroRNA-375 is down regulated in gastric carcinomas and regulates cell survival by targeting PDK1 and 14-3-3 zeta. Cancer Res. 70, 2339–2349.

Tsuruo, T., Naito, M., Tomida, A., Fujita, N., Mashima, T., Sakamoto, H., et al., 2003. Molecular targeting therapy of cancer: drug resistance, apoptosis and survival signal. Cancer Sci. 94, 15–21.

Tsuruta, F., Sunyama, J., Mori, Y., Hattori, S., Shimizu, S., Tsujimoto, Y., et al., 2004. JNK promotes Bax translocation to mitochondria through phosphorylation of 14-3-3 proteins. EMBO J. 23, 1889–1899.

Tsuruta, T., Kozaki, K.I., Uesugi, A., Furuta, M., Hirasawa, A., Imoto, I., et al., 2011. miR-152 is a tumor suppressor microRNA that is silenced by DNA hypermethylation in endometrial cancer. Cancer Res. 71, 6450–6462.

Tsutsui, S., Matsuyama, A., Yamamoto, M., Takeuchi, H., Oshiro, Y., Ishida, T., et al., 2010. The Akt expression correlates with VEGF-A and -C expression as well as microvessel and lymphatic vessel density in breast cancer. Oncol. Rep. 23, 621–630.

Turley, R.S., Finger, E.C., Hempel, N., How, T., Fields, T.A., Blobe, G.C., 2007. The type III transforming growth factor-beta receptor as a novel tumor suppressor gene in prostate cancer. Cancer Res. 67, 1090–1098.

Turner, N.C., Reis-Filho, J.S., Russell, A.M., Springall, R.J., Ryder, K., Steele, D., et al. 2007. BRCA1 dysfunction in sporadic basal-like breast cancer. Oncogene 26, 2126–2132.

Udd, L., Makela, T.P., 2011. LKB1 signaling in advancing cell differentiation. Familial Cancer 10 (special issue), 425–435.

Uesugi, A., Kozaki, K.I., Tsuruta, T., Furuta, M., Morita, K.I., Imoto, I., et al., 2011. The tumor suppressive microRNA miR-218 targets the mTOR component rictor and inhibits AKT phosphorylation in oral cancer. Cancer Res. 71, 5765–5778.

Uhm, K.O., Lee, E.S., Lee, Y.M., Park, J.S., Kim, S.J., Kim, B.S., et al. 2009. Differential methylation pattern of ID4, SFRP1, and SHP1 between acute myeloid leukemia and chronic myeloid leukemia. J. Korean Med. Sci. 24, 493–497 [seen Abstract form only].

Ullmannova, V., Popescu, N.C., 2006. Expression profile of the tumor suppressor genes DLC-1 and DLC-2 in solid tumors. Int. J. Oncol. 29, 1127–1132.

Ullmannova, V., Popescu, N.C., 2007. Inhibition of cell proliferation, induction of apoptosis, reactivation of DLC1, and modulation of other gene expression by dietary flavone in breast cancer cell lines. Cancer Detect. Prevent. 31, 110–118.

Umekita, Y., Ohi, Y., Souda, M., Rai, Y., Sagara, Y., Sagara, Y., et al., 2011. Maspin expression is frequent and correlates with basal markers in triple-negative breast cancer. Diagn. Pathol. 6, 36.

Umetani, N., Takeuchi, H., Fujimoto, A., Shinozaki, M., Bilchik, A.J., Hoon, D.S.B., 2004. Epigenetic inactivation of ID4 in colorectal carcinomas correlates with poor differentiation and unfavorable prognosis. Clin. Cancer Res. 10, 7475–7483.

Umetani, N., Mori, T., Koyanagi, K., Shinozaki, M., Kim, J., Giuliano, A.E., et al. 2005. Aberrant hypermethylation of ID4 gene promoter region increases risk of lymph node metastasis in T1 breast cancer. Oncogene 24, 4721–4727.

Ungefroren, H., Sebens, S., Groth, S., Gieseler, F., Faendrich, F., 2011. The Src Family kinase inhibitors PP2 and PP1 block TGF-beta1-mediated cellular responses by direct and differential inhibition of type I and type II TGF-beta receptors. Curr. Cancer Drug Targets 11, 524–535.

Unoki, M., Nakamura, Y., 2001. Growth-suppressive effects of BPOZ and EGR2, two genes involved in the PTEN signaling pathway. Oncogene 20, 4457–4465.

Unoki, M., Kumamoto, K., Harris, C.C., 2009. ING proteins as potential anticancer drug targets. Curr. Drug Targets 10, 442–454.

Urbich, C., Kuehbacher, A., Dimmeler, S., 2008. Role of microRNAs in vascular diseases, inflammation, and angiogenesis. Cardiovasc. Res. 79, 581–588.

Urtasun, R., Latasa, M.U., Demartis, M.I., Balzani, S., Goni, S., Garcia-Irigoyen, O., et al., 2011. Connective tissue growth factor autocriny in human hepatocellular carcinoma: oncogenic role and regulation by epidermal growth factor receptor/Yes-associated protein-mediated activation. Hepatology 54, 2149–2158.

Valencia-Sanchez, M.A., Liu, J., Hannon, G.J., Parker, R., 2006. Control of translation and mRNA degradation by miRNAs and siRNAs. Genes. Dev. 20, 515–524.

Van der Weyden, L., Adams, D.J., 2007. The Ras-association domain family (RASSF) members and their role in tumorigenesis. Biochim. Biophys. Acta 1776, 58–85.

Van Dieck, J., Fernandez-Fernandez, M.R., Veprintsev, D.B., Fersht, A.R., 2009. Modulation of the oligomerization state of p53 by differential binding of proteins of the S100 family to p53 monomers and tetramers. J. Biol. Chem. 284, 13804–13811.

Van Dieck, J., Brandt, T., Teufel, D.P., Veprintsev, D.B., Joerger, A.C., Fersht, A.R., 2010. Molecular basis of S100 proteins interacting with the p53 homologs p63 and p73. Oncogene 29, 2024–2035.

Van Jaarsveld, M.T.M., Helleman, J., Berns EMJJ, Wiemer E.A.C., 2010. MicroRNAs in ovarian cancer biology and therapy resistance. Int. J. Biochem. Cell. Biol. 42, 1282–1290.

Varelas, X., Samavarchi-Tehrani, P., Narimatsu, M., Weiss, A., Cockburn, K., Larsen, B.G., et al., 2010. The Crumbs complex couples cell density sensing to Hippo-dependent control of the TGF-beta-SMAD pathway. Develop. Cell 19, 831–844.

Vazques-Martin, A., Oliveras-Ferraros, C., Menendez, J.A., 2009. The antidiabetic drug metformin suppresses HER2 (erbB2) oncoprotein over expression via inhibition of the mTOR effector p70S6K1 human breast carcinoma cells. Cell Cycle 8, 88–96.

Vazquez-Martin, A., Oliveras-Ferraros, C., Cufi, S., Del Barco, S., Martin-Castillo, B., Lopez-Bonet, E., et al., 2011. The anti-diabetic drug metformin suppresses the metastasis-associated protein CD24 in MDA-MB-468 triple-negative breast cancer cells. Oncol. Rep. 25, 135–140.

Vazquez-Martin, A., Vellon, L., Quiros, P.M., Cufi, S., de Galarreta, E.R., Oliveras-Ferraros, C., et al., 2012. Activation of AMP-activated protein kinase (AMPK) provides a metabolic barrier to reprogramming somatic cells into stem cells. Cell Cycle 11, 974–989.

Veerapandiyan, A., Abdul-Rahman, O.A., Adam, M.P., Lyons, M.J., Manning, M., Coleman, K., et al., 2011. Chromosome 22q11.2 deletion syndrome in African-American patients: a diagnostic challenge. Am. J. Med. Genet. 155A, 2186–2195.

Vega, M.I., Martinez-Paniagua, M., Huerta-Yepez, S., Gonzalez-Bonilla, C., Uematsu, N., Bonavida, B., 2009. Dysregulation of the cell survival/anti-apoptotic NF-kappa B pathway by the novel humanized BM-ca anti-CD20 mAb: implication in chemosensitization. Int. J. Oncol. 35, 1289–1296.

Vergara, D., Valente, C.M., Tinelli, A., Siciliano, C., Lorusso, V., Acierno, R., et al., 2011. Resveratrol inhibits the epidermal growth factor-induced epithelial mesenchymal transition in MCF-7 cells. Cancer Lett. 310, 1–8.

Vetter, G., Saumet, A., Moes, M., Vallar, L., Le Bechec, A., Laurini, C., et al., 2010. miR-661 expression in SNAI1-induced epithelial to mesenchymal transition contributes to breast cancer cell invasion by targeting Nectin-1 and StarD10 messengers. Oncogene 29, 4436–4448.

Vinarskaja, A., Goering, W., Ingenwerth, M., Schulz, W.A., 2012. ID4 is frequently downregulated and partially hypermethylated in prostate cancer. World J. Urol. 30, 319–325.

Visser-Grieve, S., Hao, Y., Yang, X., 2012. Human homolog of Drosophila expanded, hEx, functions as a putative tumor suppressor in human cancer cell lines independently of the Hippo pathway. Oncogene 31, 1189–1195.

Vitolo, M.I., Anglin, I.E., Mahoney, W.M., Renoud, K.J., Gartenhaus, R.B., Bachman, K.E., et al., 2007. The Runx2 transciption factor cooperates with the YES-associated protein YAP65 to promote cell trnsformation. Cancer Biol. Ther. 6, 856–863.

Volinia, S., Calin, G.A., Liu, C.G., Ambs, S., Cimmino, A., Petrocca, F.A., 2006. microRNA expression signature of human solid tumors defines cancer gene targets. Proc. Natl. Acad. Sci. U.S.A. 103, 2257–2261.

Volpe, T.A., Kidner, C., Hall, I.M., Teng, G., Grewal, S.I., Martienssen, R.A., 2002. Regulation of heterochromatic silencing and histone H3 lysine-9 methylation by RNAi. Science 297, 1833–1837.

Voorhoeve, P.M., Le Sage, C., Schrier, M., Gillis, A.J.M., Stoop, H., Nagel, R.R., et al., 2006. A genetic screen implicates miRNA-372 and miRNA-373 as oncogenes in testicular germ cell tumors. Cell 124, 1169–1181.

Vultur, A., Buettner, R., Kowolik, C., Liang, W., Smith, D., Boschelli, F., et al., 2008. SKI-606 (bosutinib), a novel Src kinase inhibitor, suppresses migration and invasion of human breast cancer cells. Mol. Cancer Ther. 7, 1185–1194.

Wales, M.M., Biel, M.A., Eldeiry, W., Nelkibn, B.D., Issa, P., Cavenee, W.K., et al., 1995. p53 activates expression of HIC-1, a new candidate tumor suppressor gene on 17p13.3. Nat. Med. 1, 570–577.

Wang, B., Matsuoka, S., Carpenter, P.B., Elledge, S.J., 2002. 53BP1, a mediator of thee DNA damage checkpoint. Science 298, 1435–1438.

Wang, B.C., Yang, H.Z., Liu, Y.C., Jelinek, T., Zhang, L.X., Ruoslahti, E., et al., 1999. Isolation of high-affinity peptide antagonists of 14-3-3 proteins by phage display. Biochemistry 38, 12499–12504.

Wang, D.S., Su, L., Huang, D.H., Zhang, H.Z., Shin, D.M., Chen, Z.G., 2011a. Downregulation of E-cadherin enhances proliferation of head and neck cancer through transcriptional regulation of EGFR. Mol. Cancer 10, 116.

Wang, H., Hertlein, E., Bakkar, N., Sun, H., Acharyya, S., Wang, J., et al., 2007. NF-kappaB regulation of YY1 inhibits skeletal myogenesis through transcriptional silencing of myofibrillar genes. Mol. Cell Biol. 27, 4374–4387.

Wang, H., Wang, X.Q., Xu, X.P., Lin, G.W., 2010a. ID4 methylation predicts high risk of leukemic transformation in patients with myelodysplastic syndrome. Leukemia Res. 34, 598–604.

Wang, H., Wu, J., Meng, X., Ying, X., Zuo, Y., Liu, R., et al., 2011b. MicroRNA-342 inhibits colorectal cancer cell proliferation and invasion by directly targeting DNA methyltransferase 1. Carcinogenesis 32, 1033–1042.

Wang, L., Pan, Y., Dai, J.L., 2004a. Evidence of Mkk4 pro-oncogenic activity in breast and pancreatic tumors. Oncogene 23, 5978–5985.

Wang, K., Pan, L., Che, X., Cui, D., Li, C., 2010b. Sonic Hedgehog/GLI1 signaling pathway inhibition restricts cell migration and invasion in human gliomas. Neurol. Res. 32, 975–980.

Wang, L., Zhang, Q., Zhang, J., Sun, S., Guo, H., Jia, Z., et al., 2011c. PI3K pathway activation results in low efficacy of both trastuzumab and lapatinib. BMC Cancer 11, 248.

Wang, Q., Zhang, Y., Yang, H.S., 2012a. Pdcd4 knockdown up-regulates MAP4K1 expression and activation of AP-1 dependent transcription through c-Myc. Biochim. Biophys. Acta 1823, 1807–1814.

Wang, S., Wilkesd, M.C., Leof, E.B., Hirschberg, R., 2005. Imatinib mesylate blocks a non-Smad TGF-beta pathway and reduces renal fibrogenesis in vivo. FASEB J. 19, 1–11.

Wang, S., Song, P., Zou, M.H., 2012b. AMP-activated protein kinase, stress responses and cardiovascular diseases. Clin. Sci. 122, 555–573.

Wang, Y., Wang, X., Zhang, J., Sun, G., Luo, H., Kang, C., et al., 2012c. MicroRNAs involved in the EGFR/PTEN/AKT pathway in gliomas. J. Neuro-Oncol. 106, 217–224.

Wang, Y.F., Kuramitsu, Y., Ueno, T., Suzuki, N., Yoshino, S., Iizuka, N., et al., 2012d. Glyoxalase I (GLO1) is up-regulated in pancreatic cancerous tissues compared with related non-cancerous tissues. Anticancer Res. 32, 3219–3222.

Wang, Z., Wang, F., Wang, W.Q., Gao, Q., Wei, W.L., Yang, Y., et al., 2004b. Correlation of N-myc downstream-regulated gene 1 overexpression with progressive growth of colorectal neoplasm. World J. Gastroenterol. 10, 550–554.

Wang, Z., Rong, Y.P., Malone, M.H., Davis, M.C., Zhong, F., Distelhorst, C.W., 2006. Thioredoxin-interacting protein (txnip) is a glucocorticoid-regulated primary response gene involved in mediating glucocorticoid-induced apoptosis. Oncogene 25, 1903–1913.

Wang, Z., Nesland, J.M., Suo, Z., Trope, C.G., Holm, R., 2011d. The prognostic value of 14-3-3 isoforms in vulvar squamous cell carcinoma cases: 14-3-3 beta and epsilon are independent prognostic factors for these tumors. Plos ONE 6, e24843.

Wanshura, L.E.C., Galvin, K.E., Ye, H., Fernandez-Zapico, M.E., Wetmore, C., 2011. Sequential activation of snail1 and N-myc modulates sonic hedgehog-induced transformation of neural cells. Cancer Res. 71, 5336–5345.

Warfel, J.M., Steele, A.D., D'Agnillo, F., 2005. Anthrax lethal toxin induces endothelial barrier dysfunction. Am. J. Pathol. 166, 1871–1881.

Wassenegger, M., 2000. RNA-directed DNA methylation. Plant Mol. Biol. 43, 203–220.

Wassenegger, M., 2005. The role of the RNAi machinery in heterochromatin formation. Cell 122, 13–16.

Watanabe, K., Hasegawa, Y., Yamashita, H., Shimizu, K., Ding, Y.Y., Abe, M., et al., 2004. Vasohibin as an endothelium-derived negative feedback regulator of angiogenesis. J. Clin. Invest. 114, 898–907.

Watanabe, T., Totoki, Y., Toyoda, A., Kaneda, M., Kuramochi-Miyagawa, S., Obata, Y., et al., 2008. Endogenous siRNAs from naturally formed dsRNAs regulate transcripts in mouse oocytes. Nature 453, 539–543.

Watanabe, T., Sato, K., Kaibuchi, K., 2009. Cadherin-mediated intercellular adhesion and signaling cascades involving small GTPases. Cold Spring Harb. Perspect. Biol. 1, a003020.

Waters, L.C., Veverka, V., Bohm, M., Schmedt, T., Choong, P.T., Muskett, F.W., et al., 2007. Structure of the C-terminal MA-3 domain of the tumour suppressor protein Pdcd4 and characterization of its interaction with eIF4A. Oncogene 26, 4941–4950.

Webster, J.I., Moayeri, M., Sternberg, E.M., 2004. Novel repression of the glucocorticoid receptor by anthrax lethal toxin. Ann. NY Acad. Sci. 1024, 9–23.

Wee-Yong, V., Power, C., Forsyth, P., Edwards, D.R., 2001. Metalloproteinases in biology and pathology of the nervous system. Nat. Rev. Neurosci. 2, 502–511.

Wei, H.J., Yin, T., Zhu, Z., Shi, P.F., Tian, Y., Wang, C., 2011. Expression of CD44, CD24 and ESA in pancreatic adenocarcinoma cell lines varies with local microenvironment. Hepatobiliary Pancreatic Dis. Int. 10, 428–434.

Wei, N., Liu, S.S., Chan, K.K.L., Ngan, H.Y.S., 2012a. Tumour suppressive function and modulation of programmed cell death 4 (PDCD4) in ovarian cancer. Plos ONE 7, e30311.

Wei, Q.J., He, W., Lu, Y.J., Yao, J., Cao, X., 2012b. Effect of the tumor suppressor gene ING4 on the proliferation of MCF-7 human breast cancer cells. Oncol. Lett. 4, 438–442.

Weir, B.A., Woo, M.S., Getz, G., Perner, S., Ding, L., Beroukhim, R., et al., 2007. Characterizing the cancer genome in lung adenocarcinoma. Nature 450, 893–898.

Welsh, S., Williams, R., Kirkpatrick, L., Paine-Murrieta, G., Powis, G., 2004. Antitumor activity and pharmacodynamic properties of PX-478, an inhibitor of hypoxia-inducible factor-1alpha. Mol. Cancer Ther. 3, 233–244.

Wen, Y., Wang, Q., Zhou, C., Yan, D., Qiu, G., Yang, C., et al., 2011. Decreased expression of RASSF6 is a novel independent prognostic marker of a worse outcome in gastric cancer patients after curative surgery. Ann. Surg. Oncol. 18, 3858–3867.

Wen, Y.H., Shi, X., Chiriboga, L., Matsahash, S., Yee, H., Afonja, O., 2007. Alterations in the expression of PDCD4 in ductal carcinoma of the breast. Oncol. Res. 18, 1387–1393.

Wen, Y.H., Ho, A., Patil, S., Akram, M., Catalano, J., Eaton, A., et al., 2012. Id4 protein is highly expressed in triple negative breast carcimnomas: possible implications for BRCA1 downregulation. Breast Cancer Res. Treat. 135, 93–102.

West, A., Vojta, P.J., Welch, D.R., Weissman, B.E., 1998. Chromosome localization and genomic structure of the KiSS-1 metastasis suppressor gene (KISS1). Genomics 54, 145–148.

Will, M.A., Palaniappan, M., Peegel, H., Kayampilly, P., K.M.J., Menon, 2012. Metformin: direct inhibition of rat ovarian theca-interstitial cell proliferation. Fertility Sterility 98, 207–214.

Williams, J.A., Guicherit, O.M., Zaharian, B.I., Xu, Y., Chai, L., Wichterle, H., et al., 2003. Identification of a small molecule inhibitor of the hedgehog signaling pathway: effects on basal cell carcinoma-like lesions. Proc. Natl. Acad. Sci. U.S.A. 100, 4616–4621.

Williams, J.A., Thomas, A.M., Li, G., Kong, B., Zhan, L., Inaba, Y., et al., 2012. Tissue specific induction of p62/Sqstm1 by farnesoid X receptor. PloS ONE 7, e43961.

Williams, S.A., Maecker, H.L., French, D.M., Liu, J., Gregg, A., Silverstein, L.B., et al. 2011. USP1 deubiquitinates ID proteins to preserve a mesenchymal stem cell program in osteosarcoma. Cell 146, 917–929.

Wingo, S.N., Gallardo, T.D., Akbay, E.A., Liang, M.C., Contreras, C.M., Boren, T., et al., 2009. Somatic LKB1 mutations promote cervical cancer progression. PLoS ONE 4, e5137.

Winter, J., Jung, S., Keller, S., Gregory, R.I., Sven Diederichs, S., 2009. Many roads to maturity: microRNA biogenesis pathways and their regulation. Nat. Cell Biol. 11, 228–234.

Wong, C.M., Yam, J.W., Ching, Y.P., et al., 2005. Rho GTPase-activating protein deleted in liver cancer suppresses cell proliferation and invasion in hepatocellular carcinoma. Cancer Res. 65, 8861–8868.

Wong, H., Chen, J.Z., Chou, B., Halladay, J.S., Kenny, J.R., La, H., et al., 2009. Preclinical assessment of the absorption, distribution, metabolism and excretion of GDC-0449 (2-chloro-*N*-(4-chloro-3-(pyridin-2-yl)phenyl)-4-(methylsulfonyl)benzamide), an orally bioavailable systemic Hedgehog signalling pathway inhibitor. Xenobiotica 39, 850–861.

Wong, I.H.N., Chan, J., Wong, J., Tam, P.K.H., 2004. Ubiquitous aberrant RASSF1A promoter methylation in childhood neoplasia. Clin. Cancer Res. 10, 994–1002.

Wong, K.Y., So, C.C., Loong, F., Chung, L.P., Lam, W.W.L., Liang, R., et al., 2011. Epigenetic inactivation of the miR-124-1 in haematological malignancies. Plos ONE 6, e19027.

Woodcock, J.M., Ma, Y., Coolen, C., Pham, D., Jones, C., Lopez, A.F., et al., 2010. Sphingosine and FTY720 directly bind pro-survival 14-3-3 proteins to regulate their function. Cell Signalling 22, 1291–1299.

Woods, K.M., Grewal, N.K., Markwart, S.M., Vellaichamy, A., Chinnaiyan, A.M., Yeung, K., et al., 2008. Loss of Raf kinase inhibitory protein induces radioresistance in prostate cancer. Int. J. Radiat. Oncol. Biol. Phys. 72, 153–160.

Woodward, W.A., Sulman, E.P., 2008. Cancer stem cells: markers or biomarkers? Cancer Metastasis Rev. 27, 459–470.

Wormstone, I.M., Tamiya, S., Anderson, I., Duncan, G., 2002. TGF-beta 2-induced matrix modification and cell transdifferentiation in the human lens capsular bag. Invest. Ophthalmol. Vis. Sci. 43, 2301–2308.

Wu, K., Bonavida, B., 2009. The activated NF-kappa B-Snail-RKIP circuitry in cancer regulates both the metastatic cascade and resistance to apoptosis by cytotoxic drugs. Crit. Rev. Immunol. 29, 241–254.

Wu, J., Omene, C., Karkoszka, J., Bosland, M., Eckard, J., Klein, C.B., et al., 2011a. Caffeic acid phenethyl ester (CAPE), derived from a honeybee product propolis, exhibits a diversity of anti-tumor effects in pre-clinical models of human breast cancer. Cancer Lett. 308, 43–53.

Wu, N., Liu, X., Xu, X., Fan, X., Liu, M., Li, X., et al., 2011b. MicroRNA-373, a new regulator of protein phosphatase 6, functions as an oncogene in hepatocellular carcinoma. FEBS J. 278, 2044–2054.

Wu, Q., Liu, C.Z., Tao, L.Y., Yu, L., Liu, W., Chen, S.S., et al., 2012a. Clinicopathological and prognostic impact of 14-3-3 protein isoforms expression in human cholangiocarcinoma by immunohistochemistry. Asian Pacific J. Cancer Prevent. 13, 1253–1259.

Wu, Y., Jiang, W., Wang, Y., Wu, J., Saiyin, H., Qiao, X., et al., 2012b. Breast cancer metastasis suppressor 1 regulates hepatocellular carcinoma cell apoptosis via suppressing osteopontin expression. PloS ONE 7, e42976.

Wu, Y., Richard, J.P., Wang, S.D., Rath, P., Laterra, J., Xia, S., 2012c. Regulation of glioblastoma multiforme stem like cells by inhibitor of DNA binding proteions and oligodendroglial lineage-associted transcription factors. Cancer Sci. 103, 1028–1037.

Wu, Z., Yang, S., Weng, X., Liu, X., 2011c. MicroRNA-21 is involved in osteosarcoma cell invasion and migration. Med. Oncol. 28, 1469–1474.

Wullschleger, S., Loewith, R., Hall, M.N., 2006. TOR signaling in growth and metabolism. Cell 124, 471–484.

Xia, L., Zhang, D., Du, R, Pan, Y, Zhao, L, Sun, S, et al., 2008. miR-15b and miR-16 modulate multidrug resistance by targeting BCL2 in human gastric cancer cells. Int. J. Cancer 123, 372–379.

Xiang, T., Li, L., Fan, Y., Jiang, Y., Ying, Y., Putti, T.C., et al., 2010. PLCD1 is a functional tumor suppressor inducing G(2)/M arrest and frequently methylated in breast cancer. Cancer Biol. Ther. 10, 520–527.

Xiao, L., Chen, Y., Ji, M., Volle, D.J., Lewis, R.E., Tsai, M.Y., et al., 2011. KIBRA protein phosphorylation is regulated by mitotic kinase Aurora and protein phosphatase 1. J. Biol. Chem. 286, 36304–36315.

Xiao, X., He, Q., Lu, C., Werle, K.D., Zhao, R.X., Chen, J., et al., 2012. Metformin impairs the growth of liver kinase B1-intact cervical cancer cells. Gynaecol. Oncol. 127, 249–255.

Xie, L., Ushmorov, A., Leithaeuser, F., Guan, H., Steidl, C., Faerbinger, J., et al., 2012. FOXO1 is a tumor suppressor in classical Hodgkin lymphoma. Blood 119, 3503–3511.

Xin M., Kim Y., Sutherland L.B., Qi X., McAnally J., Schwartz R.J., et al., (2011). Regulation of insulin-like growth factor signaling by Yap governs cardiomyocyte proliferation and embryonic heart size. Sci. Signal 4, ra70.

Xinzhou, H., Ning, Y., Ou, W., Xiaodan, L., Fumin, Y., Huitu, L., et al., 2011. RKIP inhibits the migration and invasion of human prostate cancer PC-3M cells through regulation of extracellular matrix. Mol. Biol. 45, 921–928.

Xu, B., Niu, X., Zhang, X., Tao, J., Wu, D., Wang, Z., et al., 2011a. miR-143 decreases prostate cancer cells proliferation and migration and enhances their sensitivity to docetaxel through suppression of KRAS. Mol. Cell Biochem. 350, 207–213.

Xu, C.R., Lee, S., Ho, C., Bommi, P., Huang, S.A., Cheung, S.T., et al., 2009. Bmi1 functions as an oncogene independent of Ink4A/Arf repression in hepatic carcinogenesis. Mol Cancer Res. 7, 1937–1945.

Xu, H., Fu, J., Ha, S.W., Ju, D., Zheng, J., Li, L., et al., 2012. The CCAAT box-binding transcription factor NF-Y regulates basal expression of human proteasome genes. Biochim. Biophys. Acta Mol. Cell Res. 1823, 818–825.

Xu, X., Omelchenko, T., Hall, A., 2010. LKB1 tumor suppressor protein regulates actin filament assembly through Rho and its exchange factor Dbl independently of kinase activity. BMC Cell Biol. 77, 11.

Xu, Y, Xia, F, Ma, L, Shan, J, Shen, J, Yang, Z, et al., 2011b. MicroRNA-122 sensitizes HCC cancer cells to adriamycin and vincristine through modulating expression of MDR and inducing cell cycle arrest. Cancer Lett. 310, 160–169.

Yageta, M, Kuramochi, M, Masuda, M., Fukami, T., Fukuhara, H., Maruyama, T., et al., 2002. Direct association of TSLC1 and DAL-1, two distinct tumor suppressor proteins in lung cancer. Cancer Res. 62, 5129–5133.

Yam, J.W., Ko, F.C., Chan, C.Y., Jin, D.Y., Ng, I.O., 2006. Interaction of deleted in liver cancer 1 with tensin2 in caveolae and implications in tumor suppression. Cancer Res. 66, 8367–8372.

Yamagata, K., Fujiyama, S., Ito, S., Ueda, T., Murata, T., Naitou, M., et al., 2009. Maturation of microRNA is hormonally regulated by a nuclear receptor. Mol. Cell. 36, 340–347.

Yamamoto, Y., Kato, I., Doi, T., Yonekura, H., Ohashi, S., Takeuchi, M., et al., 2001. Development and prevention of advanced diabetic nephropathy in RAGE-overexpressing mice. J. Clin. Invest. 108, 261–268.

Yan, C., Wang, H., Boyd, D.D., 2001. KiSS-1 represses 92-kDa type IV collagenase expression by down-regulating NF-kappa B binding to the promoter as a consequence of Ikappa Balpha -induced block of p65/p50 nuclear translocation. J. Biol. Chem. 276, 1164–1172.

Yan, G.R., Xu, S.H., Tan, Z.L., Liu, L., He, Q.Y., 2011. Global identification of miR-373-regulated genes in breast cancer by quantitative proteomics. Proteomics 11, 9912–9920.

Yan, Y.M., Dai, H.Q., Du, Y., Schneider, B., Guo, H., Li, D.P., et al., 2012. Identification of blapsins A and B as potent small-molecule 14-3-3 inhibitors from the insect Blaps japanensis. Bioorg. Med. Chem. Lett. 22, 4179–4181.

Yang, C.H., Yue, J., Pfeffer, S.R., Handorf, C.R., Pfeffer, L.M., 2011a. MicroRNA miR-21 regulates the metastatic behavior of B16 melanoma cells. J. Biol. Chem. 286, 39172–39178.

Yang, G., He, W., Cai, M., Luo, F., Kung, H., Guan, X., et al., 2011b. Loss/down-regulation of tumor suppressor in lung cancer 1 expression is associated with tumor progression and is a biomarker of poor prognosis in ovarian carcinoma. Int. J. Gynecol. Cancer 21, 486–493.

Yang, H., Jansen, A., Komar, A., Zheng, X., Merrick, W., Costes, S., et al., 2003a. The transformation suppressor Pdcd4 is a novel eukaryotic translation initiation factor 4A binding protein that inhibits translation. Mol. Cell. Biol. 23, 26–37.

Yang, H., Xiang, J., Wang, N., Zhao, Y., Hyman, J., Li, S., et al., 2009a. Converse conformational control of Smoothened activity by structurally related small molecules. J. Biol. Chem. 284, 20876–20884.

Yang, H.S., Matthews, C.P., Clair, T., Wang, Q., Baker, A.R., Li, C.C.H., et al., 2006. Tumorigenesis suppressor Pdcd4 down-regulates mitogen-activated protein kinase kinase kinase kinase 1 expression to suppress colon carcinoma cell Iinvasion. Mol. Cell Biol. 26, 1297–1306.

Yang, H.Y., Wen, Y.Y., Chen, C.H., Lozano, G., Lee, M.H., 2003b. 14-3-3σ positively regulates p53 and suppresses tumor growth. Mol. Cell. Biol. 23, 7096–7107.

Yang, H.Y., Liu, H.L., Liu, G.Y., Zhu, H., Meng, Q.W., Qu, L.D., et al., 2011c. Expression and prognostic value of ID-1 and ID3 in gastric adenocarcinoma. J. Surg. Res. 167, 258–266.

Yang, J., Mani, S.A., Donaher, J.L., Ramaswamy, S., Itzykson, R.A., Come, C., et al., 2004. Twist, a master regulator of morphogenesis, plays an essential role in tumor metastasis. Cell 117, 927–939.

Yang, K., Handorean, A.M., Iczkowski, K.A., 2009b. MicroRNAs 373 and 520c are downregulated in prostate cancer, suppress CD44 translation and enhance invasion of prostate cancer cells in vitro. Int. J. Clin. Oncol. Exp. Pathol. 2, 361–369.

Yang, L., Wu, X., Wang, Y., Zhang, K., Wu, J., Yuan, Y.C., et al., 2011d. FZD7 has a critical role in cell proliferation in triple negative breast cancer. Oncogene 30, 4437–4446.

Yang, X., Yin, J., Yu, J., Xiang, Q., Liu, Y., Tang, S., et al., 2012. miRNA-195 sensitizes human hepatocellular carcinoma cells to 5-FU by targeting BCL-w. Oncol. Res. 27, 250–257.

Yao, J., Liang, L., Huang, S., Ding, J., Tan, N., Zhao, Y., et al., 2010. MicroRNA-30d promotes tumor invasion and metastasis by targeting Galphai2 in hepatocellular carcinoma. Hepatology 51, 846–856.

Yao, M., Wang, Y., Shen, H., Ning, W., Tang, J., Wang, X., et al., 2008. Identification of the core promoter of STK11 gene and its transcriptional regulation by p53. Progr. Nat. Sci. Mater. Int. 18, 273–280.

Yatabe, Y., Mitsudomi, T., Takahashi, T., 2004. Maspin expression in normal lung and non-small-cell lung cancers: cellular property-associated expression under the control of promoter DNA methylation. Oncogene 23, 4041–4049.

Yau, T.O., Leung, T.H.Y., Lam, S., Cheung, O.F., Tung, E.K.K.K., Khong, P.L., et al., 2009. Deleted in liver cancer 2 (DLC2) was dispensable for development and its deficiency did not aggravate hepatocarcinogenesis. Plos ONE 4, e6566.

Ye, K., 2007. Phosphorylation of Merlin regulates its stability and tumor suppressive activity. Cell Adhes. Migrat. 1, 196–198.

Yeh, Y.C., Wei, W.C., Wang, Y.K., Lin, S.C., Sung, J., Tang, M.J., 2010. Transforming growth factor-beta 1 induces Smad3-dependent beta(1) integrin gene expression in epithelial-to-mesenchymal transition during chronic tubulointerstitial fibrosis. Am. J. Pathol. 177, 1743–1754.

Yeung, K., Seitz, T., Li, S., Janosch, P., McFerran, B., Kaiser, C., et al., 1999. Suppression of Raf-1 kinase activity and MAP kinase signalling by RKIP. Nature 401, 173–177.

Yeung, K.C., Rose, D.W., Dhillon, A.S., Yaros, D., Gustafsson, M., Chatterjee, D., et al., 2001. Raf kinase inhibitor protein interacts with NF-kappa B-inducing kinase and TAK1 and inhibits NF-kappa B activation. Mol. Cell Biol. 21, 7207–7217.

Yi, B., Tan, S.X., Tang, C.E., Huang, W.G., Cheng, A.L., Li, C., et al., 2009. Inactivation of 14-3-3 sigma by promoter methylation correlates with metastasis in nasopharyngeal carcinoma. J. Cell Biochem. 106, 858–866.

Yi, M., Yang, J., Chen, X., Li, J., Li, X., Wang, L., et al., 2011. RASSF1A suppresses melanoma development by modulating apoptosis and cell-cycle progression. J. Cell Physiol. 226, 2360–2369.

Ying, H., Shen, Y.H., Godlewski, J., Bronisz, A., Zhu, J., Comb, M.J., et al., 2003. Significance of 14-3-3 self-dimerization for phosphorylation-dependent target binding. Mol. Biol. Cell 14, 4721–4733.

Ying, S., Hamdy, F.C., Helleday, T., 2012. Mre11-dependent degradation of stalled DNA replication forks is prevented by BRCA2 and PARP1. Cancer Res. 72, 2814–2821.

Yokosaki, Y., Tanaka, K., Higashikawa, E., Yamashita, K., Eboshida, A., 2005. Distinct structural requirement for binding fo the integrins alphavbeeta6, alphavbeeta5, alpha5beta1 and alpha9beta1 to osteopontin. Matrix Biol. 24, 418–427.

Yoo, B.K., Gredler, R., Chen, D., Santhekadur, P.K., Fisher, P.B., Sarkar, D., 2011. c-Met activation through a novel pathway involving osteopontin mediates oncogenesis by the transcription factor LSF. J. Hepatol. 55, 1317–1324.

Yook, J.I., Li, X.Y., Ota, I., Fearon, E.R., Weiss, S.J., 2005. Wnt-dependent regulation of the E-cadherin repressor snail. J. Biol. Chem. 280, 11740–11748.

Yook, J.I., Li, X.Y., Ota, I., Hu, C., Kim, H.S., Kim, N.H., et al., 2006. A Wnt-Axin2-GSK3beta cascade regulates Snail1 activity in breast cancer cells. Nat. Cell Biol. 8, 1398–1406.

Yoon, B.C., Zivraj, K.H., Strochlic, L., Holt, C.E., 2012. 14-3-3 proteins regulate retinal axon growth by modulating ADF/cofilin activity. Develop. Neurobiol. 72, 600–614.

Yoon, C.S., Hyung, W.J., Lee, J.H., Chae, Y.S., Won, N.H., Yeom, B.W., et al., 2008. Expression of S100A4, E-cadherin, alpha- and beta-catenin in gastric adenocarcinoma. Hepatogastroenterology 55, 1916–1920.

Yoshida, B.A., Dubauskas, Z., Chekmareva, M.A., Christiano, T.R., Stadler, W.M., Rinker-Schaeffer, C.W., 1999. Mitogen-activated protein kinase kinase 4/stress-activated protein/Erk kinase 1 (MKK4/SEK1), a prostate cancer metastasis suppressor gene encoded by human chromosome 17. Cancer Res. 59, 5483–5487.

Yoshizawa, K., Nozaki, S., Kitahara, H., Kato, K., Noguchi, N., Kawashiri, S., et al., 2011. Expression of urokinase-type plasminogen activator/urokinase-type plasminogen activator receptor and maspin in oral squamous cell carcinoma: association with mode of invasion and clinicopathological factors. Oncol. Rep. 26, 1555–1560.

Young, J.A., Collier, R.J., 2007. Anthrax toxin: receptor binding, internalization, pore formation, and translocation. Annu. Rev. Biochem. 76, 243–265.

Yu, F., Jiao, Y., Zhu, Y., Wang, Y., Zhu, J., Cui, X., et al., 2012a. MicroRNA 34c gene down-regulation via DNA methylation promotes self-renewal and epithelial–mesenchymal transition in breast tumor-initiating cells. J. Biol. Chem. 287, 465–473.

Yu, J., Zheng, Y., Dong, J., Klusza, S., Deng, W.M., Pan, D., 2010. Kibra functions as a tumor suppressor protein that regulates Hippo signalling in conjunction with Merlin and Expanded. Dev. Cell 18, 288–299.

Yu, L., Li, H.Z., Lu, S.M., Tian, J.J., Ma, J.K., Wang, H.B., et al., 2011. Down-regulation of TWIST decreases migration and invasion of laryngeal carcinoma Hep-2 cells by regulating the E-cadherin, N-cadherin expression. J. Cancer Res. Clin. Oncol. 137, 1487–1493.

Yu, L., Lu, S., Tian, J., Ma, J., Li, J., Wang, H., et al., 2012b. TWIST expression in hypopharyngeal cancer and the mechanism of TWIST-induced promotion of metastasis. Oncol. Rep. 27, 416–422.

Yu, S.W., Wang, H., Poitras, M.F., et al., 2002. Mediation of poly(ADP-ribose) polymerase-1-dependent cell death by apoptosis-inducing factor. Science 297, 259–263.

Yuan, B.Z., Zhou, X.L., Durkin, M.E., Zimonjic, D.B., Gumundsdottir, K., Eyfjord, J.E., et al., 2003a. DLC-1 gene inhibits human breast cancer cell growth and in vivo tumorigenicity. Oncogene 22, 445–450.

Yuan, B.Z., Durkin, M.E., Popescu, N.C., 2003b. Promoter hypermethylation of DLC-1, a candidate tumor suppressor gene, in several common human cancers. Cancer Genet. Cytogenet. 140, 113–117.

Yuan, P., Wang, L., Wei, D., Zhang, J., Jia, Z., Li, Q., et al., 2007. Therapeutic inhibition of Sp1 expression in growing tumors by mithramycin a correlates directly with potent antiangiogenic effects on human pancreatic cancer. Cancer 110, 2682–2690.

Yuen, H.F., Chua, C.W., Chan, Y.P., Wong, Y.C., Wang, X.H., Chan, K.W., 2006. Id proteins expression in prostate cancer: high-level expression of Id-4 in primary prostate cancer is associated with development of metastases. Modern Pathol. 19, 931–941.

Yun, J., Frankenberger, C.A., Kuo, W.L., Boelens, M.C., Eves, E.M., Cheng, N., et al., 2011. Signalling pathway for RKIP and Let-7 regulates and predicts metastatic breast cancer. EMBO J. 30, 4500–4514.

Zajkowicz, A., Rusin, M., 2011. The activation of the p53 pathway by the AMP mimetic AICAR is reduced by inhibitors of the ATM or mTOR kinases. Mech. Ageing Dev. 132, 543–551.

Zaman, M.S., Chen, Y., Deng, G., Shahryari, V., Suh, S.O., Saini, S., et al., 2010. The functional significance of microRNA-145 in prostate cancer. Br. J. Cancer 103, 256–264.

Zang, D., Li, X., Zhang, L., 2010. 14-3-3 zeta overexpression and abnormal beta-catenin expression are associated with poor differentiation and progression in stage I non-small cell lung cancer. Clin. Exp. Med. 10, 221–228.

Zang, Z.J., Cutcutache, I., Poon, S.L., Zhang, S.L., McPherson, J.R., Tao, J., et al., 2012. Exome sequencing of gastric adenocarcinoma identifies recurrent somatic mutations in cell adhesion and chromatin remodeling genes. Nat. Genet. 44, 570–574.

Zaratiegui, M., Irvine, D.V., Martienssen, R.A., 2007. Noncoding RNAs and gene silencing. Cell 128, 763–776.

Zavadil, J., Bitzer, M., Liang, D., Yang, Y.C., Massimi, A., Kneitz, S., et al., 2001. Genetic programs of epithelial cell plasticity directed by transforming growth factor-beta. Proc. Natl. Acad. Sci. U.S.A. 98, 6686–6691.

Zawacka-Pankau, J., Kostecka, A., Sznarkowska, A., Hedstrom, E., Kawiak, A., 2010. p73 tumor suppressor protein A close relative of p53 not only in structure but also in anticancer approach? Cell Cycle 9, 720–728.

Zeng, P.Y., Berger, S.L., 2006. LKB1 is recruited to the p21/WAF1 promoter by p53 to mediate transcriptional activation. Cancer Res. 66, 10701–10708.

Zeng, Y., Kotake, Y., Pei, X.H., Smith, M.D., Xiong, Y., 2011. P53 binds to and is required for the repression of Arf tumor suppressor by HDAC and polycomb. Cancer Res. 71, 2781–2792.

Zhan, Y., Modi, N., Stewart, A.M., Hieronimus, R.I., Liu, J., Gutmann, D.H., et al., 2011. Regulation of mixed lineage kinase 3 is required for Neurofibromatosis-2-mediated growth suppression in human cancer. Oncogene 30, 781–789.

Zhang, A.H., Rao, J.N., Zou, T., Liu, L., Marasa, B.S., Xiao, L., et al., 2007. p53-Dependent NDRG1 expression induces inhibition of intestinal epithelial cell proliferation but not apoptosis after polyamine depletion. Am. J. Physiol. Cell Physiol. 293, C379–C389.

Zhang, B.G., Li, J.F., Yu, B.Q., Zhu, Z.G., Liu, B.Y., Yan, M., 2012a. MicroRNA-21 promotes tumor proliferation and invasion in gastric cancer by targeting PTEN. Oncol. Rep. 27, 1019–1026.

Zhang, C., Li, F., Zhou, L., Li, X., Wu, X.X., Hoffman, R.M., 2009a. Effect of verapamil on the expression of EGFR and NM23 in A549 human lung cancer cells. Anticancer Res. 29, 27–32.

Zhang, C.Z., Han, L., Zhang, A.L., Fu, Y.C., Yue, X.A., Wang, G.X., et al., 2010a. MicroRNA-221 and microRNA-222 regulate gastric carcinoma cell proliferation and radioresistance by targeting PTEN. BMC Cancer 10, 367.

Zhang, H., Li, M., Han, Y., Hong, L., Gong, T., Sun, L., et al., 2010b. Down-regulation of miR-27a might reverse multidrug resistance of esophageal squamous cell carcinoma. Dig. Dis. Sci. 55, 2545–2551.

Zhang, H., Singh, R.R., Talukder, A.H., Kumar, R., 2006a. Metastatic tumor antigen 3 is a direct co-repressor of the Wnt4 pathway. Genes Dev. 20, 2943–2948.

Zhang, H.K., Pan, K., Wang, H., Weng, D.S., Song, H.F., Zhou, J., et al., 2008a. Decreased expression of ING2 gene and its clinicopathological significance in hepatocellular carcinoma. Cancer Lett. 261, 183–192.

Zhang, H.Y., Rumilla, K.M., Jin, L., Nakamura, N., Stilling, G.A., Ruebel, K.H., et al., 2006b. Association of DNA methylation and epigenetic inactivation of RASSF1A and beta-catenin with metastasis in small bowel carcinoid tumors. Endocrine 30, 299–306.

Zhang, J., Smolen, G.A., Haber, D.A., 2008b. Negative regulation of YAP by LATS1 underscores evolutionary conservation of the Drosophila Hippo pathway. Cancer Res. 68, 2789–2794.

Zhang, J., Ji, J.Y., Yu, M., Overholtzer, M., Smolen, G.A., Wang, R., et al., 2009b. YAP-dependent induction of amphiregulin identifies a non-cell-autonomous component of the Hippo pathway. Nat. Cell Biol. 11, 1444–1450.

Zhang, M., Volpert, O., Shi, Y.H., Bouck, N., 2000. Maspin is an angiogenesis inhibitor. Nat. Med. 6, 196–199.

Zhang, R., He, Y., Zhang, X., Xing, B., Sheng, Y., Lu, H., et al., 2012b. Estrogen receptor-regulated microRNAs contribute to the BCL2/BAX imbalance in endometrial adenocarcinoma and precancerous lesions. Cancer Lett. 314, 155–165.

Zhang, S., Hao, J., Xie, F., Hu, X., Liu, C., Tong, J., et al., 2011a. Downregulation of miR-132 by promoter methylation contributes to pancreatic cancer development. Carcinogenesis 32, 1183–1189.

Zhang, S., Kim, K., Jin, U.H., Pfent, C., Cao, H.J., Amendt, B., et al., 2012c. Aryl hydrocarbon receptor agonists induce microRNA-335 expression and inhibit lung metastasis of estrogen receptor negative breast cancer cells. Mol. Cancer Ther. 11, 108–118.

Zhang, W., Zhang, M., 2002. Tissue microarray analysis of maspin expression and its reverse correlation with mutant p53 in various tumors. Int. J. Oncol. 20, 1145–1150.

Zhang, W., Cheng, G.Z., Gong, J., Hermanto, U., Zong, C.S., Chan, J., et al., 2008c. RACK1 and CIS mediate the degradation of BimEL in cancer cells. J. Biol. Chem. 283, 16416–16426.

Zhang, X., Xu, L.S., Wang, Z.Q., Wang, K.S., Li, N., Cheng, Z.H., et al., 2004. ING4 induces G2/M cell cycle arrest and enhances the chemosensitivity to DNA-damage agents in HepG2 cells. FEBS Lett. 570, 7–12.

Zhang, X., Wang, K.S., Wang, Z.Q., Xu, L.S., Wang, Q.W., Chen, F., et al., 2005. Nuclear localization signal of ING4 plays a key role in its binding to p53. Biochem. Biophys. Res. Commun. 331, 1032–1038.

Zhang, X.M., Gu, H., Yan, L., Zhang, G.Y., 2013. RKIP inhibits the malignant phenotypes of gastric cancer cells. Neoplasma 60, 196–202.

Zhang, Y., Laterra, J., Pomper, M.G., 2009c. Hedgehog pathway inhibitor HhAntag691 is a potent inhibitor of ABCG2/BCRP and ABCB1/Pgp. Neoplasia 11, 96–101.

Zhang, Y., Yan, L.X., Wu, Q.N., Du, Z.M., Chen, J., Liao, D.Z., et al., 2011b. miR-125b is methylated and functions as a tumor suppressor by regulating the ETS1 proto-oncogene in human invasive breast cancer. Cancer Res. 71, 3552–3562.

Zhang, Y., Xu, R., Li, G., Xie, X., Long, J., Wang, H., 2012d. Loss of expression of the differentially expressed in adenocarcinoma of the lung (DAL-1) protein is associated with metastasis of non-small cell ljung carcinoma cells. Tumour Biol. 33, 1915–1925.

Zhang, Y., Wang, Y., Bao, C., Xu, Y., Shen, H., Chen, J., et al., 2012e. Metformin interacts with AMPK through binding to gamma subunit. Mol. Cell. Biochem. 368, 69–76.

Zhang, Z., Liu, Z.B., Ren, W.M., Ye, X.G., Zhang, Y.Y., 2012f. The miR-200 family regulates the epithelial–mesenchymal transition induced by EGF/EGFR in anaplastic thyroid cancer cells. Int. J. Mol. Med. 30, 856–862.

Zhao, B., Wei, X., Li, W., Udan, R.S., Yaang, Q., Xie, J., et al., 2007a. Inactivation of YAP oncoprotein by the Hippo pathway is involved in cell contact inhibition and tissue growth control. Genes Dev. 21, 2747–2761.

Zhao, C., Matthews, J., Tujague, M., Wan, J., Ström, A., Toresson, G., et al., 2007b. Estrogen receptor beta2 negatively regulates the transactivation of estrogen receptor alpha in human breast cancer cells. Cancer Res. 67, 3955–3962.

Zhao, H.B., Wang, C., Li, R.X., Tang, C.L., Li, M.Q., Du, M.R., et al., 2010. E-Cadherin, as a negative regulator of invasive behavior of human trophoblast cells, is down-regulated by cyclosporin A via epidermal growth factor/extracellular signal-regulated protein kinase signaling pathway. Biol. Reprod. 83, 370–376.

Zhao, J., Du, Y., Horton, J.R., Upadhyay, A.K., Lou, B., Bai, Y., et al. 2011. Discovery and structural charactersization of a small molecule 14-3-3 protein-protein interaction inhibitor. Proc. Natl. Acad. Sci. USA 108, 16212–16216.

Zheng, G.P., Xiong, Y., Yi, S., Zhang, W., Peng, B., Zhang, Q., et al., 2012. 14-3-3 sigma regulation by p53 mediates a chemotherapy response to 5-fluorouracil in MCF-7 breast cancer cells via Akt inactivation. FEBS Lett. 586, 163–168.

Zheng, H., Ying, H., Yan, H., Kimmelman, A.C., Hiller, D.J., Chen, A.J., et al., 2008. p53 and Pten control neural and glioma stem/progenitor cell renewal and differentiation. Nature 455 1129–U13

Zheng, J., Xue, H., Wang, T., Jiang, Y., Liu, B., Li, J., et al., 2011. miR-21 downregulates the tumor suppressor P12CDK2AP1 and stimulates cell proliferation and invasion. J. Cell. Biochem. 112, 872–880.

Zheng, X., Yao, Y., Xu, Q., Tu, K., Liu, Q., 2010. Evaluation of glioma-associated oncogene 1 expression and its correlation with the expression of sonic hedgehog, E-cadherin and S100a4 in human hepatocellular carcinoma. Mol. Med. Rep. 3, 965–970.

Zhong, D., Liu, X., Khuri, F.R., Sun, S.Y., Vertino, P.M., Zhou, W., 2008. LKB1 is necessary for Akt-mediated phosphorylation of proapoptotic proteins. Cancer Res. 68, 7270–7277.

Zhou, A.D., Diao, L.T., Xu, H., Xiao, Z.D., Li, J.H., Zhou, H., et al., 2012a. Beta-Catenin/LEF1 transactivates the microRNA-371-373 cluster that modulates the Wnt/beta-catenin-signaling pathway. Oncogene 31, 2968–2978.

Zhou, J., Huang, W., Tao, R., Ibaragi, S., Lan, F., Ido, Y., et al., 2009. Inactivation of AMPK alters gene expression and promotes growth of prostate cancer cells. Oncogene 28, 1993–2002.

Zhou, L., Ercolano, E., Ammoun, S., Schmid, M.C., Barczyk, M.A., Hanemann, C.O., 2011a. Merlin-Deficient human tumors show loss of contact inhibition and activation of Wnt/beta-catenin signaling linked to the PDGFR/Src and Rac/PAK pathways. Neoplasia 13, 1101–1112.

Zhou, M., Liu, Z., Zhao, Y., Ding, Y., Liu, H., Xi, Y., et al., 2010a. MicroRNA-125b confers the resistance of breast cancer cells to Paclitaxel through suppression of pro-apoptotic Bcl-2 antagonist killer 1 (Bak1) expression. J. Biol. Chem. 285, 21496–21507.

Zhou, Q., Lui, V.W.Y., Lau, C.P.Y., Cheng, S.H., Ng, M.H.L., Cai, Y., et al., 2012b. Sustained antitumor activity by co-targeting mTOR and the microtubule with temsirolimus/vinblastine combination in hepatocellular carcinoma. Biochem. Pharmacol. 83, 1146–1158.

Zhou, R.H., Kokame, K., Tsukamoto, Y., Yutani, C., Kato, H., Miyata, T., 2001. Characterization of the human NDRG gene family: a newly identified member, NDRG4, is specifically expressed in brain and heart. Genomics 73, 86–97.

Zhou, W.H., Tang, F., Xu, J., Wu, X., Feng, Z.Y., Li, H.G., et al., 2011b. Aberrant upregulation of 14-3-3o' expression serves as an inferior prognostic biomarker for gastric cancer. BMC Cancer 11, 397.

Zhou, X., Thorgeirsson, S.S., Popescu, N.C., 2004. Restoration of DLC-1 gene expression induces apoptosis and inhibits both cell growth and tumorigenicity in human hepatocellular carcinoma cells. Oncogene 23, 1308–1313.

Zhou, X., Zhang, J., Jia, Q., Ren, Y., Wang, Y., Shi, L., et al., 2010b. Reduction of miR-21 induces glioma cell apoptosis via activating caspase 9 and 3. Oncol. Rep. 24, 195–201.

Zhou, Z., Hao, Y., Liu, N., Raptis, L., Tsao, M.S., Yang, X., 2011c. TAZ is a novel oncogene in non-small cell lung cancer. Oncogene 30, 2181–2186.

Zhu, C.H., Cheng, K.W., Ouyang, N.T., Huang, L.Q., Sun, Y., Constantinides, P., et al., 2012. Phosphosulindac (OXT-328) selectively targets breast cancer stem cells in vitro and in human breast cancer xenografts. Stem Cells 30, 2065–2075.

Zhu, H., Wu, H., Liu, X., Evans, B.R., Medina, D.J., Liu, C.G., et al., 2008a. Role of MicroRNA miR-27a and miR-451 in the regulation of MDR1/P glyglycoprotein expression in human cancer cells. Biochem. Pharmacol. 76, 582–588.

Zhu, J., Shimizu, E., Zhang, X., Partridge, N.C., Qin, L., 2011. EGFR signaling suppresses osteoblast differentiation and inhibits expression of master osteoblastic transcription factors Runx2 and Osterix. J. Cell. Biochem. 112, 1749–1760.

Zhu, S., Wu, H., Wu, F., Nie, D., Sheng, S., Mo, Y.Y., 2008b. MicroRNA-21 targets tumor suppressor genes in invasion and metastasis. Cell Res. 18, 350–359.

Zhu, Z., Luo, Z., Li, Y., Ni, C., Li, H., Zhu, M., 2009. Human inhibitor of growth 1 inhibits hepatoma cell growth and influences p53 stability in a variant-dependent manner. Hepatology 49, 504–512.

Zhuang, Z.G., Di, G.H., Shen, Z.Z., Ding, J., Shao, Z.M., 2006. Enhanced expression of LKB1 in breast cancer cells attenuates angiogenesis, invasion, and metastatic potential. Mol. Cancer Res. 4, 843–849.

Zou, Z.Q., Gao, C.L., Nagaich, A.K., Connell, T., Saito, S., Moul, J.W., et al., 2000. p53 regulates the expression of the tumor suppressor gene maspin. J. Biol. Chem. 275, 6051–6054.

Zuo, J.H., Zhu, W., Li, M.Y., Li, X.H., Yi, H., Zeng, G.Q., et al., 2011. Activation of EGFR promotes squamous carcinoma SCC10A cell migration and invasion via inducing EMT-Like phenotype change and MMP-9-mediated degradation of E-cadherin. J. Cell. Biochem. 112, 2508–2517.

Printed in the United States
By Bookmasters